the BESS BOOK

A Cell to Grid Guide to Utility-scale Battery Energy Storage Systems

Drew Lebowitz, P.E.
Sean Daly
Swetha Sundaram

ISBN: 979-8-218-44798-4

Illustrations by Sidra Aslam unless otherwise noted.
Cover design by Casey Ramsey, with contributions from Dall-E 3
Book interior and e-book design by Amit Dey.

First edition.

BOOK DESCRIPTION

Recent advances in lithium-ion battery technology and rapidly falling prices have ushered in an era of unprecedented growth in utility-scale energy storage projects. Propelled by ample government incentives, markets around the world have seen a swift rise in the number and size of battery energy storage systems (BESS) connected to the power grid. This expansion has spurred massive advancements and interest in the design, construction, finance, and operation of these plants.

The BESS Book is intended for the reader who is interested in learning about the fundamentals of lithium-ion battery systems. All key concepts of BESS technology, engineering, finance, construction, and operation and maintenance (O&M) are covered, as is hands-on learning from the field. Later chapters discuss battery controls, finance, development, and safety. The final section presents an outlook on the future of energy storage technology, and how anyone can start or advance a career in the battery industry. The book features examples from actual projects, market trends, lists of leading firms, photographs, and original graphics throughout.

The content has been written in a way to help present ideas clearly for newcomers to the field, while including comprehensive technical content for energy storage professionals. The authors have over 5 GWh of combined energy storage project experience, including work with some of the largest projects and leading names in energy storage development and finance. Their hands-on experience combines theory with practice to give a clear picture of the current state of the industry, where it is going, and how the reader can get involved.

ABOUT THE AUTHORS

Drew Lebowitz, P.E., is Principal and Managing Director at PowerSwitch, an energy storage advisory firm based in Portland, Oregon, where he is a licensed professional engineer. Drew manages the firm, providing owner's engineering, technical due diligence, and training services around utility-scale energy storage projects. Prior to joining PowerSwitch, Drew worked as Principal Consultant on DNV's Energy Storage Engineering team, where he led desktop and field projects for some of the world's largest battery plants. He holds an engineering degree from Cornell University. Drew started his career in Panama and Haiti, where he lived for a combined 12 years. These experiences continue to inspire his involvement in Energy Access, an industry working to attract sustainable investment to connect the 500m+ people in the world living without reliable access to electricity.

Sean Daly is an energy storage engineer based in Portland, Oregon. After a decade of wind energy consulting, Sean shifted to energy storage to be part of the next transition needed to accomplish a net-zero carbon future. Currently a Project Design Engineer at Akaysha Energy, Sean provides technical support for utility-scale energy storage systems, specializing in utility interconnection. Sean holds bachelor's and master's degrees from the University of Massachusetts Amherst and previously acted in several roles for DNV's wind turbine testing accredited laboratory, including Project Manager, Engineer, and Quality Manager. When not engineering, Sean records and performs original music at venues in Portland.

Swetha Meenakshi Sundaram is an experienced energy professional with expertise covering diverse aspects of the energy industry. She currently serves as Director of Energy Storage Project Design at leading renewables developer RWE Clean Energy. Swetha is based in Austin, Texas and has niche, hands-on experience for close to ten years in the rapidly growing energy storage space as both project developer and independent engineer. Prior to RWE, Swetha worked at leading independent engineering firm DNV, and utilities PG&E and AES Indiana. Her experience is backed by robust knowledge of energy market fundamentals and techno-economic analysis, and a Master's degree from Purdue University.

DEDICATION

Drew dedicates this book to his parents Mark and Lisa, who instilled in him a love of learning, the value of service and the tenacity to do big things, and to Jess, who has been a towering source of love and patience without whom this book never would have been written.

Sean dedicates this book to his parents Kathy and Tom, who supported him through all his wild projects, and Erik Brown, for his friendship.

Swetha dedicates this book to her husband Avinash and daughter Akshara for their unconditional support through this journey, her parents Kavitha and Sundar for backing all her life choices, and Andrea Hu-Bianco for his thoughtful mentorship.

Lastly, this book is dedicated to the more than 500 million people living around the world without access to reliable electricity.

ACKNOWLEDGEMENTS

This book has been a labor of love more than four years in the making. We could not have completed this project without the encouragement and kindness from this cast of hundreds. To name only a few of the many:

Our patient and knowledgeable team of technical reviewers improved the text of this book in thousands of ways, large and small. Thanks to Michael Kleinberg, Ali Ghorashi, Andrea Hu-Bianco, Ken Elser, Frank Bergh, Henry Louie, John Ruiz and Jamie Daggett for all of your valuable additions. Juan Pablo Zagorodny and Christian Sanchez were especially helpful and diligent reviewers. Stephan Williams was an early and fervent supporter of this project.

All three authors are former employees of DNV, the Norwegian energy consulting company that has been a vocal leader in the energy transition. This book owes a great debt to the firm and its passionate and brilliant team of DNVers who were supportive of this project from its earliest days.

Daniel Crotzer provided valuable input on battery controls. Euromina Thevenin performed thorough research from Port-au-Prince. Anita Roberts was a thorough and relentless technical reviewer of the book. Craig Lebowitz lent a helping hand on several fronts. Nico Johnson was helpful in promoting this book.

DISCLAIMER

The work is an original work composed by the authors, separate and apart from their ongoing employers. Any and all views expressed in this book reflect the views of the authors alone. All references made in the book are either publicly available or have been cited with express permission of the original authors.

TABLE OF CONTENTS

1. INTRODUCTION

Energy blogs, newsletters, and major newspapers are bristling with news about energy storage. These stories tout massive battery announcements, game-changing technological breakthroughs, and land deals for project development, lithium resources, and new battery products. It often seems that every few months new battery energy storage system (BESS) projects break the previous size records for the biggest system. In 2014, battery energy storage was a fringe topic within the energy industry, seen as a promising but expensive and underutilized technology. Ten years later, it has become a sector that all major utilities, project developers, banks, and investors are scrambling to participate in. The projects are driven by the ability to balance increasing renewable resources, a task for which energy storage projects are uniquely suited. These developments are driven by breakthroughs in both the physics of batteries themselves as well as advances in software, financing, and government incentives in the industry. In the United States, the Inflation Reduction Act (IRA), signed into law in August 2022, included large incentives for energy storage projects, technology development, and manufacturing, which promise to further accelerate growth in the sector. All major forecasts predict batteries playing a central role in power grids of the 21st century.

The numbers tell an intriguing story: looking back to the end of 2019, the total collective power output from all lithium-ion (Li-ion) battery projects installed worldwide was approximately 4 gigawatts (GW), according to Bloomberg New Energy Finance (BNEF), a leading industry reporter [1]. For context, this is approximately 0.3% of the total US Energy Generation capacity of 1,300 GW. Collectively, these plants occupied roughly 160 hectares (400 acres) of land, or approximately 4,000 40-foot shipping containers (at 2019 densities).

Fast forward to today, and BNEF estimates that the world added 42 GW in 2023 alone [1]. This value is predicted to rise to reach 110 GW by 2030. This meteoric growth can be seen graphically in Figure 1-1, which also shows the breakdown between geographic regions.

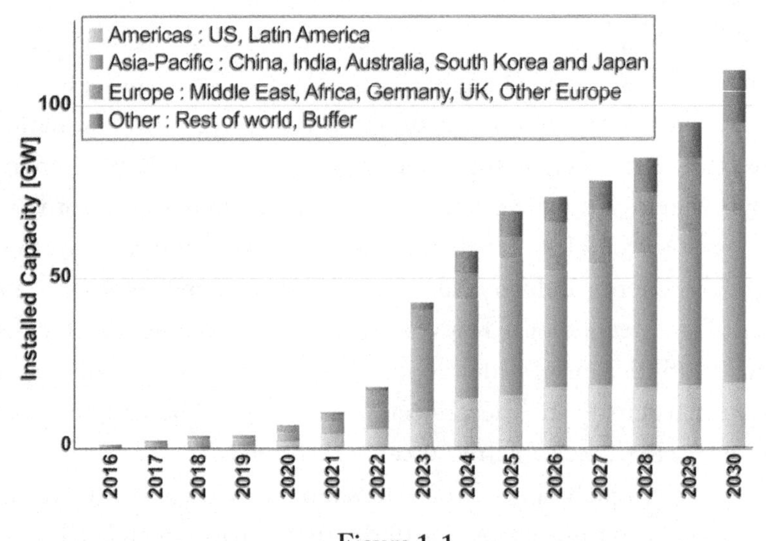

Figure 1-1
Historic and projected global energy storage capacity additions [2]

In economic terms, storage has also become a juggernaut. As of 2023, it was a $37 billion annual business [3], and one in which many different parts of the economy have become involved: manufacturers who make the cells, modules, and racks that go into systems; software providers who build the controls; contractors who install the systems; developers who plan and execute projects; and the banks and investors who finance them. Additionally, the story of batteries is a global one – one recent BESS project installed in California featured cells made in

China using lithium from Chile, assembled into containers in Mexico, and shipped for final installation in the US with a Spanish-made PCS and Korean controls system!

Despite their growth, batteries still represent a small portion of the global power generation – for example, if the roughly 90 GW of batteries in operation in 2023 were cycled once every day, they would collectively discharge only 0.4% of the world's annual energy consumption! However, by 2030 the expected total storage capacity will rise to over 600 GW, while energy demand is expected to rise around 8% [4], so by that point BESS projects can supply 2.6% of all energy demand. Although these numbers seem small, having this flexibility in the grid enables the grid to be versatile in ways that were never possible before batteries became available.

Energy storage projects are being deployed in nearly every country in the world. The US and China were the sites of many of the first deployments, but in recent years South Korea, Germany, the United Kingdom, and Japan have all seen large upticks in projects. The deployment of energy storage is primarily driven by the regulatory environment, as will be discussed in Chapter Safety and Environmental Considerations. This book is focused on the energy storage boom in the US, which has been one of the leaders in the deployment of BESS technology, as well as in innovations in battery markets and financial incentives.

One note on terminology: the term BESS is the most common, although some manufacturers prefer to use the term **ESS**, or **Energy Storage System**, particularly when distinguishing from electric vehicle applications. Some others refer to these projects as stationary storage, to distinguish it from batteries used in electric vehicles. Technically ESS is the broadest term, since it encompasses non-battery energy storage technologies, some of which are covered in Section 11.2 on the future of energy storage. For the purposes of this book, we will use the term BESS, since we are exclusively concerned with energy storage using batteries.

1.1 Front or Behind the Meter

The growth in battery projects includes a wide range of project sizes, from individual units hanging in residential garages, up to large battery facilities with hundreds of containers of batteries. One major distinction among types of battery projects is whether they are **front-of-the-meter (FTM)** or **behind-the-meter (BTM)** installations. This naming convention refers to the location of the project, from the customer's point of view.

BTM projects are energy storage products installed behind a customer's electrical meter, meaning that they are installed in parallel with a commercial, industrial, or residential installation. When a BTM battery charges or discharges, its power may go onto the grid, but it may offset the consumption or production of the business or residence it is working alongside. For example, during periods of peak demand, a BTM BESS at a factory might discharge to reduce the power draw required in peak periods, reducing the electrical bill (this is known as **peak-shaving** and is covered in depth in Section 6.2).

In contrast, FTM projects have their own interconnection to the project, and hence sit in front of the meter of any business or residence. When an FTM project charges or discharges, its power is going directly into or out of the grid. These projects still have meters installed to measure the power flowing into or out of the grid, but they exist to meter only the battery facility, not any associated industrial or commercial loads.

The scale of BTM installations varies from a few kilowatts (kW) of power in residential projects, up to 50 MW for the largest BTM installations at industrial facilities. In contrast, the smallest FTM projects are typically around 1 MW, and the largest range up to 500 MW or more. As of this writing, world's battery facility under with the largest power rating is the Waratah under construction in Australia at 850 MW, which is covered in depth in a Case Study in Section 13.1.

This book is focused on utility-scale projects, which are almost exclusively FTM because they provide services directly to the grid at high voltage. Some of the larger industrial BTM installations may be considered utility-scale, but they are rare compared with utility-scale FTM projects. While the technologies powering utility-scale systems are similar to those found in home and commercial / industrial battery systems, the revenue streams and value propositions of utility-scale projects are distinct. These topics are discussed in greater depth in Chapter Project Development on BESS project development. Figure 1-2 from McKinsey [5] shows the deployment of energy storage systems and the divide between BTM and FTM assets.

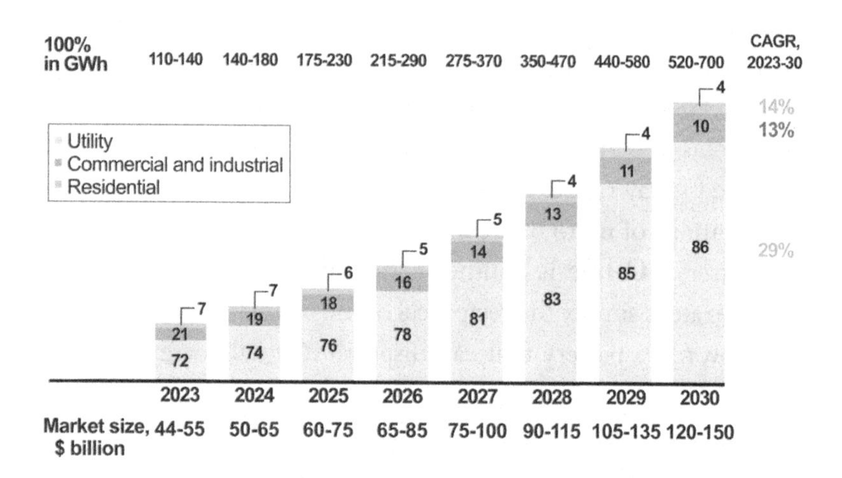

Figure 1-2
Global BTM and FTM installations [5]

1.2 Power, Energy and C-rate

A note on nomenclature: throughout this book batteries are referred to by their **power rating** (kW or MW), their **energy rating** (kWh or MWh), and their **C-rate**, or charge rate, which are all defined below. Most power plants are typically referred to by their maximum power rating. This rating may be dispatchable, in the case of a natural gas plant, or intermittent, in the case of a wind turbine. Batteries are closer to dispatchable assets, since they can send power into the grid, or absorb power from the grid, at a moment's notice, provided their **state-of-charge (SOC)** allows them to do so. All batteries have a maximum power output, which is typically limited by the maximum current the battery cells can tolerate. The capacity of most batteries is measured by the amount of energy that can be dispatched at the battery's full power output. For example, if a battery has a rating of 1 MW / 4 MWh, that means that if it is fully charged, it can output 1 MW of power into the grid for 4 hours. Alternatively, if totally empty, the same battery can charge 1 MW of power from the grid for 4 hours. The time it takes to discharge a battery at its full power rating is known as a battery's **duration.** The battery from the prior example would be referred to as a 1 MW battery with a 4-hour duration (although it may be more simply referred to as a "4-hour battery"). The duration of most Li-ion batteries installed today ranges from 1 to 4 hours, though falling costs have made durations of up to 6 hours economically feasible in some cases.

A battery's **C-rate** is another common term used to quantify its output. C-rate is simply the reciprocal of the battery's duration in hours. In other words, a battery that can dispatch its total energy capacity in 1 hour is a 1C system, while a 2-hour battery is a 0.5C system. Batteries that have lower C-rates (less than 0.5C) are often referred to as "high-energy" batteries, since they allow for larger stored energy relative to the power output. Conversely, batteries with C-rates over 0.5 are "high-power" batteries, since they discharge their energy quickly at higher power. Some batteries have different C-rates for charging than they do for discharging, but most offer the same power rating. This is discussed in depth in Chapter 3 Battery Basics. These concepts are critical to understanding how batteries, both large and small, function.

One common way to visualize the concepts of power and energy in a battery is to imagine the battery as a tank of water, with a valve at the bottom for filling and emptying. The amount of energy in the battery is analogous to the amount of water in the tank – every tank has a specific capacity, and can accept no more water when full, or provide no more water when empty. The flow rate into or out of the tank corresponds to the power of the system. A battery that was able to output a high amount of power would be like a tank with a large pipe allowing water to move into or out of the system quickly. In contrast, a tank with a narrow pipe would be analogous to a battery with a low power output (or input).

1.3 The Goal of this Book

This book aims to help the reader gain an understanding of the BESS industry as a whole, with a focus on the practical aspects of how systems are being built today. After reading this book, the reader will have a basic understanding of:

- The key concepts of Li-ion battery energy storage technology
- How that technology is being deployed in utility-scale systems
- The main economic drivers of energy storage projects, and how they are financed
- How energy storage projects are built and operated
- The key players in the global energy storage industry

Lastly, the book will cover how anyone can get involved in the industry. Energy storage is a tremendously diverse field, with its topics ranging from the atomic structure of lithium cells to the selection of proper personal protective equipment (PPE) to battery maintenance schedules to complex financial instruments used to price battery services. As of 2023, approximately 73,000 people worked full-time jobs in energy storage in the US across a vast array of different careers [6]. The content in this book is presented as a broad overview of the industry, its major trends, and principles. It is intended to be useful for experienced energy storage professionals and newcomers alike.

Batteries have a crucial role to play in the global effort to reduce carbon emissions to combat climate change. As we rise to the challenge of decarbonizing our economy, energy storage's biggest role will be enabling the grid to transition from fossil fuels to renewables. While there is some debate about whether batteries are a green technology (Section 10.5 discusses this in depth), their key role to store and dispatch energy as needed; they are a crucial partner to large-scale solar and wind power, which provide cost-effective but intermittent energy sources. Environmental impacts of batteries are discussed throughout the book, especially as they relate to the reduction of carbon emissions.

While performing research for this book, the authors noted that while there are many specific resources on different aspects of the battery industry, there was no centralized, holistic knowledge base for Li-ion battery systems in real-world applications. Each of the authors has been on the front lines of battery storage, advising on some of the world's most prominent transactions, technologies, and projects to date. The intention of this book is to share the knowledge of the field and how it has evolved, encouraging more study, observation, and participation in the industry as it continues to grow.

While the text covers many abstract topics, it is intended to be a useful guide to how projects are really being built, drawing heavily on real-world examples the authors have encountered in the field. The more everyone in the energy industry understands energy storage, the better prepared we will be for the upcoming grid revolution. The goal of this book is to help spread that knowledge.

Throughout the book we have relied on facts, figures, and data to help make the concepts more easily understood, and to ground the theory in the real happenings in this industry. While we have done our best to document sources and attribute our work, the fast-changing nature of the battery world means that there may be some aspects of the book which become out of date. We would recommend that the information in this book be combined with the latest battery news and data.

1.4 Structure of the Book

Following this introduction, Chapter 2 provides a brief history of battery energy storage, starting from the invention of the battery cell in 1799 to the 2020s, what Bloomberg has called the "decade of energy storage."[1] While the focus of this book is on practical applications, the path of previous development is critical for understanding the future path of batteries.

Chapter 3 introduces battery building blocks, or the basic elements contained within all batteries. Li-ion cells and the chemistries powering them are covered first, including their parts, types, and manufacture. The chapter discusses how cells are built up into battery modules and racks that work together to power a battery system. Non-lithium chemistries are covered in brief, along with alternative energy storage technologies.

Chapter 4 gives an overview of the other main components of a fully functional BESS project. This includes the integrated BESS product, a power conversion system (PCS, sometimes referred to as 'inverter', its main component) along with the enclosures, collection systems, thermal management, and metering systems that let the BESS project operate safely and effectively.

Chapter 5 covers BESS controls, the hardware and software systems that allow the battery to safely and efficiently charge or discharge their power onto the grid. This includes the battery management system (BMS), energy management system (EMS), plant controllers, and fleet-level controls. Cybersecurity and virtual power plants are also covered in this section.

Chapter 6 is on BESS project development, exploring the economic drivers powering the new wave of investment in battery projects. This includes covering the main BESS use cases, how battery plants earn profits, and the major BESS development business models. We review the process used by developers to select sites, technology, and markets, and how projects are contracted, permitted, and executed.

Chapter 7 evaluates energy storage finance. Typical finance structures are reviewed, along with how projects are modeled, lenders and investors are selected, and how projects progress from planning to eventual execution. The project finance model is reviewed in depth, along with some of the key government incentives for energy storage and how they affect battery financing.

Chapter 8 focuses on the engineering, procurement, and construction (EPC) of a utility-scale BESS project. This includes system design fundamentals, procurement strategies, and building processes from groundbreaking to commercial operation. Typical codes and standards are covered here along with the standard procedures for testing and commissioning of a BESS plant.

Chapter 9 covers the operations and maintenance (O&M) of a typical battery facility after it has been connected to the power grid. This includes both preventative and corrective work required to keep a BESS running smoothly. This chapter covers the typical contracts that govern the operation of a BESS, along with battery warranties and guarantees.

Chapter 10 discusses safety and environmental concerns about batteries. This covers the dangers and safety incidents that come with BESS operations, along with how system hardware, software, and procedures have evolved to mitigate these risks. This section reviews all current and forthcoming codes and standards that apply to the battery industry. Lastly, the socio-political and environmental effects of battery projects are covered.

Chapter 11 is on the future of energy storage and considers forthcoming trends in both Li-ion batteries and other forms of energy storage, such as how storage will work with electric vehicles (EVs), and how regulatory markets may adapt to incentivize battery storage.

Chapter 12 discusses the growing battery workforce, including all the key jobs that comprise the sector, covering manufacturing, development, construction, O&M, and finance. This section discusses the more common pathways for anyone to get involved in the energy storage industry.

Throughout the book, there are terms **in bold** – these indicate key terminology that is used in the BESS industry.

1.5 The Usual Suspects

Every chapter concludes with a section called "The Usual Suspects," detailing the leading players in that section of the industry. While the authors have worked closely with many players in the industry, we are committed to writing this as a technology-agnostic resource: we have no agenda and have received no financial incentive to promote any one product, technology, or firm over any other. The lists are all ordered alphabetically and include a summary of the firm and their relative size or specialization within the industry. This information is intended to be current as of the publication of this book, but may change rapidly as firms arrive, merge, and disband.

Unless otherwise noted, the facts and figures provided in the Usual Suspects are those that have been publicized by the firms themselves. We have made every effort to include data around these firms as best as possible, and in all cases based on publicly available information.

1.6 Case Studies

Chapter 13 is a special section highlighting real-world case studies of three energy storage systems that are being successfully deployed in the field. For each, we have detailed the technology, application, location, and process of financing, construction, and operation, to the extent that information can be publicly shared. The goal is to help demonstrate how the concepts in the text are deployed in a real-world context. The projects are in various stages of development and construction and cover the most common current applications for energy storage. They are:

- **Standalone BESS**: An 850 MW / 1680 MWh outdoor BESS to the north of Sydney, Australia. As of this publication, this project is the world's largest plant by power rating. The project is being developed by Akaysha Energy, an Australia-based energy storage developer.
- **PV+Storage**: A 150 MW solar + 137 MW / 548 MWh BESS located south of Fresno, CA. The project is currently operational and was developed by RWE Clean Energy, the US division of the German multinational energy company RWE.
- **Commercial / Industrial Storage**: A grid-connected PV+Storage plant, developed by the utility Seattle City & Light. This is an operational system that was built in 2021 consists of a 200 kW roof-mount solar PV plant coupled with a 200 kW / 800 kWh BESS, located in Seattle, WA. This system typically follows the power grid but can also operate as a microgrid in the case of a power outage.

Whether the reader comes to energy storage as an investor, hobbyist, student, or climate warrior – this book has something to share. By the end, the reader will understand the benefits, applications, and limitations of energy storage. While the book is structured to be read straight-through, it may also be useful to focus on the sections, case-studies, and principles that are most applicable to the reader's work or interests. However the text is approached, we are glad that the reader's interests have brought them to these pages, which will cover a fast-paced, fascinating, high-tech, and high-dollar industry that is core to the green energy transition.

And now – we charge on!

2. A BRIEF HISTORY OF ENERGY STORAGE

Batteries seem to be ubiquitous – from our mobile phones to our cars, our cordless drills and satellites, this technology has changed the lives of nearly everyone on the planet. But why was it that this technology, which has been around for hundreds of years, has so recently come into our lives in such a big way? What changed, or what was holding it back?

This chapter reviews the backstory that led to the recent burst of innovation. Although the book is focused on the cutting edge of energy storage systems, and their future, understanding the history of batteries is critical to understanding its strengths, weaknesses, and future.

We will start by summarizing the key innovations that led to the creation of the first battery cells, from the late 1700s to 1900. The chapter then turns to how battery cells evolved throughout the 20th century, which will provide the technical and economic conditions that set the stage for today's boom. Lastly, the chapter covers how lithium-based cells have evolved since their discovery in the 1980s to today.

2.1 Volta and His Cell

The modern battery was invented in 1799 by Italian physicist and chemist Alessandro Volta (1745-1827). Volta was one of Italy's greatest scientists, revered by both the public and academia, and a leading light in the era's discoveries in energy and power.

In Volta's time, electricity was poorly understood and had not yet been channeled to any practical use. It streaked across the sky as lightning, swam in electric fishes, and sputtered and sparked in parlor room experiments. Some previous physicists had begun to understand its properties, such as Georg von Kleist and Pieter van Musschenbroek, who independently invented the Leyden jar (1745), in which a spinning chain stored static electricity in a bottle. In that same era, Benjamin Franklin's famous kite experiment (1747) showed that lightning was composed of electricity. In 1791, Italian Luigi Galvani discovered that electricity was passed through the body in neurons.

Building on this previous work, Volta discovered that discs of different metals, bathed in brine or acid would result in an "electromotive force" between the two metals. After an argument with Galvani about the source of the force, Volta built what he called a "voltaic pile," a predecessor to the modern battery, shown in Figure 2-1.

Figure 2-1
Voltaic Pile

The pile consists of sheets of zinc and copper in a tube, separated by a cloth soaked in a simple saltwater brine. Volta demonstrated that terminals connected to either end of the pile would create an electrical current when touched together. The current was weak, the piles degraded quickly, and there was no way to recharge the battery, but Volta had demonstrated a key concept – that an electrochemical reaction could be used to store electricity. Up until that point, only static electricity had been recreated in a lab.

Today, we know the voltaic pile as a battery cell. Although modern batteries use far more complex materials, hold vastly more energy, and can be cycled thousands of times, the fundamental principle and components are the same: a cell composed of different metals, an electrolyte, and two terminals. This discovery, soon shared across Europe, provided the foundation for the exploration of electrochemical batteries. Volta's name is preserved in our name for electromotive force: voltage, which is measured in the SI unit "Volt."

As the 19th century progressed, other scientists built on Volta's discovery. Hans Christian Ørsted first discovered there was a connection between electricity and magnetism (Denmark, 1820). Later, Michael Faraday discovered electro-magnetism (England, 1821), Georg Ohm defined the relationship between power, voltage, and resistance (Germany, 1826) and J.C. Maxwell defined the mathematical laws relating magnetism and electricity (Scotland, 1860s).

These discoveries paved the way for the first electrical grid, built by Thomas Edison in 1878 in New York City, followed soon thereafter by Nikola Tesla's discovery of alternating current (AC) in 1888. These technologies form the basis of our power grids today.

2.2 Cell Progress

The principles of direct current (DC) battery technology discovered by Volta progressed gradually throughout the 19th century. In 1836, John Frederic Daniell created a cell with two electrolytes. In 1859

French physicist Gaston Planté connected lead plates and sulfuric acid in lieu of the original copper, zinc, and brine solution, creating the world's first lead-acid battery. By sandwiching flexible plates and rolling them together, the battery could be made more compact: an arrangement known as the **jelly roll** that still exists today in both lead-acid and Li-ion batteries. In 1866, French engineer Georges Leclanché patented a battery cell that used both zinc and carbon as anode materials. A dry cell battery, the first to use a paste, rather than liquid, electrolyte was created by Carl Gassner in 1886. This cell was later the first to be mass produced by the National Carbon Company of Cleveland, Ohio, in 1896.

American physicist Gilbert Lewis performed the first experiments with lithium batteries in 1912 at the University of California, Berkeley, but the technology was not explored until the 1960s by Robert Huggins and Carl Wagner in their studies on ions. In the 1970s, Huggins and Wagner discovered the principle of **intercalation**, which allows insertion of an ion into a layered sheet of molecules. Building on their research, M. Stanley Whittingham, in research at ExxonMobil in 1976, observed a charge-discharge cycle with a rechargeable battery cell using a lithium metal cathode and titanium sulfide anode. Although the research was discontinued due to the high cost and noxious materials, the core concept discovered by Whittingham was the first example of a Li-ion battery.

2.3 The Lithium Revolution

In the 1980s, Wittingham and the researchers John Goodenough, and Akira Yoshino began independently experimenting with new types of lithium electrodes. These batteries showed great promise in that they have higher energy density than their predecessors, meaning there is more energy stored per unit weight or volume. Table 2-1 shows the relative densities of different types of energy storage technologies. While Li-ion batteries cannot hold the energy of oil or coal, they are significantly more energy-dense than lead-acid and nickel cadmium batteries.

Table 2-1
Comparison of energy density by technology [7]

Category	Energy storage	Typical energy densities [MJ/m³]
Conventional fuel	Crude oil	37,000
	Coal	42,000
	Dry wood	10,000 – 16,000
	Ethanol	22,000
Electrochemical energy storage	Lead-acid batteries	100 – 900
	Nickel-cadmium batteries	350
	Lithium-ion batteries	1,400

The new Li-ion battery cells appeared to have less of a memory effect, or a loss of maximum energy capacity caused by partial discharge, than cells with other chemistries. This meant that they could be charged and discharged many more times before losing their effectiveness. As with lead-acid, Li-ion batteries also gradually lose their effectiveness in a process known as degradation, which is discussed at length in Chapter 3.9. However, the number of cycles available from Li-ion batteries in their useful life appeared to be potentially in the thousands, rather than the hundreds.

Sony produced the first commercially available Li-ion battery in 1991, and they soon became widespread in consumer electronics, including remote controls, cordless drills, and the two rising titans of the 1990s: laptop computers and cell phones. Li-ion cells have continued to evolve, as new breakthroughs in material science and manufacturing processes have driven down the cost of batteries and the energy density values. In 2019 Whittingham, Goodenough, and Yoshino were awarded the Nobel Prize in Chemistry for their pioneering work in Li-ion batteries.

In the early 2000s, EVs were one of the first large-scale markets that began to take advantage of the falling costs and rising densities of battery cells. Although the dream of electric cars had been around since the beginning of the 20th century, the low densities, difficulties in building charging infrastructure, and resistance from entrenched industry groups limited their growth. Starting in approximately 2005, public companies such as Tesla began to make serious moves to bring EVs to market with range and pricing that had the potential to rival internal-combustion vehicles. Other manufacturers followed suit, driven by the continued trend of falling costs and rising cell density. Today nearly all major vehicle manufacturers have some form of EV in their fleet, led by BYD, Tesla, and Volkswagen, along with corporate alliances such as Renault-Nissan-Mitsubishi. Tesla had an outsize role in rapidly advancing EV technology, and popularizing EVs in a way that had never been done before (and which propelled Elon Musk to become the wealthiest person on the planet – at least for now). The technology powering Tesla's success in the EV market was and is very similar to the tech at the heart of BESS projects.

One of the first grid-connected Li-ion battery projects was installed by AES in the US in 2008 in Lyons, Pennsylvania: a 1 MW system using Altairnano batteries. Similar projects were being built in China around this time. Government incentives led to a large uptick in such projects around the world, and the US Department of Energy described 2009 to 2014 as a "period of ferment, in which diverse chemistries were tried out in diverse applications, often with risk-sharing support from government agencies and incentives from regulators" [7]. In 2015, Li-ion became the dominant technology for grid-connected energy storage, driven by its use in frequency regulation (Chapter has more information about these value streams).

While there are many other types of batteries still in use today, Li-ion has eclipsed all other rivals in terms of MWh of deployed projects and, more importantly, in lenders' acceptability of the BESS project risks. Since 2015 the market for Li-ion BESS has grown exponentially and is expected to continue to grow as prices fall, economies of scale are achieved, and demand continues to increase.

It is notable that as of today, batteries are not the leading energy storage technology on the grid. That prize goes to **pumped hydro energy storage (PHES)**, a technology that uses pumps to store energy in the water as potential energy. However, the gap between PHES and battery storage is shrinking: in 2020 the BESS capacity was approximately 11% of the PHES capacity. By 2023 it was 48%, by the end of 2024 it is expected to be approximately 80%, and by 2025 batteries will almost certainly overtake PHES as the leading form of energy storage.

Battery technology is the focus of research and development in thousands of academia, government, and industry laboratories. There will continue to be breakthroughs in these areas, some of which may soon present safer, competitive alternatives that overtake Li-ion. Chapter 11, the Future of Energy Storage, presents some of the most promising candidate technologies discovered to date. For the time being, Li-ion batteries are the leading stationary energy storage technology. The next chapter takes a closer look at the key components involved in today's Li-ion BESS projects.

3. BATTERY BASICS

Upon visiting the site of a new utility-scale battery system, there will be a series of white metal boxes, neatly arranged on concrete pads. There will be a faint hum coming from some of these boxes, and the occasional click or spinning fan. Underground conduits typically connect the major equipment, so there are few to no visible wires, except for where the cables emerge to connect to the substation or where overhead wiring is required. There may be a logo on the sides, or plates to indicate the numbering of the equipment, but from the outside there are usually no screens or dials on which to see operational data. For all the hype about battery systems, this collection of boxes may even seem a bit boring! Despite its appearance, this plain-looking installation may be moving hundreds of megawatts of power onto and from the grid, and as it hums and clicks it may be earning thousands of dollars in revenue every hour for the project owners. Figure 3-1 shows an aerial view of a utility-scale battery system, from the 185 MW / 565 MWh Kapolei BESS in Oahu County, Hawai'i. This photo was taken by Plus Power, the developer of the project.

What is happening inside these boxes, and what are the core components that make up the system? Battery systems come in different shapes, sizes, makes, and models, but share several similar components. On the project site they are hidden within layers of enclosures, linked by miles of cabling and ductwork. The next three chapters focus on explaining these building blocks, and how they work together to create a BESS:

Figure 3-1
Battery system layout [8]

- **Chapter 3** covers the basics of battery technology, and it is built up into cells, modules, and eventually racks –the components at the heart of every battery system.
- **Chapter 4** covers a fully functional BESS project – the systems that allow the battery and its controls to function as a cohesive unit.
- **Chapter 5** describes battery controls – the key hardware and software systems that allow a DC battery to connect to an AC power grid safely and effectively.

After reviewing the basics, we will describe the more complex systems that allow these components to work together to perform their functions in energy storage markets and other applications.

3.1 Battery Basics

Batteries are a technology that is familiar to most people. From the batteries in our phones to our cars to our cordless drills, batteries are a common ingredient in modern life. But what is happening inside the cylindrical cells or pouches hidden in our devices? Although they come in different shapes and sizes, most cells used in large-scale energy storage have the same principles at work. This section explains those key concepts from the molecular level on up to the modules and racks that make up a modern energy storage system. While the focus will be on utility-scale systems, the components discussed are also present in any residential, commercial, or industrial Li-ion BESS.

A brief note on battery types: this book is focused on Li-ion batteries, because that is the most commercially viable BESS technology today. There are many other types of cells available in the marketplace, and promising new developments in battery technology seem to be announced on a daily basis. However, this book focuses on the technologies that are most commercially viable, meaning tech that can attract large-scale financing from commercial banks and institutional investors. By that measure, Li-ion is, by far, the most commercially viable battery energy storage technology being deployed today. If interested in other forms of energy storage, Chapter 3.7 explores non-lithium battery technologies. Some of these are rising quickly and may one day overtake Li-ion.

While reading these sections, it is important to remember that the battery world is constantly changing. Given the exponential growth in the sector, large investments in research and development are being made by laboratories, universities, and governments around the world. So far this decade, the industry has grown faster than even the most aggressive predictors forecasted. This chapter discusses the latest technology as of publication of the book, but we encourage anyone interested in the sector to continue their research using the many reference sources, industry publications, and conferences to stay abreast of the latest developments.

3.2 The Battery Cell

At their core, batteries are a technology used to turn electrical energy into chemical energy and back again. While a beaker full of chemicals storing energy may sound odd, it is a concept that all of us are familiar with. One common example is in liquid fuels such as gasoline and diesel, but alcohol, hydrogen, and all fossil fuels retain energy the same way: in the form of chemical bonds holding their molecules together. When these chemicals are burned, they release their energy, powering the engines in our lawnmowers, tractors, and convertibles. Once the energy is released it is gone—converted into heat, light, or motion. A barrel of oil cannot be un-burned; this is because the chemical energy released by oil is an irreversible process; the energy can be released from the molecular bonds but not returned to the bonds without requiring energy input. The byproducts of this reaction, such as carbon dioxide (CO_2), are some of the primary drivers of climate change, and are also known carcinogens.

By contrast, a rechargeable battery cell holds energy in a process that is reversible. We can put electricity into the cell by inserting electrons, charging the battery, or we can remove electrons to discharge the cell. Each of these processes is the result of a chemical reaction happening inside the battery cell. Many of the leading cell chemistries are trade secrets that comprise the heart of the intellectual property of cell manufacturers, and have been patented by private companies, governments, or individuals.

The cells of most Li-ion batteries consist of a thin-layer structure with four main components:

- A **cathode**, which corresponds to the positive terminal of the battery
- An **anode**, which corresponds to the negative terminal of the battery
- An **electrolyte**, which allows ions to flow between cathode and anode
- A **separator**, which is a thin layer of polymer preventing electrical contact between electrodes, while allowing passage of the Li-ions.

The structure of a typical Li-ion cell is laminar, in the sense that the electrodes and separator are thin layers of few micrometers – as opposed to lead-acid batteries where the plates are a few millimeters thick. The layers are made of a metal sheet, typically copper or aluminum, coated with active materials. Figure 3-2 shows a close-up view of a typical Li-ion laminar structure with the anode, cathode, and separator layers.

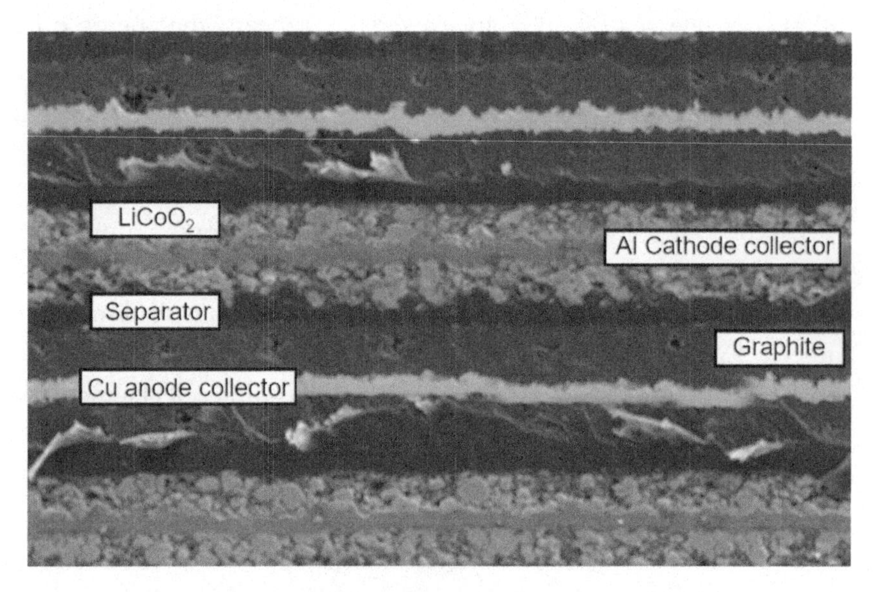

Figure 3-2
Close-up of laminar Li-ion battery [9]

The key principle in the individual layers of the cell is intercalation, which is the process by which Li-ions can fit within the lattice structure of the anode and cathode. Figure 3-3 shows a schematic version of a typical Li-ion cell.

When the battery is **discharging** and providing an electric current, the anode releases the ions, shown in red, to the cathode, shown in green. This generates a flow of electrons between the positive and negative terminals.

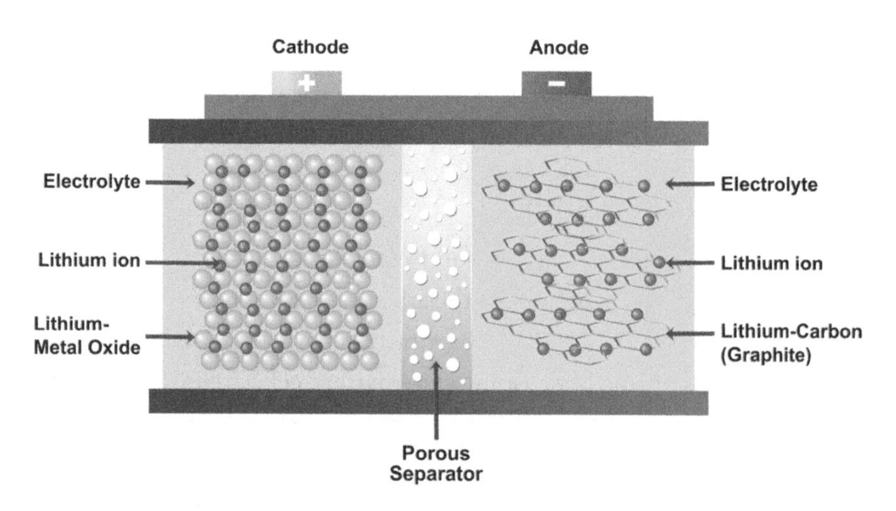

Figure 3-3
Li-ion cell structure [10]

When the cell is **charging**, the opposite happens: ions are released by the cathode and received by the anode. This absorbs a flow of electrons from between the positive and negative terminals.

These charging and discharging chemistries are the heart of the Li-ion technology – whether within a 10 watt-hour cell phone battery or cells that make up a 5 MWh containerized BESS.

When we talk about the **chemistry** of a battery, this is generally referring to the type of cathodes that are used, which are made of advanced materials designed to form efficient structures in which lithium ions can be efficiently stored. This is further explored in Section 3.3. Some are composed of thin layers, some winding helixes, and others of even more complex three-dimensional shapes.

The electrolyte is a liquid or gel which facilitates the electrochemical reaction by allowing ions to flow between the active materials of the two electrodes. Electrolytes are usually made of a lithium salt dissolved in an organic solvent. These differ from batteries which use other chemistries, like lead-acid batteries, which use an aqueous acid solution, usually sulfuric acid (commonly known as **battery acid**).

The anode of a Li-ion battery is a thin sheet of metal, usually copper, which is coated by graphite, and in some cases silica. Many energy breakthrough announcements tout novel structures of the anode, into which Li-ions can be more tightly packed.

Recently there have been developments in a **solid-state battery cell**, which is a cell that uses a solid electrolyte rather than a liquid one. This type of cell would have the advantage of avoiding flammable electrolytes, as well as allowing for a faster charge or discharge. As of the publication of this book, solid-state battery cells are just beginning to be deployed commercially in EVs or stationary storage.

Figure 3-4 shows the processes in greater depth – the electrons moving between cathode and anode on the external circuit, and Li-ions moving between their internal structures.

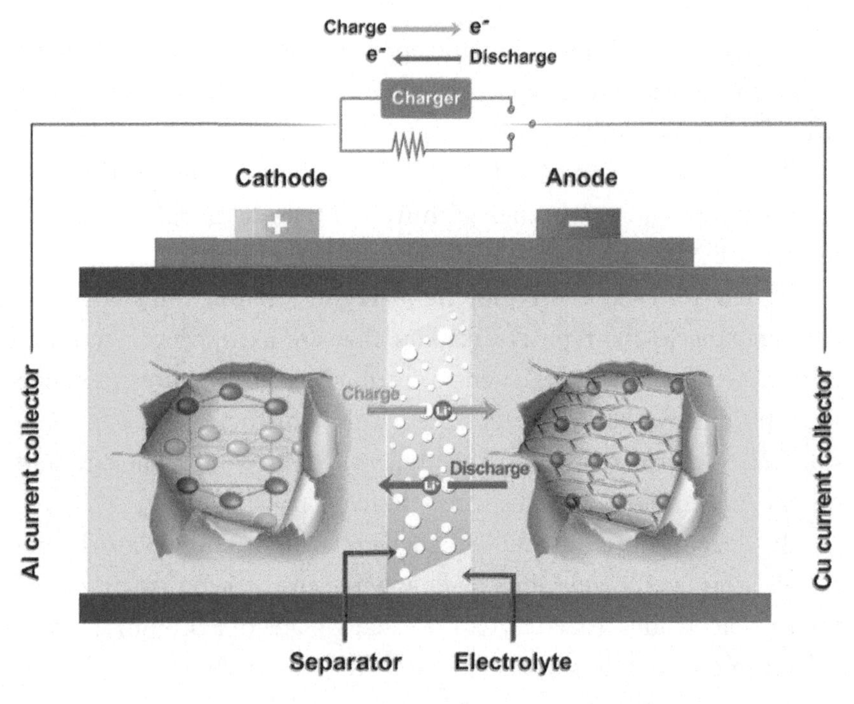

Figure 3-4
Charging and discharging of Li-ion battery [11]

There are a few common battery cell shapes that are used to make the flat sheets of the chemical reaction into a useable unit of energy storage. These shapes are referred to as the battery's **form-factor**. The most common form-factors are:

- **Cylindrical**: This resembles a large AA battery. This is often referred to as a jelly-roll construction since inside the metal shell there is tight coil of anode, cathode, and separator layers.
- **Pouch**: As the name implies, this is a flat pouch that houses layers of electrodes and separators. This is the battery type used in most mobile phones and laptops.
- **Prismatic**: The prismatic cell is a boxy enclosure that contains a pouch with many stacked layers. These cells may include clamping, which compresses the cells to avoid swelling. This is the preferred form-factor for many large-scale lithium iron phosphate (LFP) battery cells.
- **Blade:** A newer type of cell, blade cells are long and narrow forms of prismatic cells. Certain manufacturers, such as BYD, have preferred this architecture and used it to increase their densities, in some cases exceeding 6 MWhDC in a 20-foot container.

Many cylindrical cell shapes follow a standard form factor, such as 18650 for cylindrical cells (referring to their 18 mm diameter and 65 mm height), or typical dimensions for prismatic cells. Some other common sizes are 21700 (22 mm diameter x 70 mm height), or Tesla's proprietary 4680 cells, which are 46 mm diameter and 80 mm height. Figure 3-5 shows common sizes of cylindrical cells, drawn to scale. AA and AAA batteries are uncommon for Li-ion cells but are shown for size reference only.

Figure 3-5
Examples of different cylindrical battery sizes [12]

Most utility-scale BESS products are built around prismatic cells. These cells vary widely in their size, since their amp-hour ratings (more on this in Section 3.4) may vary anywhere from 100 to 600 amp-hours. Figure 3-6 shows some typical prismatic cell sizes for reference.

Across various form factors, standardized sizing enables battery integrators to become **cell-agnostic**, meaning they can utilize different battery manufacturers, or even chemistries, with minimal headaches, as required based on market trends. In contrast, some cell manufacturers

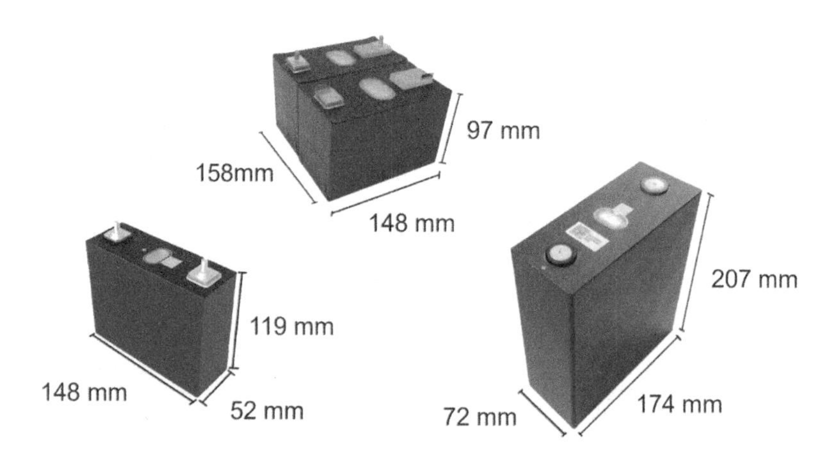

Figure 3-6
Typical sizes of prismatic cells

have produced cells with unique dimensions. While this can provide a certain cachet or an edge in technical capabilities, it can also complicate the efforts of integrators who wish to offer multiple cells for a product without varying the size. The largest integrators may partner with a manufacturer to produce a proprietary cell – for example, the Megapack 2 XL product uses a specially-produced CATL battery cell that is manufactured specifically for this product.

3.3 Cell Voltages, Currents, and States of Charge

A battery is designed to operate in an ideal voltage range. When it is fully charged, the battery will show a higher voltage, while when it is fully discharged, it will show a lower voltage. On the cell level, the most common voltages are a fully charged cell from 4.0 to 4.5 VDC, and a fully discharged cell from 2.0 to 2.5 VDC. The amount of voltage typically correlates to a percentage value, indicating what fraction of its total stored energy a Li-ion cell has available for discharge. This is referred to as a cell's **state of charge**, commonly abbreviated as SoC.

The relationship between voltage and SoC is not linear – a typical curve showing voltage vs. SoC is shown in Figure 3-7. This shows the common S-shaped relationship – a steep voltage drop from 100% to around 75%, a slow voltage drop from 75% to 25% and a steep voltage drop again from 25% to 0%.

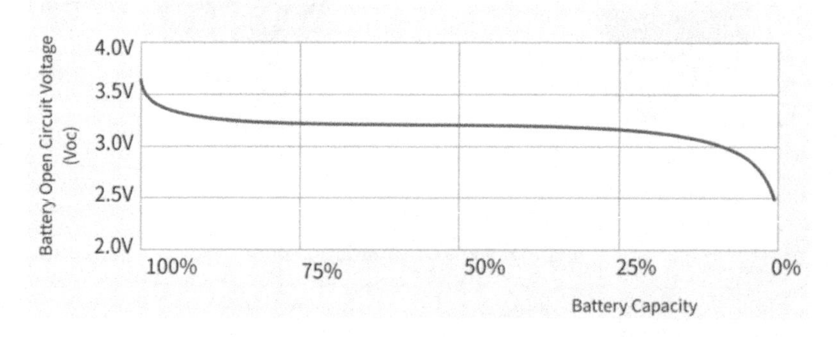

Figure 3-7
Typical Li-ion voltage vs. SoC curve [13]

The average voltage of the cell from 75% to 25% SoC is often referred to as the cell **nominal voltage**. For most utility-scale energy storage cells, the cell nominal voltage is 3.2 V, while some other cells have nominal voltages of approximately 3.6 – 3.7 V. For comparison, most lead-acid car batteries are composed of six 2V cells in series, giving a voltage of 12 V. Every new cell will list its voltage range and nominal voltage, which may have slightly different values.

If a cell is pushed beyond its maximum or minimum voltage, it may be damaged irreversibly. This is one of the reasons most battery cells have a battery management system, or BMS, which protects and keeps the cells inside a safe operating regime and is further discussed in Section 5.1. The manufacturer defines the voltages for each cell which correspond to 0% and 100%, depending on the area in which the cell can operate safely.

Sometimes, instead of SoC, a manufacturer may refer to a battery's **depth of discharge (DoD)**. This is simply 100% minus the SoC, when expressed as a percentage: An SoC of 100% corresponds to 0% DoD, while an SoC of 20% corresponds to a DoD of 80%.

3.4 Cell Currents, Power, and Energy

To charge or discharge a battery, a current must flow into or out of the cell. Since batteries are operating on DC, this means that physical electrons are flowing into or out of the cell. The unit of current is the ampere, or amp (A), and it is common to measure the current flows in the unit of amps over a period of hours, commonly referred to as **amp-hours (Ah)**. One amp-hour is equivalent one amp flowing over one hour. The amp-hour rating can be obtained by multiplying current by time – 0.5 amps over 2 hours is also 1 Ah, as is 100 amps for 0.01 hours (36 seconds). One helpful way to think of an amp-hour is as a "packet of charge," representing a current applied over a unit of time. An amp-hour is a basic unit of charge that is seen across the world.

Since many cells have similar nominal voltages (3.2 to 3.7 V), it is common to list the amp-hour rating of cells, rather than the Watt-hour rating – even though Watt-hour is a better representation of how much energy is contained in the cell. This is why portable electronic battery banks and cells alike are often sold in Ah or mAh ratings.

To calculate the amount of **energy** each battery cell can store, we use two simplified relationships for DC power and energy, with units shown in parentheses for clarity:

$$Power\ (W) = Voltage\ (V) \times Current(A)$$

$$Energy\ (Wh) = Power\ (W) \times Time(h)$$

Combining these gives:

$$Energy\ (Wh) = Voltage\ (V) \times Current(A) \times Time(h)$$

Since, as we saw above, the voltage changes throughout the charging and discharging of a cell, calculating the exact value of the energy involves calculating the power received at each instant throughout the cell's life. A simple approximation of this value can be obtained by using the nominal voltage, since that is, approximately, the average voltage.

Cells in energy storage systems typically range from 100 to 700 Ah. As an example, a cell with a 320 Ah rating and a nominal voltage of 3.2 V would have an energy of:

$$3.2\ V\ X\ 320\ Ah = 1{,}024\ Wh = 1.024\ kWh$$

In this case, the battery has just over one kilowatt-hour (abbreviated as kWh) of energy. For comparison, this is around the same amount of energy contained in a traditional lead-acid battery found in a gasoline-powered vehicle, which commonly has 75 Ah at 12 V, or around 0.9 kWh.

There is some subtlety in these ratings – the amp-hour rating of the battery depends on how much power is drawn. In general, charging or discharging a cell more slowly results in a greater amp-hour rating, while drawing it more quickly decreases the amp-hour rating. In general, the amp-hour rating for Li-ion battery cells is typically given at the rated power. In other words, if the same 320 Ah cell from the prior example was a 0.25C battery, it would allow for charging and discharging at a maximum of 80 A, meaning that it would take four hours to discharge the battery at this power. In case the reader missed the discussion on what a C-rate is, it can be found in Section 1.2. In general, it is more stressful on the cell to be charged, rather than discharged, since charging is putting energy into the cell, while discharging is extracting that energy (and hence going into a more relaxed state). However, the specifications of most utility-scale BESS products allow for power charging and discharging, known as **symmetrical** charging / discharging.

As stationary Li-ion energy storage systems have gotten bigger, so have their cells. Where in the past the standard was 100-200 Ah, prismatic cells, nowadays 500 - 600 Ah cells have become common. In 2023, leading LFP cell manufacturer EVE introduced a cell called "Mr. Big," which had 628 Ah – meaning that each cell has over 2 kWh of energy. While these larger systems mean fewer cells, they also have brought new safety concerns – the individual energy in a cell has no way to be cut off by a fuse or other device, since these devices can only be installed on the lines connecting cells. As we will see, modern BESS products consist of hundreds of cells in series, resulting in voltages over 1000 VDC.

As with all batteries, we can combine cells in series and parallel to produce higher voltages and/or currents. When we put two batteries in **series**, the voltage is additive – so two 3.2V cells connected this way give an overall voltage of 6.4V. When we put two cells in **parallel**, those cells have the same voltage, but can supply twice the current. As we will see in Chapter 8 when looking at battery engineering, most BESS projects have racks consisting of hundreds of cells in series, with the racks and enclosures later connected in parallel.

The currents produced if BESS projects are subjected to a short-circuit can reach into many thousands of amperes, enough to melt the battery terminals or vaporize the insulation around the wires. Because of this, large diameter conductors are used to connect the battery cells and racks to one another. BESS project safety is covered in full in Chapter 10.

The above examples might seem to imply that the charging and discharging happens at a constant amperage. It is important to note that there are several ways to charge or discharge a battery. Some of the more common methods are listed below. Exploring this topic in depth is beyond the scope of this book, but these terms describe three of the most common charging / discharging methods used in BESS projects:

- **Constant power (CP)** charge or discharge: In this mode, a cell is either charged or discharged with constant power. This is achieved by varying the amount of current flowing into or out of the cell, to maintain a constant power rating. This is the most common method for BESS charge and discharge, since it will keep the power (typically in MW) at a steady level.
- **Constant current (CC)** charge or discharge: In this mode, a cell is charged with a constant current, regardless of voltage. Since the voltage changes throughout the charge process, this means that the amount of power will vary throughout the charge or discharge cycle. This is often done early in the charge cycle.
- **Constant** voltage (CV) charge or discharge: This mode is often used at the end of a charge or discharge cycle. In this mode, the system's discharge happens at a constant voltage, meaning that the amperage (and hence power) will vary. CV charging is often used for topping up a battery cell at the end of charging, often referred to as a **trickle charge**.

It is common to see a charge method described as a combination of these methods, such as CPCV – this would mean that the system charges (or discharges) at constant power, until the cell is nearly full (or empty), at which point the system switches to constant voltage to finish the cycle.

3.5 Li-ion Chemistries

The chemistry of a Li-ion battery, which generally refers to the chemistry of the cell´s cathode, is the technology at the heart of the battery system. The cell´s chemistry is one of the main parameters that will determine if a battery is an appropriate fit for any given application. The choice of chemistry has a large effect on the size, cost, and safety aspects of a given battery. Many viable chemistries have been deployed in commercial applications, and there are many subcategories of chemistry within each of these groupings.

Every year millions of dollars and thousands of researchers seek out novel and more effective battery materials. Some of this research is done by private companies, such as CATL, the world's largest battery manufacturer, whose campus features a massive R&D center in Ningde, China. Other research is done by state actors, such as the US research labs Argonne National Laboratory and the National Renewable Energy Laboratory (NREL), both of whom have large battery research departments. Still other research is conducted by universities, many of which are supported by government funding.

The key factors differentiating battery chemistries are their energy density, safety, price, and available C-rate.

- **Energy density**, is a measure of how much power can be packed into a given mass of cell. Energy density is typically expressed in Wh / kg. This metric has been a driving factor for EV batteries, since their acceleration is tied to the weight of the battery. For stationary applications, a bigger driver has been **volumetric energy density**, commonly expressed in Wh / L. This is because while stationary cells can be heavy, it is preferable for them to be compact, resulting in smaller footprints of energy storage products and products.
- Safety is another key differentiator among chemistries. **Thermal runaway** is when batteries go into uncontrollable combustion, with catastrophic effects. NMC (as defined below) shows this effect at a lower temperature than some other Li-ion cells. This makes safety precautions, such as cooling, monitoring, and gas detection key concerns for all Li-ion products. Products with a

higher thermal runaway threshold are preferable, since they can withstand more heat before entering this condition; products with a lower threshold are more dangerous, since they enter this condition more readily. Battery safety, including thermal runaway, is covered in depth in Chapter 10.

- Price is, of course, one of the key drivers in batteries. The most common price metric is in $ / kWh, which measures the value per unit of energy. This value can be given at the cell, rack, enclosure, or project level, so it is important to note where it is being measured. Additionally, the price may be quoted at a specific point in the supply chain, ranging from **ex-works**, meaning quoted at the factory gates, or **DDP**, meaning delivered duty paid to the location where the product is delivered. The prices shown in this section are all quoted ex-works at the factory producing the cells. Prices have fallen dramatically in recent years from over $500 / kWh in 2013 to under $100 / kWh in 2023.
- Lastly, C-rate is a key driver. As covered in Section 1.2, C-rate is a measure of how quickly the energy can be inserted or drained from a cell. High C-rates (0.5C or higher) indicate that a cell can be charged or discharged quickly, while lower C-rates (those less than 0.5C) indicate that a cell must be charged or discharged more slowly. Another way to think about this is that cells that allow for high C-rates (typically 0.5C or higher) are **high-power** cells, while those that allow for lower C-rates (typically 0.25C and lower) are **high-energy** cells.

There are dozens of Li-ion chemistries which have been researched and commercialized. The following sections outline four of the most common Li-ion chemistries used in commercial utility-scale BESS projects today.

NMC - Nickel Manganese Cobalt

The chemistry lithium-nickel-manganese cobalt ($LiNiMnCoO_2$ in molecular terms) is commonly referred to as **NMC**, or sometimes NCM. NMC was one of the first large-scale commercialized chemistries for both EVs and stationary storage and is thus considered the original king of Li-ion.

NMC cells have a cathode composed of nickel, manganese, and cobalt. Cell manufacturers can alter the ratios of these components to increase or decrease specific energy and power, each at the expense of the other.

NMC was originally described by Christopher Johnson, et. al. in 1998 at the Argonne National Laboratory, a leading energy storage research laboratory run by the US Department of Energy (DOE) [6]. Since then, they have made great strides in increasing performance, scaling up manufacturing, and improving safety concerns.

NMC cells offer a relatively high C-rate, when compared to other chemistries. This means they are good candidates in applications where a large burst of power is required (typically < 1 hour duration, that is, having a C-rate > 1C), such as frequency regulation, rather than a longer duration backup battery (typically 2- to 4-hour duration, that is, having a C-rate < 1C). That is one reason why NMC was, and remains, popular in EV applications – the rapid acceleration required in a vehicle needs a cell that can withstand a high C-rate. NMC is the chemistry of choice in passenger EV applications, although in 2021 LFP started appearing in this market as well, primarily driven by falling prices.

Two prominent downsides to NMC cells are their relatively low thermal runaway temperature, and their cobalt content.

Cobalt content is critical because cobalt is a rare mineral that is abundant in the Democratic Republic of the Congo (DRC), where determining the source of its extraction is difficult or impossible. As described at length in Section 10.5, this means that some cobalt is the product of child labor, environmental disasters, and horrific conditions and low pay for miners. Because of this, the industry has tried to move away from chemistries containing cobalt, and to minimize the cobalt content for those that still rely on it.

Key characteristics of NMC cells:

- High relative energy density: 200-275 Wh / kg
- Low thermal runaway temperature: ~ 210 °C, meaning it is a higher safety concern
- Medium relative cost: $80-90/kWh as of October 2023 [14]
- Ethical sourcing concerns related to the use of cobalt

Prior to approximately 2021, NMC was still the leading chemistry for stationary storage, but since then it has ceded its crown to the next chemistry we will look at: LFP.

LFP - Lithium Iron Phosphate

Lithium iron phosphate, or LFP, sometimes referred to by its molecular structure $LiFePO_4$, is the leading chemistry in stationary energy storage systems. LFP is not an all-star in any single performance category but is more of an all-around performer with competitive energy density and degradation, along with leading pricing and safety performance.

LFP cells use iron-phosphate as their cathode, which was discovered and developed by Akshaya Padhi, Kirakoda Nanjundaswamy, and John Goodenough, working at the University of Texas in 1996. Goodenough was one of three awardees of the 2019 Nobel Prize in Chemistry for his work with Li-ion batteries. This chemistry was one of the early frontrunners in battery technology, featured in 18650 form-factor cells produced by companies like A123 and across many fleets of buses in China.

Due to its lower energy density relative to NMC and NCA chemistries, LFP cells were, prior to 2021, favored for stationary applications and heavy transportation like electric buses, rather than in passenger vehicles.

According to the 2024 DNV Battery Energy Scorecard [15], LFP is currently the market-leader, returning after a domination by NMC and NCA chemistries between 2016-2020.

LFP chemistry has a reputation for safety, as its thermal runaway temperatures are typically higher than those of NMC and NCA cells, meaning it can withstand more heat before a catastrophic fire occurs. Given the increasing focus on safety and evolving codes, this feature has given another boost to LFP adoption. Safety is explored in greater depth in Chapter 10, which covers both battery safety and environmental issues.

LFP is notable for a letter it lacks: no C, which means it is not made with any cobalt. The use of cobalt has concerns regarding ethical sourcing of materials, as discussed in depth in Section 10.6.

Key characteristics of LFP cells:

- Low relative energy density: 160 – 200 Wh / kg
- Higher thermal runaway temperature: ~ 270 °C, meaning it is a lower safety concern
- Low relative cost: $70-75 / kWh as of October 2023 [14]
- Less ethical sourcing concern due to the absence of cobalt

NCA - Nickel Cobalt Aluminum

The NCA chemistry ($LiNiCoAlO_2$) chemistry is closely related to NMC, using aluminum as the metal in place of manganese. NCA is notable for being the chemistry used in Panasonic cells, which are deployed in Tesla EVs and stationary energy storage products, although in 2021 Tesla switched their utility-scale battery product, the Megapack, to an LFP chemistry. The relatively high energy density and allowable power is balanced by higher costs and similar safety concerns to NMC.

- High relative energy density: 200 – 260 Wh / kg
- Very low thermal runaway temperature: ~ 150 °C, meaning it is a high safety concern
- High relative cost: $120 / kWh as of October 2023 [14]
- Ethical sourcing concerns related to the use of cobalt

LTO - Lithium Titanate Oxide

Lithium titanate battery cells are some of the older formulations of lithium battery, being identified in the 1980s. With a chemical makeup of $Li_4Ti_5O_{12}$, LTO is a boutique chemistry that is not often seen in large-scale stationary applications. LTO is known for being able to deliver much larger instantaneous power, in some cases up to 40C, allows for fast-charging, and long cycle life. Additionally, LTO performs well at low temperatures, relative to other chemistries. This chemistry has a relatively low energy density, the lowest nominal cell voltage of Li-ion cells at 2.4 VDC, and very high cost. Titanate batteries are strong performers in low-temperature applications.

- High relative energy density: 200 – 260 Wh / kg
- Very high thermal runaway temperature: ~ 280 °C, meaning it is a low safety concern

- Very high relative cost
- Less ethical sourcing concerns due to the absence of cobalt

Trends in Cell Chemistry

Over time, utility-scale batteries have trended to cluster around specific chemistries. Figure 3-8 shows trends in cell chemistry based on data and predictions from Wood McKenzie [16]. This plot includes both stationary and EV BESS, with the EV being dominant. The clear trend is away from NMC and toward LFP. As discussed, this is driven by improving LFP density, decreasing cost, and the safety and sociopolitical drawbacks of NMC.

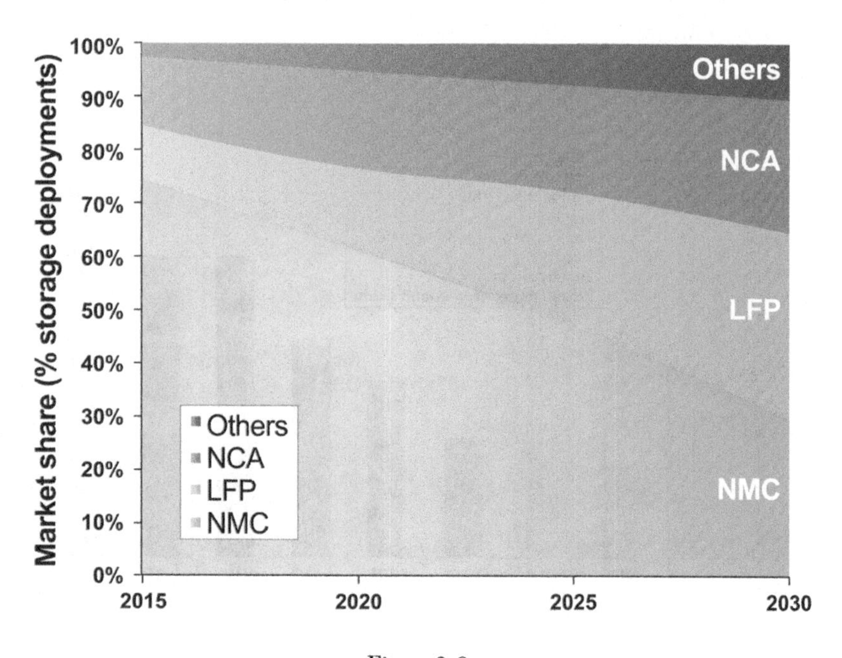

Figure 3-8
Stationary and EV BESS chemistry market share forecast [16]

Specific chemistry in a battery cell typically has a C-rate at which it prefers to operate. Section 1.2 further explains what a battery C-rate means and why it is important. Currently, most utility-scale BESS projects have a C-rate ranging from 0.25 to 0.5. In 2023, according to Energy Storage News [17], there were 7.9 GW and 20.6 GWh of new storage installed in the US, with an average C-rate of 0.4, which equates to a 2.6-hour battery.

Higher power applications are also in demand, which typically range from 0.5C to 1.0C, and often may opt for non-LFP, although these, too, may be skewing toward LFP mostly because of price competition. Figure 3-9 compares the specific energy density, which is the amount of energy (Wh) stored per unit weight (kg), across different battery chemistries: most are Li-ion and some non-lithium chemistries are included for comparison.

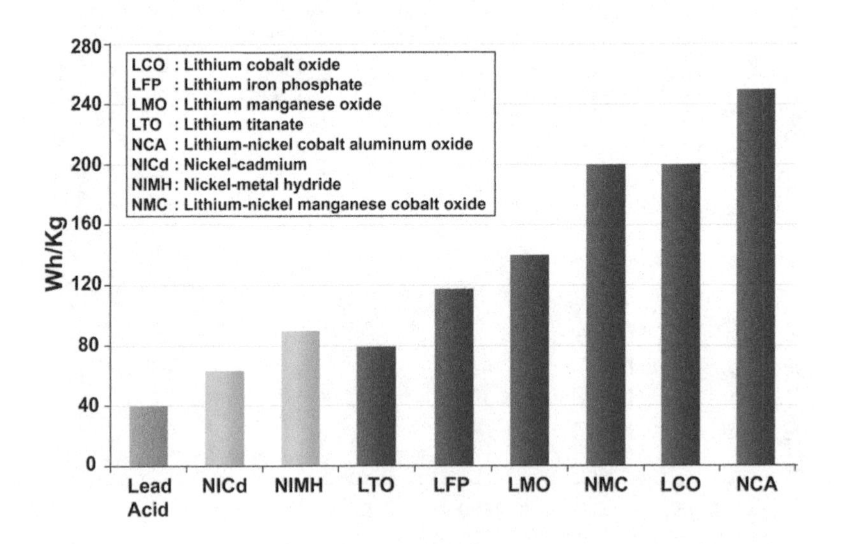

Figure 3-9
Specific density of BESS chemistries [18]

Note these acronyms use only the first letter of the chemical element abbreviations: Li, Fe, P, Ni, Mn, Co, etc.

3.6 Battery Manufacturing

Battery manufacturing is a complex and detailed topic that has many books dedicated to it. This section provides a broad overview of how cells are made and the industry trends that affect the BESS sector.

The raw materials of batteries are mined – lithium, nickel, cobalt, manganese, graphite, and copper are all common inputs, although the exact inputs will vary by each manufacturer and each cell. The most prevalent mining processes extract these elements from the ground as ore, which is later refined into a purer form of the element that can be used at the next manufacturing stage. For example, lithium is often mined as ore and then refined to lithium carbonate, a commodity which is bought and sold around the world. Lithium hydroxide is also widely used. Converting ore into usable materials is done through industrial processes that often involve hazardous chemicals, which are used to extract the desired metals or compounds. These processes may cause environmental harm if not done properly, especially since some of these materials are extracted in nations with weak regulatory bodies. The environmental impact of mining is covered in Section 10.6. A summary of the sourcing of raw materials is shown below in Figure 3-10, based on data from NREL and the Clean Energy Manufacturing Analysis Center [19].

After the precursors are refined, they are shipped from the refinery, which is usually close to the mine, to the cell manufacturer. Some elements are extracted in a relatively pure form, while others still have many steps to convert them into a usable form. Cell producers purchase lithium carbonate, nickel sulfate, and other intermediary forms of the desired compounds, which are later processed into the inputs necessary in the cell manufacturing process.

As of this writing most cells used in the BESS industry are produced in China. The US, through the 2022 Inflation Reduction Act, has begun building some cell manufacturing facilities, but these will take several years to come online and have any major effect on the

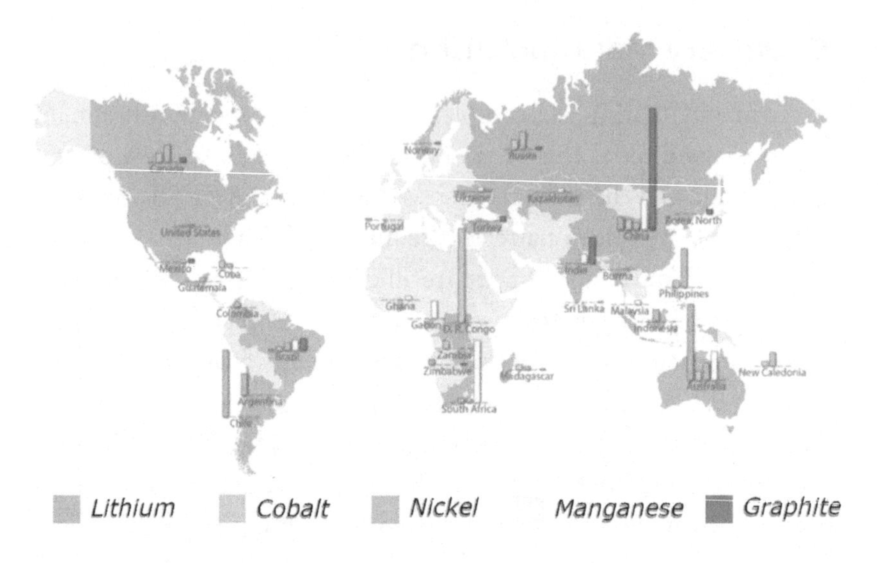

Figure 3-10
Battery raw materials [19]

industry. There are some political implications to the battery supply chain – these are discussed in depth in Section 10.6.

To make the electrodes, cell manufacturers start by mixing the active materials (say $LiFePO_4$ for the cathode or, in a separated batch, graphite for the anode), with a "binder," to make a slurry with which they coat the metal foil (called the **current collector**) in big machines that work with pressure and temperature. These machines apply the paste onto the foils to form the cathodes. This work is done either with two parallel production lines, or by using the same process twice on the same machine (with extensive cleaning between each process). The electrodes are then run through a rolling process known as **calendaring**, and slit into the desired dimensions for the cell in production, tacking layers of cathode/separator/anode/separator, and so on.

The cell is assembled by cutting and stacking layers of the electrode. The tabs, or connectors, which will drive the current to/from the current collectors to the terminals, are then welded together: anode to anode and cathode to cathode. Finally, the prismatic cell enclosure (known as the **can**) is packed with the electrode. The steps of this process are illustrated in Figure 3-11.

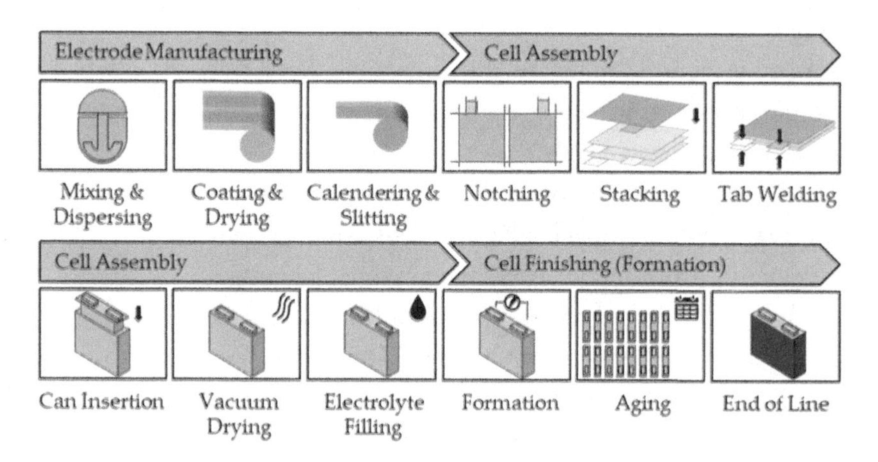

Figure 3-11
Steps in the manufacture of LFP battery cell [20]

Cell manufacturers that wish to provide a more complete product to integrators often sell battery packs or modules, which are sets of cells, often connected in series. For stationary storage products, the modules or packs are stacked into racks, which are then incorporated into enclosures by the integrator. These enclosures and connected with all the other systems that comprise the BESS. Some of the integrators using this approach are Wartsila and Fluence, both of whom have used racks manufactured by CATL, although they have since begun to use various cell suppliers. Battery manufacturer, Powin, has preferred to buy battery cells, instead of racks, and integrate them into a containerized battery product.

From 2015 through 2020, most Chinese cell manufacturers did not produce integrated products, but since 2020 many of the manufacturers have vertically integrated to offer integrated solutions. Leading firms such as CATL, REPT, Hithium, and EVE all now have containerized BESS products, where in the past they only sold cells. For the time being, it appears that these firms are content to supply third-party integrators while serving as integrators themselves. Since this means that those firms are sometimes competing with products using their own core technology, cell manufacturers may begin to focus more on integrated solutions and reduce the direct sales of cells, packs, and modules to third-party integrators. This pressures other countries to produce their own cells.

In some cases, cells are sold by a manufacturer to another supplier, who may brand and package those cells and sell them under another trade name. This practice is known as white-labeling.

As in EVs, Tesla has been a market leader, engineering and building some of the most advanced BESS products. The latest offering from Tesla, the Megapack 2 XL, uses battery modules produced by CATL, to which they add their own proprietary inverter (since the Megapack is an AC-block solution, which is discussed in Section 4.2). This novel approach allows for a standard product requires only an AC collection, rather than both a DC and AC collection like other products. Sungrow uses a similar approach with their PowerTitan 2.0 product, which incorporates a Sungrow PCS. The cell manufacturer used in the PowerTitan v2.0 has not been publicly released as of this writing.

3.7 Non-Lithium-ion Storage Technologies

Major drawbacks of Li-Ion technology include their limited power-to-energy ratio, safety risks associated with thermal runaway, limited lifetime, and the scarcity of the raw materials involved. Various commercial electrochemical technologies exist that do not use Li-ion, but none has seen the scale of deployment of Li-ion. The biggest barrier to the large-scale deployment of these products is the financing, covered further in Chapter 6.

This chapter discusses three of the most prominent non-lithium technologies that are available on the market today. Each section provides a brief overview, and resources are provided for readers who wish to learn more about these topics. Although this book is focused on Li-ion batteries, this section is included to showcase some up-and-coming technologies that may, one day, compete with and even replace lithium.

Flow Batteries

Flow batteries have gained prominence in recent years. The basic premise of these batteries is that they can store energy in a liquid, hence the 'flow' in the name. These batteries allow for longer duration storage of energy as the liquid storage can be increased independently of the power capacity, eliminating the limitation that Li-Ion has of a locked power-to-energy ratio. Additionally, these batteries offer advantages in safety, lifetime cycles, and in avoidance of supply-chain bottlenecks.

Types of Flow Batteries

There are currently three types of flow batteries in production: redox, hybrid, and membrane-less system.

Each type works in a similar fashion but utilizes different types and phases of electrolyte. Redox flow systems are currently the most popular by installed capacity, followed by hybrid, then membrane-less.

Commonly used flow battery chemistries include:

- Vanadium
- Zinc-bromine
- Polysulfide-bromine
- Iron-chromium
- Iron-iron

Mechanics of flow batteries

A flow battery works by storing positively and negatively charged electrolytes in separate containers. These electrolytes, the anolyte and catholyte (negatively and positively charged, respectively), are pumped from their respective storage tanks through an ion exchange section, and back into the storage reservoirs. When the electrolytes pass through the ion exchange section of this liquid loop, the user can leverage, through current collection electrodes, the difference in charge between the two liquids and electrode plates installed in the current collector to cause current to flow in an external circuit. When charging the flow system, energy from the grid (or an alternative source such as PV) flows into the electrolytes, creating the ionic exchange and thus charge difference between the two electrolytes, allowing for a future reversal of this process, which will discharge energy to the grid.

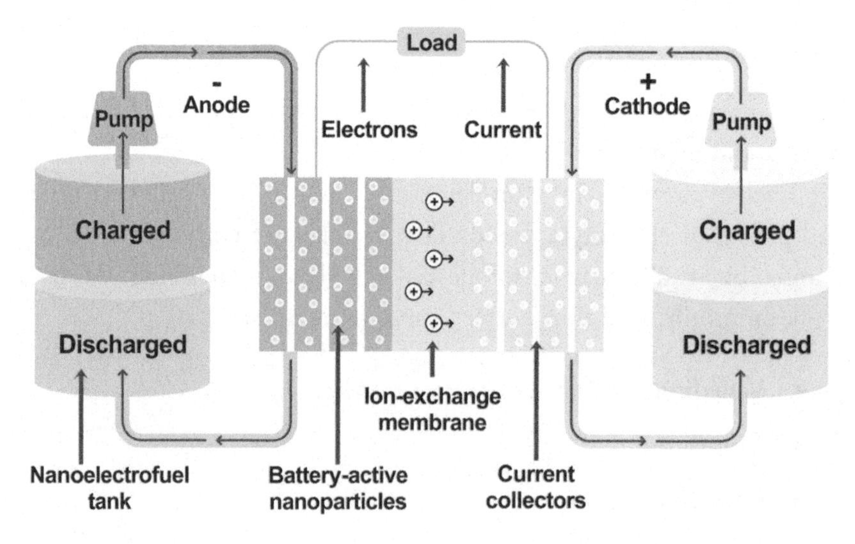

Figure 3-12
Flow battery mechanism [21]

Pros and Cons of flow batteries

The benefits of flow batteries include:

- The ability to replenish expended electrolyte liquid from an external source (like adding gasoline to a traditional internal combustion vehicle, as opposed to having to charging an EV)
- Increased lifespan
- Elimination of balancing requirements (except for hybrid flow systems)
- Decreased fire/explosion potential, access to much longer durations
- The ability to decouple power and energy.

This last point highlights that flow battery systems can be designed to meet any power and energy ratio – these parameters are not tied together and limited as they are in non-flow batteries where the electrodes are permanently integrated into the system (e.g., Li-ion cells, lead acid cells, etc.). This enables designers and developers to consider utilizing flow batteries in situations where other cell types wouldn't make engineering or economic sense.

The disadvantages of flow batteries include:

- Higher operational and maintenance costs due to pumps, piping, and other industrial plumbing hardware
- Larger footprints required for energy densities relative to Li-ion
- Higher capital expenditure costs
- Lower round-trip efficiencies
- Increased installation time
- Less institutional knowledge of the technology

Some of these disadvantages can potentially be "researched" away or mitigated at higher manufacturing volumes, such as capital expenditure (CapEx) costs and footprint sizes. Others are fundamental to the technology and can only be minimized, not removed.

Alternative Electrochemical Storage Technologies

Besides Li-ion and flow batteries, several other promising types of energy storage technologies are being developed, and on the cell-level appear to show promise as an up-and-coming technology.

Iron-air

Iron-air batteries essentially make use of a controlled and reversible oxidation or rust reaction. While there are many anode, cathode, and electrolyte combinations that can be used for this reaction, the iron-air battery is one of the most promising. In this chemistry, the iron of the battery is oxidized (rusted) to discharge energy, and the process is reversed to charge the battery. The prominent startup, Form Energy, backed by Bill Gates-backed fund, Breakthrough Energy, in 2021 unveiled their chemistry to be an iron-air type. Form Energy is in the process of building its first utility-connected project with a 1 MW / 240 MWh battery in Minnesota, connected to an electrical cooperative. The ability of this battery to feature such an extremely long-duration (240 hour) storage is promising, but the small scale shows that this technology is still in its infancy.

Aqueous Batteries

Aqueous Li-ion batteries use similar technology to traditional Li-ion, but they are distinct in that their electrolyte is a concentrated saline solution rather than the traditional Li-ion batteries, which typically use an electrolyte of lithium hexafluorophosphate (LiPF6) salt dissolved in organic carbonates. Aqueous batteries are touted as being safer and less toxic than traditional Li-ion.

However, these batteries have challenges in that they have much lower energy density, lower operational voltages, and often degrade more quickly as compared with non-aqueous Li-ion cells. Aquion energy was a high-profile manufacturer of aqueous Li-ion batteries that raised over $190 million in venture capital, only to declare bankruptcy in 2017.

One subset of aqueous batteries is **aqueous air** chemistries – this is a lithium-air battery that uses an aqueous electrolyte, which is one of several candidates for electrolyte in this type of battery. While there is still extensive research and innovation in aqueous battery technology, it is still considered experimental as of publication of this book.

Hydrogen Fuel Cells

Electrolysis and fuel cells form another method of storing and releasing chemical energy to/from a fuel, that has long claimed to be seen as a competitor with Li-ion batteries. Though a fuel cell is an electrochemical device in nature, it is quite different from a battery, in that it involves the combustion (albeit slow and controlled) of a fuel, which must be supplied from the outside of the fuel cell.

Electrolysis performs the equivalent to a battery's charge cycle; in that it stores electrical energy in the form of a fuel. Typically, this involves using electricity to split a water molecule (H_2O) into hydrogen gas (H_2) and oxygen. The hydrogen can then be stored in a variety of forms, such as compressed, liquified, or as ammonia (NH_3). In compressed form, hydrogen can be stored in tanks or naturally existing salt caverns. The density of these forms of storage is much higher than Li-ion technology in terms of energy per unit area of ground footprint), and they also have the benefit of having minimal degradation over time.

The fuel cell provides the equivalent of a battery's discharge cycle: by combining a fuel source (most often hydrogen) with oxygen, fuel cells produce energy while releasing water and heat as a byproduct.

Figure 3-7 shows the typical arrangement of a hydrogen fuel cell, with inputs and outputs of the reaction:

- Hydrogen enters the anode and air, supplying oxygen, enters the cathode.
- A membrane between the anode and cathode separates the hydrogen electrons from the protons.
- The electrons flow through an external circuit, providing DC electricity.
- As the electrons return from the external circuit, they combine with the protons to create water and heat.

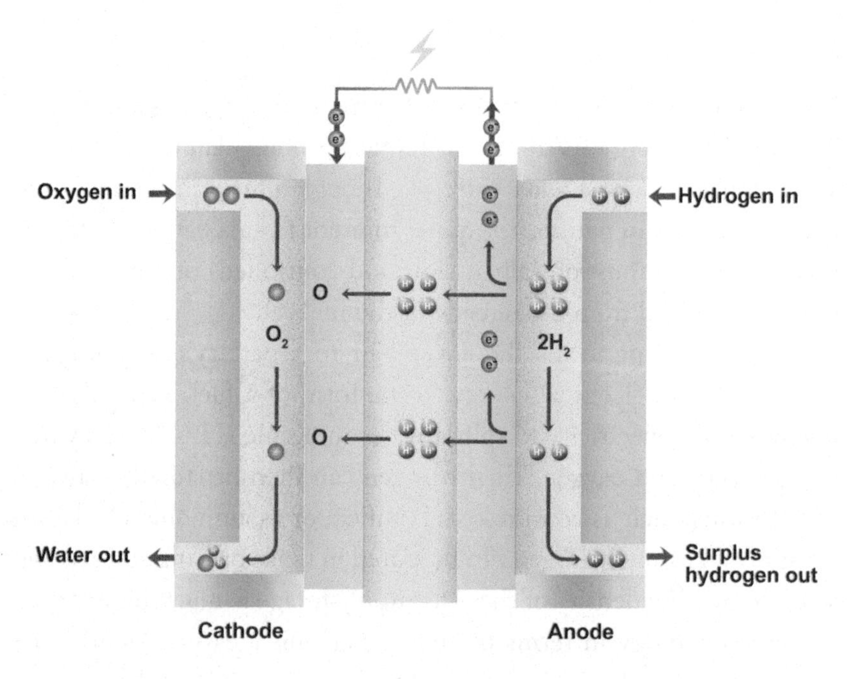

Figure 3-13
Hydrogen fuel cell [22]

At one point fuel cell EVs (FCEVs) were seen as a potential competitor with battery EVs (BEVs) for low-emission transportation.

Hydrogen can be produced by using the existing electrical grid, stored in large stationary tanks, and transferred into smaller tanks within vehicles. In the early 2000s, fuel cell and Li-ion EVs were dueling rivals, with major vehicle manufacturers aggressively pursuing research into both technologies. However, over time, Li-ion BEVs won out, primarily because of the steep decline in battery costs and EVs. Additionally, the charging infrastructure for EVs expanded more quickly than hydrogen distribution, making only certain regions feasible for FCEV ownership.

For short-duration (1-4 hour) stationary energy storage applications, Li-ion has also dominated commercialization and investment. However, hydrogen energy storage systems (HESS) have potential to serve as a contender for long-duration applications, since hydrogen can be stored at high density for long periods with very low degradation. This would make it a strong contender for storage for renewable production, such as wind, that in some cases has large seasonal variations.

According to Pacific Northwest National Laboratory (PNNL) [23], as of 2020, the capital cost of a 10-hour HESS was around $3,100 / kW or $310 / kWh but is expected to fall to $1,600 /kW or $160 / kWh by 2030. If this trajectory is accurate, HESS all-in costs will be lower than that of BESS, which are predicted to be approximately $225 / kWh in that year.

The round-trip efficiency (RTE) of HESS projects is one of its major drawbacks. According to Sandia National Laboratories (part of the US DOE), as of 2020 the RTE of hydrogen systems was around 40% [24]. This is far below the comparable value for Li-ion, which typically has an RTE of 80-90%. This value is covered in greater detail in Section 4.4. Given these values, HESS projects make strong investments in times when the cost of energy is very low, such as periods of negative real-time pricing in areas with excess wind generation.

The passage of the Inflation Reduction Act (IRA) provided a strong boost to hydrogen technology, with many provisions incentivizing electrolyzer, fuel cell, FCEVs, and HESS projects. As with the BESS provisions, these consist of extension of and increases to existing federal tax credits, as well as creation of new credits. Hydrogen supporters believe that this boost will allow hydrogen systems to see leaps in technology and declining costs, as solar and wind power have. Detractors, however, see these incentives as government support to an industry that cannot exist without it. Time will tell if hydrogen can become a viable competitor to Li-ion, but as of this publication, Li-ion is the market leader and will be for the foreseeable future.

3.8 Battery Modules and Racks

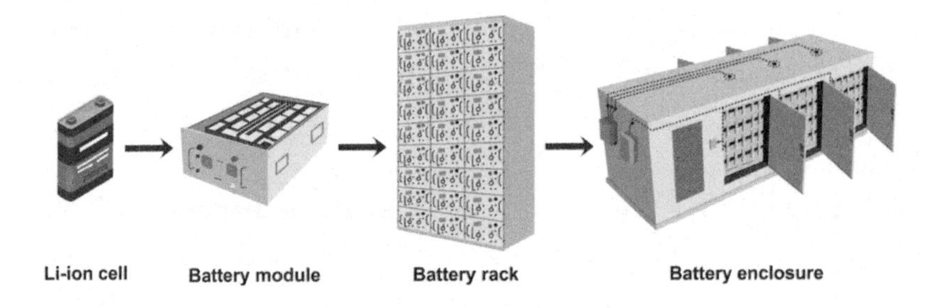

Figure 3-14
BESS construction cell to system

While the cell is the heart of any battery system, each typically holds less than 1 kWh of energy, which is roughly the energy contained in a typical lead-acid car battery. To scale cells up to form a utility-scale system, designers combine them in repeating units: larger systems, which today reach over 1 GWh, may include over one million battery cells. Cells combined in parallel and series must match voltage (V) and (Ah) levels necessary for interfacing with PCSs and meeting project energy requirements. Voltage can be thought of as electrical "pressure" and amp-hours can be thought of as "fuel-holding capacity." A 200 Ah cell will deliver one amp for 200 hours, 50 amps for 4 hours,

200 amps for one hour, or any amp and hour combination that has a product of 200.

When cells are in parallel with each other, their nominal voltages must be as close as possible, and of course they should be at almost the same SoC, to minimize circulating currents among them, that can damage and accelerate energy degradation. The easiest way to do this is to ensure that cells in parallel are from the same make, model, and production year. Even better is to ensure the cells are from the same production batch.

Some tips to remember when analyzing systems:

- Cells put in parallel will have the same voltages, but their currents (and Ah ratings) can be summed to give the overall value.
- Cells put in series sum their voltages, but their Ah levels stay the same. When cells are put in series their Ah ratings must be as close as possible to each other to avoid creating voltage deltas across the connection.
- When summing energy levels, it does not matter whether they are in series or parallel, simply multiply the size of the system by the number of cells, modules, or racks to determine the total energy storage capacity in kWh or MWh.

Most modern utility-scale energy storage systems consist of 300 to 500 cells, each at approximately 3.2 V, in series to form racks with operational voltages in the 1500-2000 VDC range. The advantage of working with higher voltages is that high amounts of power (kilowatts or megawatts) can be transmitted with less amount of current (fewer amps), resulting in smaller wires, switches, and fuses. The disadvantage of working with higher voltages is that greater insulation, spacing, and safety measures are required because higher voltage can "jump" from one terminal to the next more than lower voltage systems. This creates greater risks in the case of faults, and more safety risks for the workers building, operating, and servicing these systems, who must be trained in handling high-voltage systems.

It is worth noting that many new energy cells are unveiled often referring only to their amp-hour rating, which ranges from 20 Ah for smaller cells to over 300 Ah for larger cells. This is a bit confusing since energy is dependent on both the amp-hour and voltage ratings. The reason that amp-hour is a common metric is that nearly all battery cells operate at an average voltage of 3.2 V (when LFP cells are implied), making these ratings a mostly apples-to-apples comparison. However, it should be noted that there are slight variations in voltage ranges between cells, and this needs to be considered when determining any cell's energy storage capacity.

Various companies call these groupings different names: groups, packs, modules, strings, racks, stacks, cores, nodes, or units. This book uses the two most common groupings found in utility-scale systems:

- Modules - usually consist of 10-60 cells
- Racks - usually composed of 5-30 modules

Larger groupings such as containers and arrays are covered in the following chapter. The purpose of batteries using modules is to make the battery racks more serviceable; in other words, if there is an issue with a given cell, while that cell cannot be extracted from the rack, the module can be. This will be covered in further detail in Chapter 9, Operations and Maintenance. Additionally, constructability is a concern given the very high density of battery products.

For smaller systems, such as residential battery products, the product may be sealed as a single unit, composed of a single string of cells. Residential products are typically not field serviceable given their size. In the case of issues with a product, it is replaced, and the original unit may be sent back to the manufacturing facility for repair or decommissioning.

A battery module and rack correspond to the standard rules of electricity, as covered for cells above: when components are added in series, their voltages are sums and amperages remain the same. When components are added in parallel, their voltages remain the same and amperages are summed. For today's utility-scale battery products, it is common for battery cells to be combined in series to form modules, and modules in series to form racks, and racks connected in parallel.

Battery racks come in many shapes and sizes and are highly specific to integrator and supplier. Many modules are comprised of cells and cell groups, which are often compressed and held tightly in a plastic frame to minimize swelling of the cells during charging, which can damage the cell hardware. This aggregate hardware is then integrated into the larger battery racks and fitted with temperature, voltage, and current sensors, and in the more advanced systems, balancing circuits, which help maintain optimal voltage in the cells, relative to the system target voltage. These components are covered in Section 5.1 on Battery Management Systems.

Battery racks are also where thermal management begins. For liquid-cooled systems the fins and passages for coolant flow are integral to the rack. For air-cooled systems, fans incorporated into the rack are essential to distributing the cooled air to the battery cells. This is covered further in Section 4.3, Thermal Management.

The next level up from a module is the battery rack, a collection of battery modules, usually around 1.5-2.5 m high, and approximately 1 m by 1 m footprint. This is a standalone unit, which can be paralleled with other racks to reach a specific energy level. A rack will usually have modules, cabling, and/or bus work, a rack controller with contactors, fusing, communications, and controls electronics. Some racks may have more or less of these components, but most will have this typical configuration and are designed to be aggregated with other similar racks to build up the full system. Each rack is an independent unit and can connect to a **Power Conversion System (PCS)**, individually but is typically paralleled with many other similar racks. More to come on the PCS in Chapter 4.

Battery racks vary widely from supplier to supplier but are typically on the order of 200-400 kWh of storage. Typical options for battery racks are:

- Indoor rated vs. outdoor rated
- Liquid-cooled vs. air-cooled
- High-power vs. High-energy

As an example, consider a hypothetical battery enclosure based on cells with a 100 Ah, 3.2 V beginning-of-life (**BOL**) rating, and a 0.5C rate. This rating means that when a single cell is fully charged, it can discharge a maximum of 50 amps, and over the course of 2 hours it would provide the 100 Ah. Alternatively, it could provide 25 amps for four hours, or 12.5 amp for 8 hours, and so on. The cell's nominal, or average, voltage is 3.2 V, meaning the cell will drop to a lower voltage when discharged (~2.9 V) and climb to a higher voltage when charged (~3.5V).

If 40 of these cells are put in series to form a module, the voltage adds to be 128 VDC, and the amp-hours remain constant at 40 Ah, so the resultant module has an aggregate rating of 128 V x 40 Ah = 5120 Wh, or 5.12 kWh as measured at the DC BESS terminals.

If 8 of these modules are put in series into a rack, the rack will have a rating of 40.96 kWh, or 40 Ah with a nominal voltage of 1024 V. The typical voltage range may be +/- 5%, or 973 to 1075 V.

If an enclosure was formed from 16 of these racks in parallel, the resulting energy storage product would have an aggregate rating of 973 to 1075 V, and 1600 Ah capacity for a total of approximately 819 kWh of energy storage (or 0.819 MWh). The results of this sample calculation are shown in Table 3-1. Enclosures will be discussed in greater depth in Section 4.1.

Table 3-1
Sample battery enclosure calculation

Level	Makeup	Voltage [V]	Capacity [Ah]	Maximum continuous current [A]	Energy [kWh]
Cell	n/a	3.2	100	50	0.32
Module	40 cells	128	100	50	12.8
Rack	8 modules / 320 cells	1,024	100	50	102.4
Enclosure	16 racks / 96 modules / 3,840 cells	1,024	1,600	800	1,638.4

It is worth noting that typically the rated energy capacity of modules and racks is less than the sum of their components. This is because the standard of energy storage products is to provide a 100% usable voltage range to their customers. To ensure that the end-user has a low chance of violating the cell's operating procedures, most battery integrators reduce the stated usable energy capacity of the product in their specs or data sheets. To continue the above example, the nameplate BOL energy capacity of the hypothetical system might be marketed as 1,490 kWh rather than the 1,638.4 kWh aggregate energy capacity of all 3,840 cells.

Suppliers design their products with an integral PCS or a configuration that uses an external PCS. Some manufacturers have a preferred PCS, while some have a list of acceptable PCS to choose from.

Design based on these battery racks is covered in Chapter 8 on the EPC of battery systems.

A modular battery system can be designed by building up battery modules to match a PCS. Without the PCS, the batteries are useless, so know the target voltage range, which is determined by the

PCS, is key. In the example above, the nominal voltage is 1331 V, but this voltage changes depending on the SoC of the cells. Depending on the battery chemistry, the voltage can swing between 75% and 125% of this nominal voltage, and the PCS needs to be capable of handling the complete voltage range. If a PCS is used that could not handle the full voltage range, it would lose access to that portion of the energy in the battery and would have essentially unnecessarily oversized the system.

One important aspect of battery sizing is whether the system is quoted at the BOL, or the end of life (**EOL**). Batteries, like most products purchased for 15-20 years, show their age in gradually declining performance. However, batteries are notable since their rated energy capacity begins declining from the first day of manufacturing, even if the product is not used (known as calendar fade). Due to this phenomenon, some battery performance metrics are referred to as BOL, and sometimes as EOL. For example, the BOL energy capacity of a Powin Stack230 product is warrantied to be 230 kWh. After 15 years of use without any augmentation that same product may have an energy capacity of only 184 kWh. If the system experienced a lot of throughput, the capacity may be less than that. Other characteristics such as the RTE may also decline as the system ages. Conversely, the power that a battery can put out typically does not change, regardless of the battery's age.

Typical Duration and Configuration

Grid-scale battery systems have a range of durations and configurations, but over the last few years particular duration patterns have emerged. For Li-ion systems, regarding durations, two- to four-hour buildouts are sweet spots where the cells operate within their preferred current and power ratings, match up economically with power conversion systems, and meet particular market or other customer needs.

Battery suppliers typically provide batteries in one of three configurations:

- Battery racks that can be incorporated into indoor or outdoor containers
- Pre-packaged battery containers that come mounted with battery racks (or are populated on site)
- Pre-packaged proprietary enclosures that come fully integrated for delivery on site, which is the current dominant trend in the industry

As Li-ion technology evolves and improves in power and energy density, as well as safety attributes, we expect to see the typical configurations also evolve. The benefit that batteries bring to power systems tends to force optimal sizes and configurations in a certain direction. As renewable energy system penetration increases on our power grids, battery systems will continue to play a critical role in helping ease that transition as seamlessly as possible, and this in turn will be a big driver in how battery systems are designed and sized.

Voltage, Current, Power Ranges

Battery systems have gradually evolved toward utilizing the 1500 VDC PCSs now prevalent in the industry. The range of these PCSs typically falls between 1000 VDC and 1500 VDC and allows for higher power levels and lower currents, which translates to higher power densities and lower losses.

Typical utility-scale current levels range between 200 – 2000 A on the low voltage (LV) side before step-up transformers are required. This can be a good maximum current limit to keep in mind when sizing the AC output of the system. On the DC side currents typically range between 50 A and 200 A at the rack level and can go as high as around 4000 A at the container or enclosure level. As a reminder, current levels of this magnitude can pose extreme safety risks, and systems must be designed to appropriate required codes to ensure the safety of those working around this equipment.

Power levels for utility-scale battery systems range greatly, but at the PCS level it is currently common to see 2 to 6 MW. By paralleling multiple PCS system sizes can be as large as 1 GW. Choosing the right PCS for the job will get the right power level for the project without oversizing unnecessarily. This is covered in greater depth in Section 4.2.

3.9 Degradation

We are all aware that batteries in our daily life lose their potency over time – this applies to phone batteries, AA cells, as well as to utility-scale BESS installations. One of the main developments in battery cells over the past 30 years has been the battery's ability to maintain its capabilities over thousands of charge and discharge cycles. Even the best batteries will have a limited useful life, and planning for this both technically and commercially is a key aspect of proper design, construction, and safe operation of utility-scale battery systems.

Degradation is driven by the chemical reaction happening within each battery cell. The Li-ion material structure is designed to store and release Li-ions as part of the charge and discharge cycle. Over time, the active materials in the battery lose their ability to give and store the ions, and therefore the cells can store less energy. The net effect of this process is that year-over-year the BESS energy capacity decreases.

Different technologies and chemistries degrade at different rates. It is incumbent upon the system designer, and by extension the people doing due diligence on the system, to understand how the system will degrade over time, due to both time-based and usage-based degradation.

Engineering a 20-year battery system can be tough enough without adding in the complications of how it will lose energy capacity over time and how batteries may be added in the future, but under-estimating these requirements the last years of the project could be quite painful. Running out of physical space, anemic auxiliary power supplies, and incompatible voltages and technologies could plague the health of a system and have serious financial implications. Of course, oversizing upfront or being too conservative in the calculations can also hurt a project, so much so that the project cannot be financed and built, so striking a balance is crucial.

Batteries begin losing energy capacity the moment they are fabricated. A 280 Ah cell will begin life with the ability to discharge 280 A for one hour, and that number will go down over time. Often the energy capacity of cells will slightly increase over the first few cycles and then start to decrease, but this small effect can be ignored for now.

There are two main ways that batteries degrade: over time (calendar), and with use (throughput).

Calendar degradation, also known as calendar fade, occurs whether a battery is used or not. Everyone has bought batteries, placed them in a drawer and pulled them out years later, only to find they are worthless. This is the slow, inevitable decline in energy capacity and voltage of a battery cell. It can be slowed down in several ways, but it can never be completely avoided. For secondary (rechargeable) cells of the type used in utility scale projects, there is an additional twist since the amount of energy discharged (**throughput**, see definition below) has an additional impact on the degradation, and the two modes can interplay in subtle ways. Figure 3-15 shows calendar fade for LFP and NMC chemistries from different anonymized suppliers, as well as the temperature impact on calendar fade: higher temperatures increase the rate of calendar fade.

Operational degradation describes the way in which a battery's energy capacity declines as it is used. The mode and drivers of degradation are subtle topics, and the subject of much research. This section covers four of the key drivers of degradation: throughput, C-rate, temperature, and DoD.

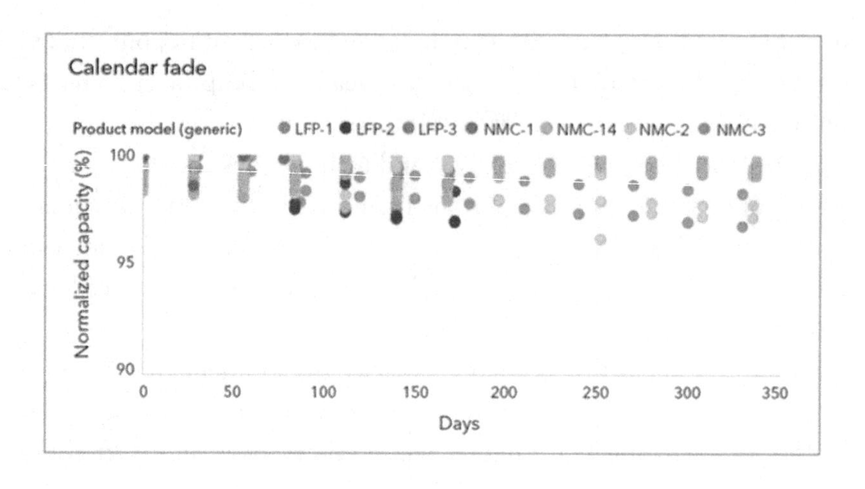

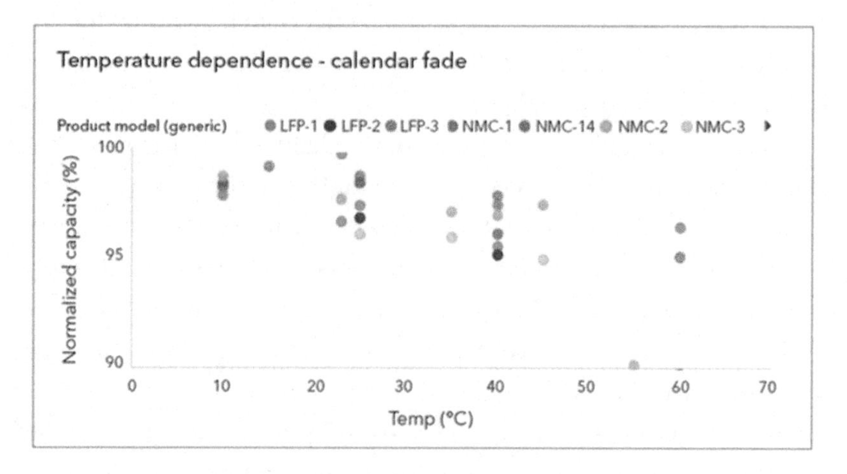

Figure 3-15
Degradation charts temperature vs. calendar [25]

Throughput

One of the biggest drivers of degradation is, unsurprisingly, how often it is used. In the battery world this is known as **throughput.** In EVs, throughput is analogous to the mileage on the vehicle – it is a measure of how much energy has been charged and discharged through the battery cell.

The most common way to measure throughput is by measuring the output energy of the battery and determining how many equivalent charge-and-discharge cycles have been used. In practice this is most often done by metering the output of the battery and measuring how much energy (kWh or MWh) passes through it. This can be done on the cell, module, rack, or unit level. By dividing the discharged energy by the initial capacity of the battery being measured, the throughput can be expressed in **cycles** (sometimes referred to as turnovers).

For example, if a 375 Ah BOL battery cell (1.2 kWh at 3.2 V nominal voltage) has its output measured over the course of one year, and it is observed to have discharged 252 kWh over this time, it could be said to have been used for 210 cycles over the course of the year. The most common battery warranties allow for one cycle per day, which is described more fully in Section 9.5 on battery warranties and guarantees.

A large BESS facility can be measured in the same way. For example, if a 150 MWh BESS meters its discharge at the point of interconnection (usually at a high-voltage transformer) and is seen to discharge at 10 MW for 4 hours in the morning, then at 60 MW in the early afternoon for 1 hour, and then at 22.5 MW for 2 hours in the evening. The net energy discharged would be (note that only real power [MW] is discussed here to simplify the example):

$$Net\ discharged\ energy = 10\ MW \times 4\ hr + 60\ MW \times 1\ hr + 22.5\ MW \times 2hrs$$

$$Net\ discharged\ energy = 40\ MWh + 60\ MWh + 45\ MW = 145\ MWh$$

Since the system has discharged 225 MWh over the course of the day, it could be said to have discharged 1.5 cycles that day (note that many warranties allow for individual days of higher cycling providing the annual average doesn't exceed 1 cycle per day). As a battery degrades, often the capacity degradation can affect the size of the cycle which is measured.

As we saw in the prior example, in practice a BESS is charged or discharged at regular intervals or at the same power level, but the throughput can be easily measured and converted into cycles.

For every cycle that the battery experiences, it will experience some degradation. This varies by chemistry and by the other degradation factors.

Figure 3-16 shows degradation for four different chemistries, as reported in the 2024 DNV Battery Energy Scorecard [15]. This document is published annually by DNV and reports the anonymous results of the testing performed on EV and stationary battery cells at the DNV laboratory. It is an excellent resource for understanding the state of the art of battery cells.

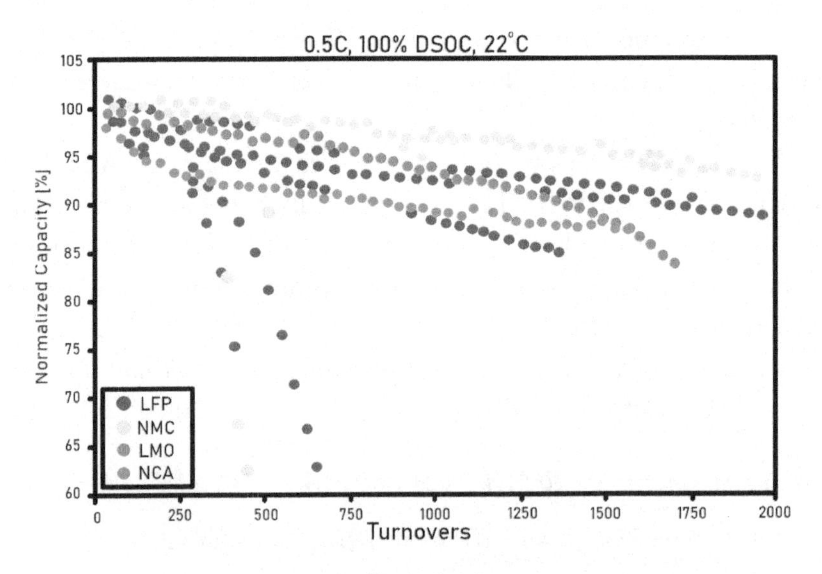

Figure 3-16
Degradation for four different chemistries at 0.5 C [15]

Plotting the data as shown above is known as a **degradation curve**, which is the most common method of measuring how a battery has aged. It is produced by stopping after the cell has been cycled to perform a capacity test – each dot in the plot represents a capacity test being performed on that cell. As shown in the curves above, there is a wide discrepancy from one manufacturer to the next, with some preserving their capacity well over thousands of cycles, and others with capacities that fall off after less than 1,000. As we will see later in the book, many battery products today are extending their life greater than ever – some of the leading products claim to retain over 80% of their original capacity at over 10,000 cycles.

C-rate

C-rate, as discussed, represents the allowed rate of power used by the battery. In operation, the C-rate defines the maximum allowable charging and discharging power. When testing, it is common to charge and discharge the battery at that maximum C-rate. For LFP cells, a C-rate of 0.25 (a four-hour battery) is considered on the lower end, while a C-rate of 0.5 (a two-hour battery) is considered on the higher end. C-rates over 1 (1-hour) and under 1/6 (6-hour) are rare in utility-scale LFP systems. In testing, it has been shown that by allowing the power to be inserted or drained by a battery more quickly, the cell will experience more degradation. This is not surprising – all EV owners know that consistent, rapid acceleration will cause the vehicle's range to decline more quickly than driving it more slowly. If a battery cell is consistently charged or discharged at high power, the cell will degrade more quickly than one treated more gently.

Wherever possible, a system, and hence its cells, should be discharged or charged at the slowest rate possible to minimize degradation. However, there are some scenarios where the end user, or the market, demands that the battery be used at high C-rates, sometimes equivalent to the maximum allowable rate. Provided that the output power of the BESS never exceeds its design limits, there should not be an issue in the cell continuing to perform, or in affecting the warranty – but wherever commercially feasible, slowing the output will result in a battery that lasts longer.

Temperature

The temperature that cells are at when they charge, discharge, or are stored is one of the key drivers of degradation. As cells receive or output power, they generate heat. This heat generation is one of the primary ways that energy is lost in a battery. Although there is some variation from cell to cell, nearly all Li-ion batteries degrade the least when they are operated between 10°C and 25°C. Lower temperatures do not typically affect the performance, but they may require the system to be pre-heated prior to discharging at full power. Higher temperatures have the effect of accelerating the cell's degradation. To keep each individual cell's temperatures low, nearly every battery system has some sort of thermal management system, which is covered in depth in Section 4.3. The most common methods are by circulating cooled air over the cells and their cooling fins (air-cooling) or by circulating a fluid throughout a plate attached to a cell or module (known as a cooling plate).

Most cell suppliers provide an upper- and lower-bound temperature, beyond which the cell warranty will be voided – this is commonly 50 to 60°C on the high end and -30 to -40°C on the low end, although this will vary between manufacturers. If the cells are purchased as part of an integrated battery product, it is more common that the end user only needs to monitor the ambient temperature, rather than the cell temperature. Some battery systems have a **de-rate** when the ambient temperature rises above a certain level.

It is important to note that even if battery cells are running very hot, say above 35°C, it is still far below the temperatures in which a cell would enter into a thermal runaway condition, a type of combustion which cannot be controlled. This is covered in full in Section 10.1, but typically thermal runaway temperatures are far over 100°C. Note that continued operation above 65°C indicates a dangerous condition and should trigger warning alarms.

Depth-of-Discharge / Average State of Charge

After a battery is in the field, we can analyze its performance around how much time it spent at each SoC. For example, we might ask: was the BESS brought down to 0% SoC and back up to 100% every day, or was it in standby mode at 100% SoC for most of the time? Was it kept mostly discharged for most of its life? Or did it float most often at 50% SoC?

These parameters have a significant effect on the way a battery degrades. Unlike some of the other parameters, this category often has counterintuitive results when the lab data is analyzed. In general, the more a battery is used in the very high or very low states of charge, the more quickly it will degrade. However, the prefered average SoC varies widely across manufacturers. As an example, Figure 3-17 shows the results of many different LFP cells cycled at three different average SoCs.

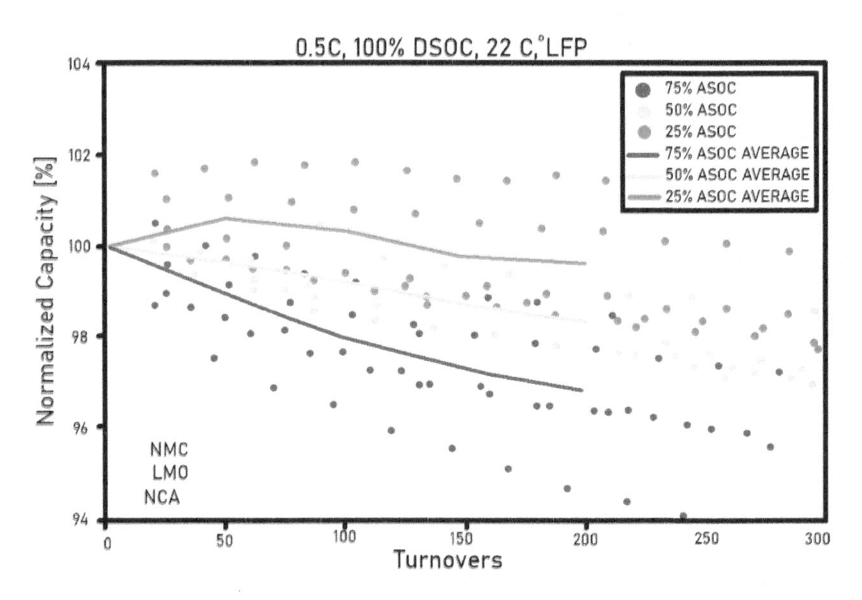

Figure 3-17
Effect of average SoC on degradation [15]

The cells degrade most slowly at an average SoC of 25%. In fact, at this average SoC, the capacity increased on average after the system began to be used! Conversely, when kept at a 75% average SoC, the systems degraded more rapidly. So, in terms of degradation, it is better to keep an LFP cell mostly empty than mostly full. However, this approach doesn't often algin with use cases and discharge strategy.

Most battery warranties typically do not limit the average SoC, or the depth of discharge, although some do. It is important to verify with the supplier of the cell, module, or rack, what the preferred SoC is, and to have the system align with these preferences, if commercially feasible.

3.10 Usual Suspects

This list represents some of the leading manufacturers of battery cells serving the stationary market. Many of the manufacturers listed here provide cells across both EV and stationary applications, and several provide integrated BESS products in addition to supplying cells, modules, and racks. Firms that also produce integrated systems are referenced again in Section 4.7.

LFP cell manufacturers:

BYD: BYD is one of the world's largest and most well-known cell manufacturers for both EV and Stationary products. Based in China, BYD began building commercially viable BESS Projects in 2011. BYD was one of the earliest entrants in the US energy storage market, and unlike its competitors began offering integrated systems when it entered the US market in 2014. In 2008 Berkshire Hathaway made a $232 million investment to obtain an approximately 10% stake in the company. In 2023, BYD deployed over 40 GWh of stationary storage projects.

CALB: China Aviation Lithium Battery, founded in 2015, has risen to be one of the top manufacturers in the LFP battery market, consistently ranked among the leading companies for installed volume and shipment of LFP batteries. The company plans to expand this capacity to 300 GWh by 2025 and further to 500 GWh by 2025 and 1 TWh by 2030.

Contemporary Amperex Technology Limited (CATL): The world's largest cell manufacturer, CATL is a China-based supplier of cells for EV and stationary applications with $55B in revenue in 2023. They have in the past been more focused on EV cells, but currently the portion of the production dedicated to stationary applications is growing. Although they started as a cell manufacturer, since 2020 CATL has been a competitor with American and European integrators as it entered the stationary storage market with its integrated battery products, the latest of which, the TENER, is one of the most advanced LFP products in the marketplace, as of 2024. According to Energy Storage News, CATL plans to have 800 GWh / yr of cell manufacturing capability online by 2030 [26]. CATL is also a leader in sodium ion cell technology.

EVE: An up-and-coming Chinese cell manufacturer, EVE Energy cells both cells to integrators and offers its own integrated BESS product. They have stood out for moving toward a larger cell format, currently at 628 Ah. In 2023, the company began constructing a 60 GWh / year gigafactory to boost its manufacturing capabilities.

Gotion: Gotion High-Tech is a prominent Chinese battery cell manufacturer. Founded in 1998, Volkswagen acquired a minority interest in Gotion in 2021. Gotion has grown significantly and now collaborates with major automakers and other industries globally. Gotion plans to reach a global production capacity of 300 GWh by 2025 between EVs and ESS. This includes new factories in Germany, Vietnam, Thailand, and potential future sites in Morocco and other regions. Gotion offers both cells and integrated BESS products.

Great Power: Great Power is a China-based cell manufacturer, as well as offering integrated BESS products since 2011. The firm has several production facilities and continues to expand globally. As of 2024 they are in the process of building a 36 GWh zero-carbon battery manufacturing plant.

Hithium: A Chinese manufacturer of storage products founded in 2019, Hithium has stated they are targeting a 70 GWh annual production capacity. In December 2023 they opened a 28 GWh / year capacity plant in Chongqing. As of April 2024, Hithium stated they had shipped 17 GWh of products. They have stated they manufacture their own cells.

LG Chem: A leading South Korean cell producer, LG Chem that also markets their own integrated energy storage products. LGChem is one of the original NMC battery providers and early entrants into stationery and EV markets. In 2023, LG Chem reported sales of approximately $40 billion, with an operating income of $1.8. LGChem was one of the early cell manufacturers whose cells were included in early-stage BESS products built in the US from 2015-2020. They also offer integrated containerized BESS products.

Li-shen: Another manufacturer that white-labels products for other brands of battery cell. Li-shen has stated their annual production capacity at 31 GWh.

REPT: Established in 2017, by the end of 2022, REPT Battero had a production capacity of 35.2 GWh. The company plans to increase this capacity to over 150 GWh / year by 2025 to meet growing demand. REPT sells modules to BESS integrators, integrates their own systems, and have partnered as OEM of an integrated product marketed by other brands.

Stationary NMC cell manufacturers:

FREYR: Founded in 2018, FREYR is a European battery firm focused on producing green battery cells to support decarbonization. They have stated plans to build 50 GWh of energy cell production by 2025 and aim to scale this up to 100 GWh by 2030. They have formed partnerships with battery firms NIDEC and have utilized 24M technology. They manufacture both NMC and LFP chemistry products.

Northvolt: A Swedish manufacturer of energy storage cells with a capacity of 60 GWh / year, Northvolt sells to both the stationary and EV market. In 2023 announced new technology in sodium-ion (Na-ion) cells. They are notable as one of the few large European cell manufacturers.

Panasonic: Part of a large Japanese electronics conglomerate, Panasonic a well-established battery cell maker, and one of Tesla's original partners for stationary storage and EV cells. Panasonic produced 2170 cylindrical cells which were notably used by Tesla in their EV production. Panasonic aims to increase its production capacity to 200 GWh by 2030.

Samsung SDI: Samsung SDI is part of Samsung, a large South Korean multi-national conglomerate. They are a leading energy storage manufacturer, offering both cells and integrated BESS solutions. The company has announced a $3 billion investment in a new EV battery cell plant in St. Joseph County, Indiana, in partnership with General Motors (GM).

Non-lithium battery providers:

EnerVenue: EnerVenue, established in 2020 and headquartered in Fremont, California, specializes in nickel-hydrogen battery technology. They are in the process of constructing a one-million-square-foot gigafactory in Shelby County, Kentucky, which will initially have an annual production capacity of 1 GWh.

EOS: EoS manufactures a zinc-hybrid chemistry cell known for safety and ease of maintenance. In 2023 they were the recipient a $400 million loan guarantee from the US Department of Energy to expand their production capabilities.

ESS: An Oregon-based flow battery company that has been a leader in flow battery technology. Their products are notable for providing long-duration storage. One notable project includes providing 2 GWh of storage capacity for the Sacramento Municipal Utility District (SMUD). As of 2024 they have a production capacity of approximately 400 MW per year.

Faradion: Based in the UK, Faradion is the leader in the sodium-ion battery technology field. In collaboration with its partner Reliance Industries, it is planning to establish an initial production capacity of 1 GWh in India.

Form Energy: A long-duration battery manufacturer backed energy fund Breakthrough Energy. Form's technology is based on an iron-air chemistry technology that offers long-duration storage. Form Energy has entered into agreements to deploy its iron-air battery systems in various locations, including a 15 MW/1,500 MWh system with Georgia Power.

QuantumScape: Founded in 2010 and headquartered in San Jose, California, QuantumScape is a notable provider solid-state lithium-metal batteries. Their technology has been more focused on EVs than on stationary storage applications. In 2024, QuantumScape began shipping Alpha-2 prototype cells to automotive customers for testing, marking a critical step towards commercial deployment.

Vizn: A zinc-iron flow battery producer with a product known for safety. Originally started in Montana, a controlling stake in Vizn was sold to a Chinese firm in 2020.

4. THE BESS PROJECT

While the battery components are the core of a BESS project, there are many other critical components required to construct a fully functional, grid-connected system. The previous chapter described battery racks, which are composed of battery cells and modules. This chapter moves on to describe each of the pieces of technology between these racks and the grid, how they operate and what they do.

Figure 4-1 provides a simplified schematic of how a typical project is built up – first from the racks to the BESS enclosure, then through a PCS, multiple transformers, a tie-line (or transmission line), and eventually on to the grid.

Outside of the battery and PCS, everything else that makes up the BESS project is known as the **balance of plant (BoP)**, or simply the BoP. This is a catch-all term for the collection systems, transformers, auxiliary power, and metering set up is required to move power between the battery and the grid. BoP typically excludes the substation and transmission lines that interconnect the project to the grid - collectively those works are known as the **high voltage (HV) scope.** Note that some projects consider transformers and/or controls to be outside of BoP. The description of what is encompassed by these terms on each project will vary, but the figure shows what the authors have found to be the most common uses of the term.

BESS controls are a key component of any system, shown as the yellow box and blue dashed lines in the diagram. While these are

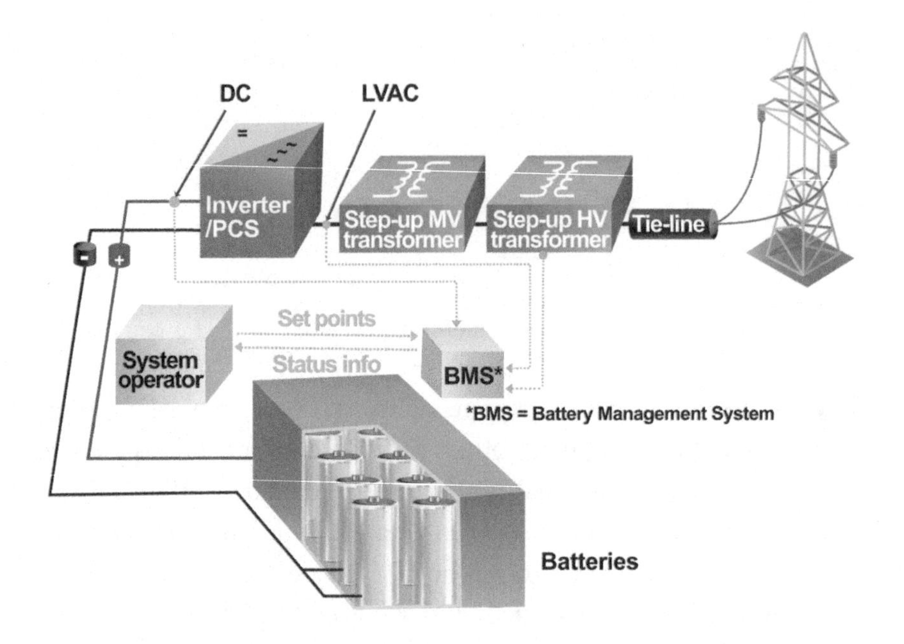

Figure 4-1
Overview of key BESS project components

certainly part of the BESS project, they are such an extensive topic that we have dedicated an entire chapter to them: Chapter 5 on BESS Controls.

When all these components come together as a single system, the result is a **BESS project.** A brief note on terminology: we use BESS project to refer to refer to every piece of hardware and software working together and connected to external systems such as the grid, communications interfaces, and first responders. Where referring exclusively to the hardware installed on site, we will use the term **BESS plant.**

When we say **integrated BESS**, this refers to the DC or AC battery enclosure product, which is sometimes called simply **the BESS**. Firms that acquire modules or racks that they later build into integrated BESS products are known as **BESS integrators**. If a battery integrator white-labels a portion of their system that is produced by others, the original equipment manufacture of the component may be referred to as the **battery OEM.** These terms reflect the evolution of the industry – as we will see later in the chapter many companies are operating at many levels of the value chain, selling cells, modules, racks, and integrated BESS.

As we will see in Chapter 6 on Project Development, BESS Project, with a capital 'P', has a distinct meaning as the legal entity which comprises the project company, holding all assets of the BESS project. This chapter is concerned exclusively with the hardware and software that comprises the project.

The chapter begins by discussing the outdoor-rated enclosures that contain the battery racks, as well as the other key components such as PCS and controls. Section 4.2 covers the PCS, the most critical piece of hardware in the project outside of the BESS itself. The following section covers thermal management, an important element that allows the system to maintain its operating temperature. Section 4.3 Balance of Plant covers all other components of the system: electrical protection, DC and AC collection systems (including substations), transformers, and auxiliary power. Section 4.5 covers metering, which is a key portion of both monetizing and monitoring the BESS project. Lastly, the Usual Suspects of this chapter is extensive, covering Integrated BESS and PCS suppliers. A special section shows images of the most prominent BESS and PCS products available in the market.

4.1 Enclosures

The enclosure is the box, typically painted steel, which houses the battery racks and the various systems that are required to help the battery product run properly. Besides the racks themselves, these include the DC busbars which connect the racks in parallel, the thermal management system (covered in the next section), and the fire protection systems (which covered in Section 10.2).

In the early days of utility-scale systems in the US, circa 2014-2015, nearly all integrated BESS products were modified shipping containers or constructed in purpose-built warehouses. Many of these systems experienced thermal management problems as they were not designed with sufficient cooling and ventilation systems to cope with the large thermal loads of the batteries. As battery integrators became more sophisticated, many designed custom enclosures to house their battery products, rather than using shipping containers or warehouses. These offered the advantage of having the size tailored to the dimensions of the battery racks and allowing for more custom solutions for a safer product with better thermal performance. While the original containerized systems were accessible by personnel, current containerized systems have moved toward exterior access only, driven by safety concerns. As many suppliers switched from air-cooled to liquid-cooled systems, the custom enclosures also allowed more space for the heating, ventilation, and air-conditioning (HVAC) gear. HVAC is used as a general term when describing any of those components). Some manufacturers, such as Tesla or Sungrow, have included the PCS as an integral part of their product enclosure, while others such as Fluence, Wartsila, and Powin opted for a DC product that could be coupled with a third-party PCS.

In the early days of batteries, there was little standardization around enclosures. Some products were enclosed in 20-foot ISO containers, some used 40-foot, and others used 53-foot containers. Between 2020 and 2023 there was a trend of using smaller "cube" products of roughly 2 x 3 m (6.6 x 9.8 ft), led by integrators like Fluence, Wartsila and Powin. As of 2024, most utility-scale BESS installations are built using pad-mounted, outdoor-rated enclosures which have tended toward a 6.1 m (20 ft) shipping container, which has a 6.1 x 2.4 m (20 x 8 ft) footprint. They tend to be "high-cube" dimensions, meaning they are 2.9 m (9.5 ft) high, rather than a standard container which is 2.6 m (8.5 ft). This design choice to move toward modular, unoccupiable units has been primarily driven by safety concerns—in the event of fire or thermal runaway, an outdoor-rated BESS unit is designed to burn on its own, with adequate spacing or firestops to prevent the flame from spreading to other units. This is covered further in Chapter 10, Safety and Environmental Considerations. Since batteries can produce flammable gases when they burn, any enclosure with large amounts of air is an explosion hazard, driving the industry toward a more tightly packed, exterior-accessible enclosure design. Shipping in standardized enclosures has also reduced shipping costs. Figure 4-2 shows a typical battery enclosure provided by FlexGen.

Figure 4-2
Typical BESS enclosure (right) and PCS + MV transformer (left) [27]

Not all suppliers use these dimensions – some, such as Tesla, with their Megapack 2XL product, and AESI, with their TeraStor product, have chosen to use unique product footprints. Some of the more common BESS products are covered in Section 4.7 at the end of this chapter.

All enclosures provide the following critical functions to the battery:

- Protection from external elements

- An envelope to contain heating and cooling, typically with thermal insulation
- Adequate space to terminate/connect – the conductors or busbars feeding the battery
- Space to connect communications cables that carry information to and from the BESS
- Space to house any electronics supporting the BESS, such as the BMS or HVAC equipment
- All battery enclosures must contain cutouts to allow equipment to enter. The most typical are:
- Openings to receive coolant carrying heat away from the batteries
- Openings to allow busbars, cables, or cable carrying conduits to deliver electrical energy to and from the battery
- Openings to allow conduits to feed the auxiliary power system, if applicable
- Doors that allow the battery to be maintained or settings to be controlled
- Ventilation ports designed to release gas buildup in the case of a safety event
- Panels designed to buckle in the case of a high-pressure explosion event, known as **deflagration panels**
- House internal smoke detectors, gas sensors, and external alarms/strobes

Dust and Moisture Protection

Most utility-scale BESS products offer various options for an enclosure's dust or moisture protection. Physically, this results in varying quality of seals, hinges, and security hardware used on the doors and ventilation ports on the enclosure. These choices are made when the battery is ordered and have varying costs based on the level of protection. The better protected from the elements the unit is, the higher the cost, as may be expected. This allows the project designers to select a product based on the local environment or operating conditions. For example, if a project is in the desert with harsh dust storms, or near the ocean with its pervasive salt fog, a higher level of ingress protection will likely be required. Table 4-1 describes the typical moisture and dust protection levels seen for batteries. NEMA ratings are more common in the US, while most other parts of the world use IP Ratings. These NEMA and IP ratings are not exactly correlated but are similar.

Table 4-1
Dust and moisture protection ratings

NEMA	IP	Protection Against	Typical applications
3	54	• Indoor or outdoor • Accidental contact, falling dirt, windblown dust • Rain, sleet, snow • External ice formation	Sites with wind and dust concerns, warm weather environments
3R	14	• Indoor or outdoor • Accidental contact, falling dirt • Rain, sleet, snow • External ice formation • Slightly less protection than NEMA 3 against windblown dust	Sites without wind and dust concerns
3S	54	• Indoor or outdoor • Accidental, falling dirt, windblown dust • Rain, sleet, snow • External mechanisms must operate with external ice formation	Cold weather environments

4	55	• Indoor or outdoor • Falling dirt, windblown dust • Rain, sleet, snow, splashing water, and hose spray • External ice formation	Sites with harsh weather
4X	55	• Indoor or outdoor • Falling dirt, windblown dust • Rain, sleet, snow, splashing water, and hose spray • Corrosion, external ice formation	Windy sites near salt water, e.g., coastal areas with harsh weather

Additional System Component Enclosures

Besides the battery, the other supporting pieces of equipment on a utility-scale project are also typically mounted nearby and designed for outdoor installation. Foundations vary by manufacturer and include concrete slabs, pylons, or piles.

The following is a list of the typical enclosures seen on BESS projects. Depending on design and manufacturer preference, some may be combined:

- AC cabinets, where the controllers, UPS, lower power switches, and sensors for the battery cells or modules
- DC cabinets, where the positive and negative cables from the battery enclosures are collected before continuing to the PCSs. Bus work and fused switches are typically found in these cabinets.
- PCS enclosures, where the power transistors do their hard work, but also where other supporting equipment is housed such as:
- DC capacitors, fuses, and contactors
- A cooling system
- An electronics control system
- A breaker and relay

- In some cases, a control power transformer (to power the electronics, relay, and cooling system)
- An auxiliary breaker
- Control rooms for outdoor installations
- Auxiliary panels, where the breakers and circuits for the system auxiliary power are found

If the system is interconnecting at a substation, there is typically a separate on-site control structure to house the required metering, switching, and communications gear. This is commonly located immediately adjacent to or within the substation.

4.2 Power Conversion System

The power conversion system (**PCS**) is one of the key pieces of equipment in all energy storage systems. The DC battery enclosure typically puts out voltage from 400-700 VDC, which is then wired in parallel with other enclosures, and then connected to the PCS.

A PCS is often referred to as an inverter in the BESS industry. This is because many people in the industry are used to working with solar PV plants, in which inverters are used to convert the DC solar power into AC grid power. This is accurate when the BESS is discharging, but when charging, the PCS must convert AC back to DC, in which the PCS is acting as a rectifier, or charger. Although PCS is more accurate, inverter is an acceptable and common name used in the BESS industry. We will use PCS throughout this book. Other common names for this component include **inverter-rectifier,** or **bi-directional inverter**.

PCSs come in many shapes and sizes, some with integrated transformers and switchgear that may be as big as 40-foot shipping containers. Regardless of the size, there are several important functions provided by the PCS in battery systems.

Critical PCS Functions

DC-AC Inverter: An inverter's main function is to take direct current (see Section 2.2) from the battery's terminals and convert it to a sine-wave output that is synchronized with the AC power grid. An inverter typically senses the frequency of the grid and uses power electronics to make thousands of rapid switches per second, resulting in a smooth coupling between the DC BESS and the AC grid. A similar function is performed by inverters used for PV power and wind power.

AC-DC Rectifier: A rectifier performs the job of the inverter in reverse. It takes AC power from the 50 or 60 Hz grid and converts it into DC power that can flow into the battery. For charging the battery the rectifier typically outputs an almost-constant voltage that is slightly higher than the battery voltage, resulting in a "flow" of current (and energy) from the AC grid into the battery. The maximum rate of this flow can be set on each PCS and must correspond to the safe operative characteristics of the battery. With DC power there is actual movement of electrons through a conductor, while with AC current the electrons do not move, but rather oscillate back and forth.

DC-DC Converter: A DC-DC converter is not present in all BESS projects, but is critical whenever DC voltages need to be stepped up or down. The power electronics in this component take a DC voltage that is variable, such as the output of PV, and adjust it to the range appropriate for the expected functionality. To cause a BESS to charge, the converter typically sets the bus voltage at slightly higher than the BESS voltage, causing current to "flow" into the BESS. To cause a discharge, the converter would set the bus voltage to be slightly less, resulting in a current gradient that causes current to flow out of the BESS. Typical applications of DC-DC converters are DC-coupled

PV+Storage systems, or when augmentations require matching DC voltages from old and new battery racks.

The PCS can be seen as a valve which allows the power to flow onto the grid when the BESS is discharging, or out of the grid when the BESS is charging. Even if the BESS is fully charged, without a functional PCS, there is no way to get the power to flow out of the BESS and onto the grid.

Each of the three above functions are conducted by the PCS by switching transistors at high frequencies, chopping the direct current of the batteries into sine waves of the appropriate frequency to match the power grid, and vice versa, in the case of AC-DC converters. In the case of DC-DC converters, there are several ways that circuits can convert a DC voltage to another DC voltage, but in general they also use transistors to manipulate currents.

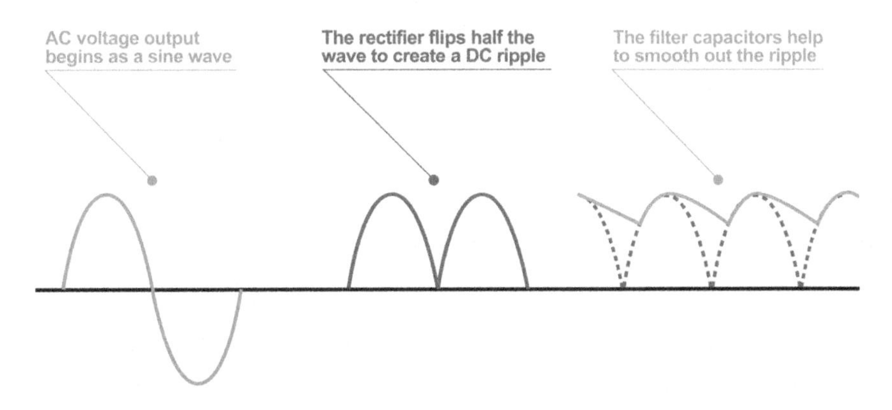

Figure 4-3
AC to DC rectification

A simple way to think of a PCS is as a black box with current and voltage minimum and maximums on both sides. For example, a 2.2 MW converter may require a minimum of 570 VDC to a maximum of 950 VDC and currents of less than 3600 A on the

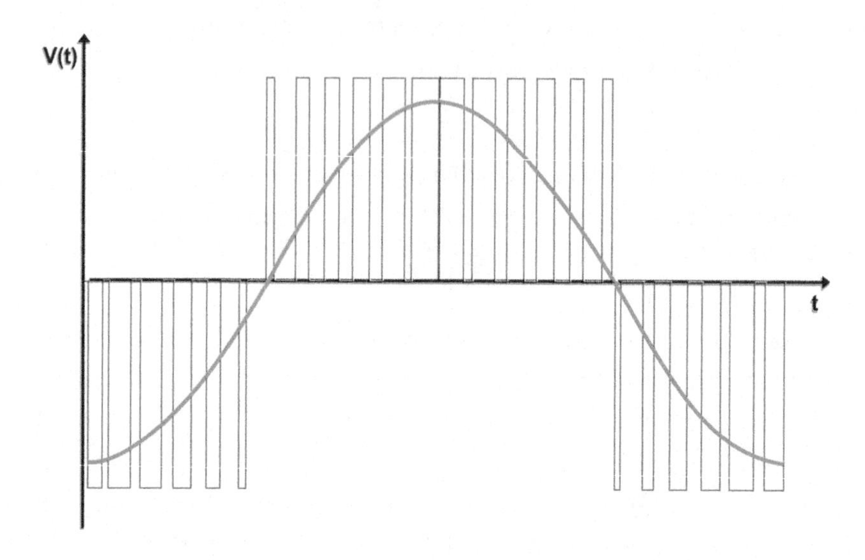

Figure 4-4
DC to AC voltage inversion

battery side and 400 VAC and currents of less than 3300 A on the grid side. With these specifications one can determine which battery models would align with this PCS by checking their voltage swing between 0% and 100% SoC. If the voltage window when the battery is between empty and full falls between 570 and 950 VDC then the battery would work with the PCS. In addition, the maximum current and power rating of the battery needs to be in-line with those required by the PCS to test compatibility. The AC side requires an appropriately sized transformer with a secondary voltage that matches the PCS output and a primary that matches the grid voltage.

DC-block versus AC-block

While all battery projects need PCSs, there are two different ways to incorporate them into the system architecture, with either a DC-blocks and AC-block product.

A **DC-block product** refers to a battery container that outputs DC voltage (typically around 1100-1400 VDC) at its terminals. DC-block products are typically pad-mounted and must be placed adjacent to a PCS to limit losses. Most DC-block products also require an auxiliary transformer to feed the BESS its power.

DC-block is the traditional approach to BESS design. Relative to an AC-block, it requires some additional BoP design, equipment, and installation. It has the advantage of offering a more customizable design. For example, in the Chilean BESS market many systems are built with a 5-hour duration. This duration is non-standard for most battery products, but with a DC-block battery product, the designer can independently select a PCS that is sized for a 5-hour solution, which can be smaller in power, and therefore cheaper – In contrast, the AC-block often comes in a fixed ratio of PCS to battery sizing, meaning it may not be able to economically be sized to a 5-hour system.

AC-block refers to a battery product that has an integrated PCS: the PCS is within the same enclosure as the batteries. These types of systems connect at 3-phase AC bus, typically from 400-800 VAC. This has the benefit of simplifying these systems, since no external PCS is needed.

Most AC-block products have many PCSs installed in repeating units throughout the BESS. This allows the system to swap out both PCS and BESS module if there is a failure.

AC-block products have the advantage that if a long-term service agreement (LTSA) is purchased for the BESS, it covers both the battery and PCS, while many DC-block products do not have warranty coverage for the PCS after an introductory period (typically 3-5 years). Although under warranty there still may be corrective maintenance costs due, AC-block products can simplify the O&M of the system. This is discussed more in Section 9.

Figure 4-5 shows an SLD highlighting the differences between a sample DC-block (red) and AC-block (blue) product. In this example, the system on the left has two enclosures, each with six racks, connected to a single PCS. The DC system has a 1500 VDC output, the PCS has a 480 VAC output, and the transformer steps the voltage up to 34.5 kV AC. In the AC block product, there is a single enclosure with six racks, with 3 internal string PCSs. The product outputs 480 VAC, which can then be fed directly into the transformer. Note that while this AC-Block product does not require external aux power feed some AC-block products still do require this circuit.

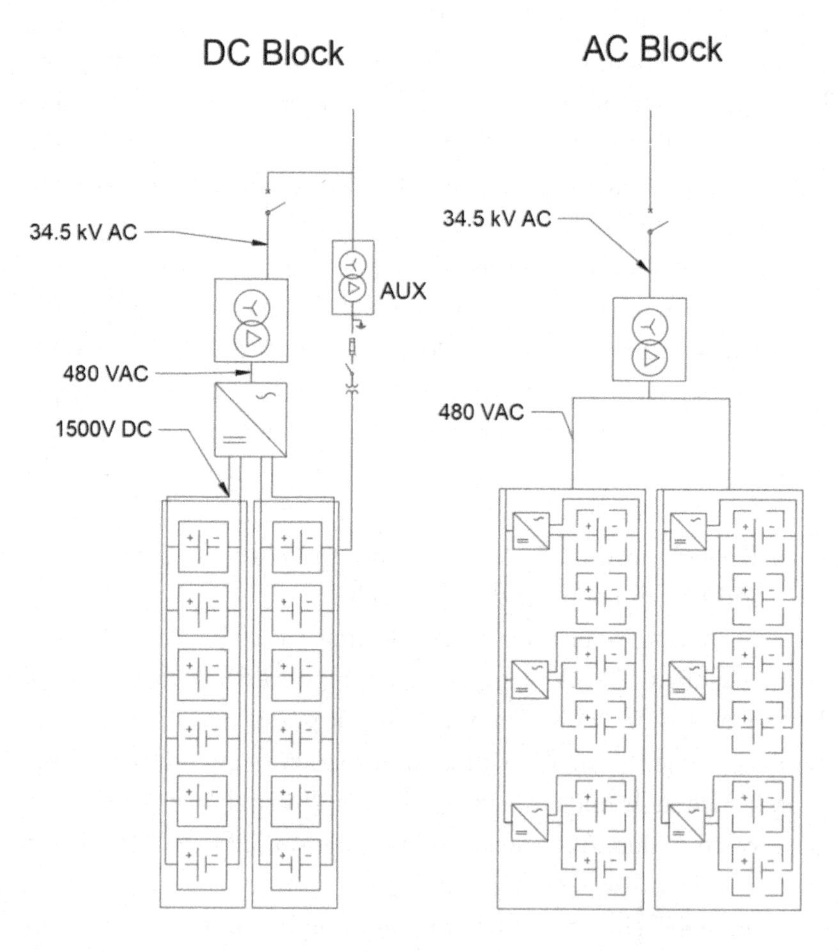

Figure 4-5
DC-Block vs. AC-Block Product SLD

Examples of both DC and AC-block products can be seen in the Usual Suspects, Section 4.7, at the end of this chapter.

PCS Safety

As the element that enables current to flow on to or off the grid, the PCS must perform a critical safety function in any BESS. Voltage and current must be continuously matched to the correct levels to avoid damage to system components and personnel. Because of the importance of this task, the initial configuration and ongoing management of the PCS is critical. This is why battery manufacturers often provide a list of acceptable PCSs that can be integrated with the battery systems, and often include the PCS in their scope. This ensures that the batteries and the PCS have been tested to work in harmony in both the electrical and communications realms.

Most PCSs utilize a circuit breaker at the AC output side and fusing at the DC input side. These components provide an extra level of protection to the system components and workers and must be sized appropriately for the system design.

More modern PCSs will also typically include an insulation resistance monitoring device, which enables ground fault monitoring. Most large-scale BESS have a floating ground voltage on the battery side, utilizing only a positive and negative leg from the batteries. This means that a single ground fault will have minimal impact on the system, but two ground faults on a system at the same time will set up a dangerous situation. To mitigate this risk, systems typically monitor for ground faults, send appropriate alerts, and alert the maintenance team to quickly address any issues.

Selecting an PCS

There are several bankable PCS manufacturers to choose from (see Usual Suspects at the end of this section). When selecting an PCS for a BESS, ensure that the PCS is UL 1741 certified, the typical standard for energy storage systems, and a requirement for the broader UL 9540 certification (which is discussed further in Chapter 10 on Safety). In California, for example, the PCS requires an additional certification known as UL 1741 SA with added technical requirements. Utility-scale PCSs produce very high currents which can be extremely dangerous, so appropriate safety certifications are not only critical, but mandatory in most jurisdictions.

The inverter chosen for an application should be well vetted to meet the project's technical requirements, at an in-market price point, and be available in the quantity needed within the project's required timeframe. Commissioning support is another concept to keep in mind, as there is often a need for a PCS representative and commissioning team to be available as the system nears its commercial operation date (COD). Elements to keep in mind when choosing a PCS include the company and PCS model's historical performance, company forecast, and the stability of the PCS manufacturer's parent organization. This is because if the manufacturer goes out of business, that would be a large risk to the project, since obtaining support or parts may be difficult.

The PCS must offer the appropriate functionality for the BESS use case, as well as meeting the sizing and interconnection requirements. Additionally, it must offer appropriate controls integration and allow for an ambient temperature range that aligns with the project site's climate and altitude. Compatibility with the chosen battery model is also a consideration, including the number of battery containers that can be connected behind a PCS can vary based on the current loading, fuse rating, voltage balancing, etc. This can lead to situations where more PCS units than the MW requirements of the project are needed, or more battery containers are needed to balance the power flow. This balance of containers to PCS must be done for every product combination carefully.

Utility Interactions

Utility companies are very cautious when it comes to battery systems and inverters, since in an outage these devices can feed power onto the grid and potentially be lethal to line workers in the vicinity or cause other electrical disturbances on the grid they are working to maintain. UL 1741 includes special "anti-islanding" provisions which require that in case of outage the inverter disconnect from the power grid. Only in cases where the BESS is used as a microgrid or to provide black start services would this safety provision need to be overridden.

For utility-scale systems these controls functions are handled either by a third-party controller or by the controller that is provided by the BESS manufacturer. PCS design and component selection is covered in greater depth in Section 8.1.

Grid-forming vs. Grid-following

Grid-following and grid-forming refer to how an inverter-based energy source interacts with the grid, specifically voltage and frequency. There could be specific requirements from the interconnecting utility/ISO on these characteristics of the PCS for the project.

Grid-following PCSs mimic current sources at their output terminals. They rely on an existing voltage source (the grid) to synchronize and inject current into the grid. They need a stiff grid voltage present to deliver active and reactive power. They cannot operate independently, black-start a system, or power a microgrid.

Grid-forming PCSs, in contrast, act as voltage sources, setting and maintaining the voltage and frequency like a synchronous generator. They can form and maintain a stable grid voltage and frequency and are capable of black-starting and operating in isolated mode. Grid-forming PCSs enable higher penetration of renewables by emulating the behavior of conventional synchronous generators.

4.3 Thermal Management

All battery systems create heat, which is generated by storing and releasing energy and having inherent losses. If 100 kWh of energy is taken from the grid, the current carrying that energy must pass through a transformer, PCS, cabling, and into the battery. And each component experiences losses, mostly in the form of heat.

Allowing batteries to overheat may shorten their life, cause permanent damage, and increase the risk of fire. There are systems built into the BESS to prevent this from happening, such as the BMS, EMS, and the thermal management system, whose purpose is to keep the BESS within its optimal temperature zone.

All thermal management systems have two main goals: maintaining the battery cells' temperatures in their optimal performance range and maintaining a uniform temperature across the system. Optimal performance ranges are typically defined in the battery energy supply agreement (BESA) or in the Long Term Service Agreement (LTSA) as conditions for warranties or performance guarantees if applicable. Typical optimal operating temperatures at the cell are 25 °C to 30 °C for LFP and 18 °C to 28 °C for NMC, although this varies across different manufacturers. Even if a system maintains all cells within the optimal range, a spread of temperatures across cells will result uneven degradation of that system, which is problematic and can create operational issues.

Note that while most BESS are concerned with overheating, BESS in cold climates, are threatened by cold temperatures. Cold temperatures reduce battery capacity and response. While charging or discharging, BESS can usually keep themselves warm, but when not operating needed to be heated.

Typical thermal management systems for batteries fall into one of two categories:

- Liquid-cooled systems, which use a closed-loop system with a refrigerant to cool the battery cells and associated hardware

- Air-cooled systems, which use a combination of ventilation and air-conditioning to ensure that the battery system maintains a steady temperature throughout

Liquid-cooled Systems

Liquid cooling works like the cooling system in traditional gas- or diesel-powered vehicles. The battery modules are designed with cooling fins integral to the system and in close contact with the battery cells. The racking has channels and tubing that allow the refrigerant to circulate and transfer heat throughout the system. After the refrigerant absorbs the heat from the cells, it returns to a type of radiator which expels the heat to the exterior of the storage unit, typically using a fan blowing across the radiator. The circulation and heat expulsion can be controlled by the HVAC system and can ramp up in high-use conditions or ramp-down when the system is being used more lightly or is inactive. Figure 4-7 shows a typical arrangement of a liquid-cooled system in a BESS.

Liquid cooling has the advantage of being an efficient and proven system. The main downside of liquid-cooled systems is the additional mechanical components, which cause more failure points. Another

Figure 4-6
Typical liquid-cooled design [28]

downside is the risk of a coolant leak, which could cause a short. For example, the August 2021 fire at the Victoria Big Battery project in Australia was caused by a system that had a coolant leak. As of June 2024, nearly all major BESS integrators use liquid-cooling for cooling the batteries. Examples of current BESS suppliers that use liquid-cooled systems are Tesla, Fluence, Wartsila, and CATL.

Air-cooled Systems

Air-cooled systems are often referred to as HVAC systems, which stands for heating, ventilation, and air-conditioning. These systems typically consist of an externally mounted air conditioner, which includes a compressor and an evaporator. The system typically relies on fan ventilation, both within the enclosure and within the battery racks themselves, to distribute the cold air around the container and cells. Figure 4-7 shows a typical air-cooled system on a BESS.

The HVAC systems that are required for system performance are some of the few moving parts in a BESS project, and they are critical its success. The maintenance and reliability of these components is covered in Chapter 9 Operations & Maintenance.

Figure 4-7
Typical air-cooled enclosure design [29]

Contractual Considerations

System thermal management design is influenced by the contractual structure used to operate the battery. The typical O&M contract structures are:

- O&M provided by owner (rare)
- O&M provided by the BESS manufacturer via a LTSA or similar (most common)
- O&M provided by a third-party via an O&M contract

Structure 1 is preferred by utility companies or by large project developers that manage their projects in-house. Under this arrangement the owner takes responsibility for maintaining the system within the required temperature range. This is a risk to the project, as the BESS warranty can be voided by a mismatch between BMS requirements and the HVAC controller. For example, if an air-conditioner breaks on a hot summer day, and the system discharges, taking average cell temperatures over the maximum temperature allowed by the contract, this could potentially void the warranty on certain cells and accelerate degradation. A properly configured controls system would avoid this condition, but it is still a risk to the owner who is running their own system. The cost for this option is the cheapest, but it carries the highest risk to the project.

Structure 2 is the preferred and most common arrangement, but also the most expensive. Under this condition, since the BESS manufacturer is maintaining their own system, the owner has less concern about potential warranty violations, since the same firm is providing both the warranty and the O&M services. Under this option typically the contract will define a minimum and maximum ambient temperature allowed, which are derived from historical weather data at the site. Under ambient temperatures which exceed the maximum or minimum temperature, the system is typically de-rated or shut off entirely. Systems that have an integral PCS and cooling system, such as the Tesla Megapack, are also typically governed by ambient temperature limitations.

Structure 3 is a hybrid approach - under this arrangement the third-party takes responsibility for the system, but typically has visibility on the monitoring system and field personnel to decrease risk. The cost for this option falls between structures 1 and 2.

Further information on the pros and cons of each of these structures is provided in Chapter 9, Operations and Maintenance.

Design Approach

The design of a thermal management system must consider a variety of factors: the expected use-case of the battery, the expected ambient temperature in each location, and external considerations such as availability of water or whether the site is remote or accessible.

Some battery products have a set design that functions in a range of ambient temperatures, while others offer various levels of customizability based on the location of the system. The design may consider the results of system testing in controlled temperature changes, as well as computational fluid dynamics (CFD) analysis used to predict the expected thermal performance of the unit.

While most projects use pad-mounted, outdoor-rated BESS products, there are some indoor energy storage projects that mount stacks inside of a purpose-built building, or even a repurposed building. These facilities require extensive thermal design to ensure the system will provide sufficient cooling for the worst-case scenario, which is typically a hot day when the system is run at its maximum capacity. This design is typically performed by a thermal management design firm.

A common approach is to design based on the expected minimum and maximum temperature on a site. Depending on location, these temperatures are available from public databases such as the American Society of Heating and Air-Conditioning Engineers (ASHRAE).

The power required for thermal management is one of the main contributors to the auxiliary load requirement for battery systems. It can range from around 1% of the total power rating of the project up to 4% depending on the product's thermal design, operational profile, and ambient temperature of the site. This power requirement tends to be variable between the products in the market, and uncertain (as of the time of writing this book) in terms of actual consumption vs. rated requirement as claimed by the manufacturer. It is starting to be one of the distinguishing factors between products as it can be a significant cost to the project.

For large projects, analysis of the expected thermal performance is a critical step that will be examined by an independent engineer (IE) as a requirement for the financing of the project. Given the potentially disastrous consequences of improper thermal management, the importance of appropriate thermal design cannot be overstated.

4.4 Balance of Plant

Balance of Plant (BoP) or ancillary components are the pieces of equipment other than the batteries and the large equipment (i.e., PCS, transformer, switchgear) that are required for the system to operate. These include switches, breakers, fuses, auxiliary transformers and switchboards, networking equipment, fire suppression systems, HVAC, UPSs, and various other components.

This section reviews some of the more common elements of the BoP, starting with the Collection Systems, both DC and AC. Substation and transmission lines are covered, including how the project is interconnected with the grid. The three types of transformers commonly found on a BESS project are covered: MV, auxiliary, and MPT. Auxiliary power is reviewed, an important consideration for DC-block products. Lastly, the section covers BESS electrical protection, which is how the BESS plant is designed to mitigate any AC or DC fault conditions that may occur.

Collection Systems

The system of conductors, busbars, and combiner boxes that connect individual units into larger flows of power and energy is known as the **collection system**. The BESS project's collection system is typically divided into the DC collection system (on the battery-side of the PCS) and the AC collection system between the PCS and the grid.

For projects using an AC-block product, there is no need for a DC collection system, since this happens within the BESS enclosure when it is delivered to the project. For projects with a DC-block product, both a DC and AC collection system is required – first to connect the BESS units to the PCS, and farther upstream to connect the many PCSs to the grid.

DC collection

The DC-collection system includes all wiring between the battery cells and the PCS. Typically, a portion of this is provided by the BESS integrator, since it is physically located in the BESS container. A DC-Block product requires an external DC collection system to connect the battery enclosures with each other and then to the PCS where the output is converted to AC. For larger utility-scale products this is typically in the form of terminations within the BESS enclosure where conductors are landed, usually with crimp-on aluminum or copper lugs that allow for a firmly bolted connection. These conductors typically exit the enclosure via conduit that is stubbed-up into the slab, and connect to the next piece of equipment, often a combiner box or PCS. Some systems use direct-bury cables that are rated for burial without a conduit. These design decisions are discussed further in Chapter 8.

DC conductors are typically a positive and a negative wire coming from each unit, although larger systems may require multiple sets of parallel conductors to handle the large amperages produced. The expected voltage and amperage determined in the design process will indicate the required size and type of conductor or busbar used. Most DC collection systems for utility-scale systems are in the range of 1000-1500 VDC.

Most containerized battery enclosures come pre-wired by the manufacturer, but if the project procures only racks that will be installed in an enclosure or building, they will require a DC-collection system to connect those racks to the DC combiner boxes.

AC collection

The AC-collection system connects the outputs of the PCS to the grid. PCS typically output at low-voltage ranging from 240 VAC split phase for smaller systems to 660 VAC three-phase for larger systems. For utility-scale battery projects it is common to put an MV-transformer immediately adjacent to the PCS, and some manufacturers even include a PCS and transformer mounted in a single skid. The connection between the PCS output and transformer on this type of system is typically busbars or large crimped conductors. The transformer outputs medium voltage (typically 10-40 kV, depending on local custom and regulation). The AC-collection system then links all the multiple PCSs, most commonly using direct-bury MV cables. Site layouts are normally done in a way that allow the AC-collection system to have a direct path through the site that passes by all PCSs enroute to the substation. The final step in a utility-scale system is interconnection with the substation, which is typically accomplished by feeders which transition from underground to aerial conductors, and feed into the system of overhead switches, breakers, and other gear within the substation. A typical AC collection system is shown in detail in Figure 4-8.

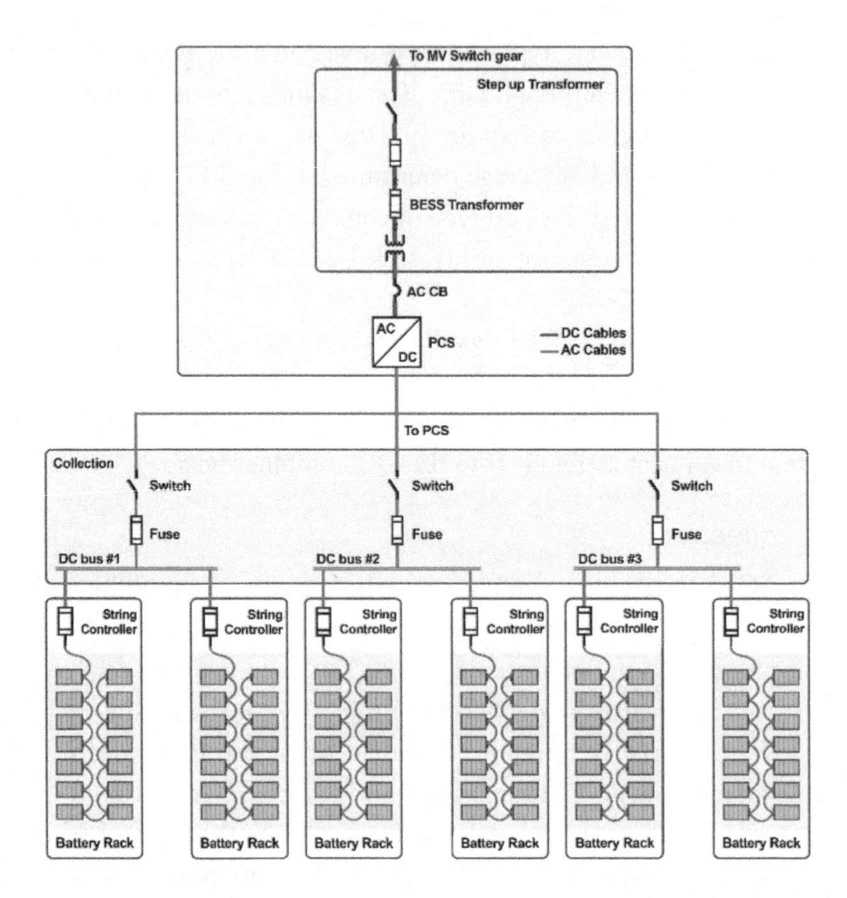

Figure 4-8
Single line diagram of DC and AC collection systems

Transformers

Utility-scale BESS are interconnected at high voltage (50 kV and higher). These systems require transformers to step-up the output of the PCSs to the interconnection voltage.

Most BESS projects have 3 types of transformers:

- **MV transformers,** which step the voltage up from the low voltage output of the PCS to a higher (medium) voltage.

- **HV transformers**, which step the voltage up from MV to high voltage for connection to a transmission line or substation. These transformers may be referred to as the **Main Power Transformer (MPT), Generator Step Up (GSU), PCS Step Up (ISU),** or **Step up Transformers (SUT)** .
- **Auxiliary transformers**, present for most DC-block products, typically have sets of MV transformers used to provide power to the auxiliary loads such as HVAC, site lighting, and communications systems.

Medium Voltage Transformers

MV transformers in battery systems step the voltage up from the output of the PCS (usually 480-690V three-phase) to the medium voltage. While they are called "step-up" transformers, this comes from traditional generation plants where power flows only in one direction – for a BESS, if the plant is discharging the transformer is stepping the voltage up, while if it is charging it is stepping the voltage down. The MV transformer is typically located immediately adjacent to the AC side of the PCS. By putting the MV transformer very close to the PCS, resistive losses can be minimized – this architecture is sometimes called "**close-coupled**". At lower voltages, high power creates high currents, and the designer will want to minimize the electrical distance and maximize the cross section of the conductors.

High Voltage Transformers

The HV transformers connects the MV (20 to 50 kV) collection system to the interconnecting HV level (69 to 500 kV), where it then goes on to connect to the utility's transmission line. These transformers tend to be very large, in some cases as big as 500 megavolt-amperes (MVA). Some projects have multiple MPTs for redundancy, while others have a single MPT. For PV+Storage systems, the MPT may be shared between the battery and solar circuits.

Projects that interconnect at HV require a substation to house the gear to step up the voltages into the power grid. Utilities typically have strict requirements for the design of these substations, and in some cases the substation may be designed and owned by the utility itself.

The determination of the required voltage of interconnection is determined by the power of the BESS, the available distribution voltage, and the amps. Table 4-2 shows typical interconnection voltages required for BESS integrations in the US, though this may vary regionally with utility codes and standards.

Table 4-2
Typical interconnection sizing

BESS size [MW]	Interconnection voltage	System size category
10-50	69-115 kV	Small utility-scale
50-100	115-230	Medium utility-scale
100+	230 kV and up	Large utility-scale

Auxiliary Transformers

Most BESS products on the market come with DC terminals that connect to a PCS, typically at 1000-1500 VDC, or to the AC circuit at 690 AC. However, a BESS requires lower voltage (LV) power to operate auxiliary service such as HVAC system, lighting, communications, and any other LV loads on the site. While some products pull this power from the DC side of the system, most have a separate auxiliary transformer that feeds power to these items. This means that each BESS enclosure requires both DC and AC electrical connections. The AC auxiliary power for utility-scale systems is typically provided via separate MV transformers, the secondary side of which distribute LV power, typically at 208 or 480 V three-phase, depending on the specific project requirements.

Depending on the manufacturer and design, auxiliary power transformers may have their own feeder directly to the MV or HV grid, or they may be fed from the MV grid that feeds the PCS. In either case, these transformers are typically required to be metered, usually as required by contractual arrangements with utilities and off-takers. This is covered further in the Metering Section.

Battery products which include an integrated PCS typically require less auxiliary power than monolithic PCS but may still require some low voltage service on site to power lighting, communications, and any other required loads. The project engineering team should be aware of all of the various auxiliary requirements and ensure that this system is adequately sized and designed.

Transformer cooling

Keeping a transformer within its design operating temperature is critical to its successful operation. There are several common methods of cooling, which vary based on the size of the transformer. Table 4-3 shows a summary of the most common methods and the sizes and types of transformers which use them.

Table 4-3
Transformer cooling methods

Abbreviation	Cooling type	Cooling method	Notes
AN	Air natural	Transformer cooled by surrounding air with no mechanical system	Used in dry-type units up to 3 MVA
ONAN	Oil natural air natural	Oil circulates by convection and is cooled by passing through fins	Typically used in units up to 5 MVA
ONAF	Oil natural air forced	Oil circulates by convection and is cooled by air blown by fans	Typically used in units from 5 – 60 MVA

OFAF	Oil forced air forced	Oil is pumped through the transformer and is cooled by blowing fans	Typically used in units from 60-100 MVA
OFWF	Oil forced water forced	Oil is pumped through the transformer and is cooled by water in a heat exchanger	Typically used in units of 100 MVA or higher

It is common for larger transformers to operate using a variety of different modes of cooling. This is because although a transformer may be designed to use a certain type of cooling, if that cooling system is damaged it may still be possible to use the transformer, but its capacity will be derated to account for the reduced cooling available. Transformer ratings often reflect this by listing multiple ratings, as shown in Figure 4-9, which shows a transformer that has a 160 MVA rating when ONAF cooled, but lower maximum power ratings when utilizing other cooling methods.

Figure 4-9
Typical transformer nameplate showing cooling methods

Auxiliary Power

Auxiliary power, commonly referred to as **aux power,** is necessary for DC-block battery products to provide all the power they need to operate. Systems that depend on aux power are thermal management system such as HVAC or chiller units, the battery control system, fire protection elements like ventilation fans and sensors and the PCS control system. Auxiliary power is an especially important requirement for battery systems that are specified by the manufacturer. It is important to ensure that auxiliary power is always enabled to keep the batteries in optimal operating conditions. The peak, average and stand-by mode auxiliary power requirements are specified by most BESS suppliers, as well as the voltage at which they must be supplied to the battery containers. These specifications must be accounted for in the design process.

Typically, the aux power has a component that is fed via an uninterruptible power supply (UPS) – this is another small battery + LV PCS set that ensures that the auxiliary system can withstand short outages (typically 2 hours or less). The aux power circuit backed up by a UPS supports critical functions such as cell performance reporting, fire safety systems, and internet connectivity.

Auxiliary power can be provided two ways: grid-supply or self-supply.

Grid-Supplied Auxiliary Power

Grid-supplied aux power is most often tapped from the MV bus, delivered via an MV transformer feeding one or several DC-block products. This aux power is metered separately and is fed into the BESS enclosure via separate terminals which are not connected to the DC output of the BESS. Benefits of grid-supplied aux power is that it is easier to track losses and it is typically more efficient to run the container loads on AC rather than DC. As a drawback, these systems are more complex, since they require a separate set of transformers and wiring to each BESS unit, as well as additional HV circuits for larger systems. There is additional metering required, since it is important to track both DC inputs/outputs of the battery and AC input of the auxiliary power. Additionally, at the time of writing, transformers of all voltage classes suffer from long-lead times in the wake of the IRA and developers rushing to secure them. Auxiliary power design and orders for transformers need to consider delivery timelines to avoid affecting the project schedule.

Self-Supplied Auxiliary Power

Self-supplied aux power means that the BESS auxiliary loads are fed from the AC side of the PCS – this is true whether the PCS is integral to the battery enclosure or is a separate pad-mounted unit. Transformers are typically needed to step down from the PCS AC output voltage (often 600-800 VAC) to the operating voltage of components (typically 480 VAC). Self-supplied aux is simpler since it doesn't require as much wiring as grid-supplied aux; however, it has the downside that if the PCS is ever de-energized, the BESS will not be able to operate at all. Alternatively, a grid-supplied aux system will be hot in the case of an offline PCS, since it is fed via a separate MV auxiliary circuit. Self-supplied aux units may require an additional UPS to prevent downtime of safety and control systems if the PCS is disconnected.

BESS Electrical Protection

The high energy density of modern batteries comes with the potential for catastrophic, or even deadly, system failures. Intelligent system design, using industry-approved standards, is the best way to ensure that any system being built is safe and durable. See Section 8.3 for more information on battery codes and standards.

A major part of safety design is **electrical protection**, which is the process of designing the fuses, switches, and breakers that allow for safe operation of an electrical system. This is its own discipline, with much literature written about it, so this section will only provide a brief introduction. Note that appropriate protection design must be performed by an experienced **professional engineer** who is authorized to work in the jurisdiction where the plant will operate.

All facilities should have a switching protocol that electricians and engineers can reference when working on or supporting work at the site. This should be included in the as-built drawings, as well as in hard copies close at hand on site. Battery systems can be especially complicated to create switching protocols for, given the bidirectionality of the systems and dual source locations, as well as potential voltage differences between the DC bus and the contactors in the racks. A reputable engineering firm with experience in switching logic is typically engaged to perform this portion of the design.

Switches

Switches, also known as **disconnects**, allow for isolation of parts of a system when work needs to be done on it, allowing other sections to remain in operation.

Switches are typically found at points in the system where isolation is required to work on the equipment safely, either for maintenance, troubleshooting, or replacements. A project owner doesn't want the entire project to shut down because one battery cell is acting up, so isolating certain circuits while keeping online is necessary. While a switch may isolate a portion of the system, the batteries are likely still charged and energizing DC buses, which is more dangerous than traditional systems. Power systems workers must be educated and diligent about working on BESS components, more so than with traditional one-way power systems.

A well-designed system will include switches at various locations:

- The rack level, which allows the cells to be disconnected from the battery DC bus
- The DC combiner, where all the battery racks come together onto the common DC bus

- The PCS, where cables from one or more DC combiner enter the input section of the PCS hardware
- At the PCS transformer
- At the step-up substation

Fuses

Fuses are components designed to be destroyed, or 'blow,' when they experience excessive current. This protects the conductors, equipment, workers, and the facility in the case of a short, fault, or other electrical overcurrent events. Fuses are installed at strategic points throughout the system, typically in the following locations on the DC side:

- High-speed semiconductor fuses and diodes at the battery module level
- Battery protection fuses, disconnect switches, and surge protection devices at the rack levels protect the cells from being overdriven by the PCS, or to protect from and open the circuit during a fault
- High-speed semiconductor fuses in the DC combiner
- High-speed semiconductor fuses and surge protection devices in the DC panel
- Arc flash relays in the container
- High-speed semiconductor fuses and DC ground fault monitors on the DC side of the PCS

Breakers

Breakers and **relays** protect equipment and workers during overcurrent and overvoltage events, while also being able to operate as switches. There are several circuit breakers in large-scale BESSs and they operate automatically based on relay settings which are determined during system design, similar fashion to switches. Breakers are found in the AC circuits of a BESS:

- Arc-flash relays and point sensors on the AC side of the PCS
- Inside of the transformers
- At the substation and/or point of interconnection

Figure 4-10 shows the typical locations of protection devices in a BESS.

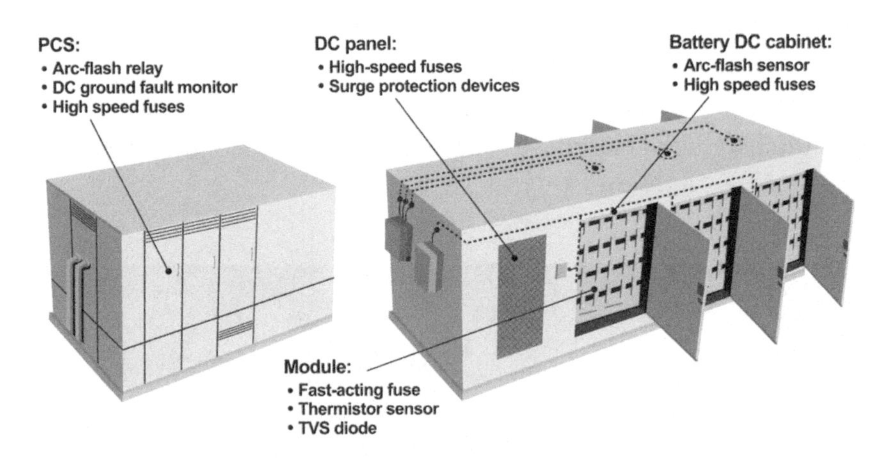

Figure 4-10
Overview of switch and fuse locations

4.5 Substations and HV Scope

The substation of a BESS plant is the interface between the batteries and PCS units collectively to the grid. There are typically two different substations on a BESS project – the substation that is part of the BESS project, and the utility substation at which the project is interconnected. Collectively the project substation, upgrades to the utility substation, and any underground or aerial transmission lines are referred to as the **HV scope** of the project.

They are typically performed by contractors specialized in HV and utility work.

The BESS project substation houses the MPT, which modifies the voltage upwards and downwards between the MV voltage and that of the transmission line the project is connected to. The substation also houses the switchgear and control systems that regulate power flow and isolate downstream faults. A BESS project substation is similar to a substation for a solar PV plant, except that BESS substation needs to be able to handle bi-directional power flow, requiring specialized equipment. In addition, BESS projects are often used for services like frequency regulation and peak shaving, which require rapid response times. Substations for BESS projects must be equipped with advanced control systems and fast-acting switches to enable the quick charging and discharging of batteries as needed. A typical BESS project substation is shown in Figure 4-11.

Figure 4-11
Typical BESS project substation [30]

Most BESS projects are constructing close, or even adjacent to, an existing utility substation. If not, the project may need a dedicated high-voltage transmission line that connects the project to the utility substation or the grid – this is called a **generator tie-line**, or **gen-tie line** for short. The gen-tie line needs to be designed and executed by the project's owner. Only very large projects tend to include transmission lines, since they are very expensive to build and require extensive permitting.

Substations are a critical part of BESS projects, but they are a portion that is typically designed, built, and maintained by firms specialized in this domain. More detail around substations can be found in the Section 8.1 of the chapter on EPC.

4.6 Metering

Most battery systems generate revenue in part by selling some type of service to a power market. This service is typically compensated by measuring the electrical power and energy that flows onto the grid (discharging) or is pulled off the grid (charging). This measurement is conducted by watt-hour meters, which are a critical component of all battery systems. All meters require measurement of current using a **current transformer (CT)** and measurement of voltage using a **voltage transformer (VT)**. Combined, these measurements allow the system to measure the amount of power flowing in either direction at the point of metering. Timestamping power levels and tracking in a register allows these measurements to be integrated over time, thereby calculating energy (kWh) production or consumption.

Metering requirements are usually stipulated by the connecting utility, and normally must be provided by a 'revenue-grade' meter, meaning that the meter has a certain level of accuracy and other requirements. Meter manufacturers typically list their products' level of certification and will work with a project to identify a meter that meets all the project requirements.

While the Project must design and supply the meter, the utility company typically performs an inspection when the meter is installed and affixes an anti-tamper seal to the meter. Data from the meter is typically provided to the utility company and the Project via internet telemetry on site.

In the case of PV+Storage, which is one of the most common configurations for BESS, metering for the PV portion and BESS portion may need to be enabled separately, especially in cases where the off-take requirement mandates it, or the Investment Tax Credit is applied to the whole project.

One of the most important functions of the metering is the ability to measure the energy capacity of the BESS. Whenever an energy capacity test is conducted a revenue-grade meter is referenced so that all parties involved can be certain of the accuracy of the test and the authenticity of the data. Testing provisions in battery contracts often include an assumption of losses of the equipment between the BESS and the testing meter, to ensure that measurement at the meter can be 'discounted' back to the BESS. For example, if the losses between the DC terminals of the BESS and the meter are described in the contract as 3%, and the revenue meter shows 26 MW of power output for 4 hours (=104 MWh), this would result in a net power of 26 x 1.03 = 26.8 MW for 4 hours, or 107.2 MWh of energy capacity, at the DC terminals.

It is notable that the current, voltage, and power may be measured by the power electronics in some components of the BESS such as the BMS, the EMS, or the PCS. While these sensors provide generally acceptable ways to check the site output and are used as inputs to the controls of the BESS, they are typically not legally auditable or calibrated by independent labs. Because of these they are not considered 'revenue-grade' and are not generally used for compensation or capacity testing of the battery.

Round-trip Efficiency

All batteries lose some energy between charging and discharging. The percent of usable energy at the POI, from the energy that entered the BESS through the POI and after all losses, is called as the **round-trip**

efficiency (RTE). Typically this includes the following losses (laid out for charging and discharging):

- HV to MV step-down transformer loss
- MV to LV step-down transformer loss
- PCS AC to DC loss
- BESS charging loss
- BESS discharging loss
- PCS DC to AC loss
- LV to MV step-up transformer loss
- MV to HV step-up transformer loss
- Line losses for all steps in the process

The values of these vary widely among different types of systems and different products, and some losses have a fixed and variable component which varies based on the loading of the system. Typical values for utility-scale BESS are around 94% efficiency for the BESS itself at its DC terminals, and around 88% for the round-trip efficiency of the system as a whole, as measured at the HV terminals of the main power transformer. These values vary substantially from one product to another.

Note that some ratings of RTE include measurement of the additional auxiliary load, while others do not. This is a critical factor in determining what an acceptable RTE is for any project. Auxiliary power varies from product to product and can vary significantly based on how the BESS is being used and the ambient temperature conditions.

4.7 Usual suspects

This list contains a sampling of the leading energy storage integrators and PCS providers. Following the descriptions of the firms is a list of some of the most common utility-scale BESS and PCS products in the marketplace, as of June 2024.

Battery Integrators

AESI: American Energy Storage Innovations (AESI) specializes in advanced energy storage solutions, offering a range of products for utility-scale, commercial, and industrial applications. Founded in 2021, the company's TeraStor product line is notable for offering a product that can be based on multiple different cell suppliers.

BYD: A leading Chinese manufacturer of electric vehicles and batteries. See the full company description in Section 3.10. BYD was an early entrant to the integrated ESS market. Their current product boasts 6.4 MWh DC storage, adhering to the Chinese battery standard GB/T 36276.

Canadian Solar: Canadian Solar is a global solar power company that manufactures solar PV modules and provides solar energy solutions, including energy storage systems. Their utility-scale BESS product is known as SolBank.

CATL: The world's largest battery manufacturer. See the full company description in Section 3.10. Contemporary Amperex Technology Co., Limited (CATL) entered the ESS market in 2021 with the EnerC, a containerized liquid-cooled DC storage product. In 2022 they released the newer version EnerOne, and in 2024 launched TENER, also known as Tianheng, 6.25 MWh DC BESS product touting 5 years with zero degradation.

EnergyVault: EnergyVault is an energy storage company that began with a focus on gravity-based energy storage solutions, lifting concrete to store and discharge energy. In 2023 they began offering Li-ion products. Their products are available with multiple cell technologies.

EVE: See the full company description in Section 3.10. In addition to offering cell manufacture, EVE Energy offers fully integrated BESS products. Their current product is a 20-foot integrated BESS known as "Mr. Giant" has been stated to have a capacity of up to 5 MWh.

FlexGen: Founded in 2009, FlexGen is a supplier of BESS projects, including system supply, EMS, and O&M of BESS plants. They integrate BESS products using components from various suppliers. Their controls product is known as HybridOS which sits on top of the controls offered by the battery OEM. In 2023, FlexGen's revenue was approximately $41 million.

Fluence: Fluence is a global energy storage technology and services provider. Originally founded as a joint venture between energy giants AES and Siemens in 2018, Fluence went public in 2021 on the NASDAQ. They are one of the leading energy storage companies and their current flagship product is the Gridstack 5000, which can be built on multiple cell chemistries.

Gotion: See the full company description in Section 3.10. Gotion's current integrated stationary utility-scale BESS is the ESD series product.

Hithium: See the full company description in Section 3.10. Hithium's integrated product offering is a DC-block with over 5 MWh of energy storage capacity, featuring cells they manufacture themselves.

Hyperstrong: A Beijing-based ESS provider established in 2011, they have stated they have built over 300 projects and 15 GWh of battery deployments. As of 2024 they are expanding into North America, Europe, and Asia Pacific regions. Their latest utility-scale BESS product is known as the HyperBlock III.

LG Energy Solution: See the full company description in Section 3.10. A subsidiary of Korean conglomerate LG, LGES offers several integrated battery products. Their utility-scale BESS product line is known as the DC-Link. As of December 2022 they stated they had deployed 21.6 GWh of energy storage worldwide.

Powin: Powin Energy is a provider of energy storage solutions based in Portland, Oregon. Powin was an early entrant in energy storage market in 2011. They were notable for building their own energy management system (EMS) around cells and modules they purchased direct from OEMs, and for having exclusively air-cooled products. Powin launched their first energy storage products in 2017 to power sites they had developed, but later that year they spun off their in-house development team to form esVolta. Powin's initial product offering was their Stack products (Stack140, Stack225, and Stack230) for indoor and outdoor projects. In 2021 they introduced their Centipede product, and in 2024 unveiled the Powin Pod, a 20-foot enclosure BESS with liquid cooling. At the end of 2023, Powin stated they had deployed 3.2 GWh of systems and had 11.9 GWh under construction.

Prevalon: In February 2024 Mitsubishi announced that it would be replacing its previous Emerald storage product with a new brand, Prevalon, of which it is part owner. Prevalon featured a rebranded product that is being provided in partnership with battery integrator REPT, with the backing of the Prevalon organization.

REPT: Find the full company description in Section 3.10. In addition to manufacturing cells, modules, and white-labeled systems for other firms, REPT also has their own utility-scale ESS product line. In April they announced a 6.9 MWh 20-foot energy storage product, which they state has zero degradation for the first five years.

SamsungSDI: Find the full company description in Section 3.10. Samsung uses their own cells in a utility-scale BESS product called the

Samsung Battery Box (SBB). They are notable for producing systems that can serve high-power applications using NMC cell chemistry.

Sungrow: Sungrow is the largest inverter manufacturer in the world. They are a Chinese company primarily known for their inverters and PCS products, but they also offer integrated BESS products. Their latest offering, Sungrow's PowerTitan 2.0 is notable for being an AC-block product with an integrated PCS.

Trina: Trina Solar is a leading provider of solar PV modules. They first entered the energy storage market in 2017, and in 2022 they began offering energy storage products when they launched their Elementa BESS product.

Wartsila: Wärtsilä is a Finnish corporation known as one of the world's leading manufacturers of large internal combustion engines for large ships and land-based power plants. In 2015 they acquired Greensmith EMS, a leading American storage controls provider. They launched their first BESS product in 2018, which later evolved into the GridSolv Quantum product line. The latest offering is Quantum2, a 20-foot enclosure DC-block product.

Leading BESS Products

Figure 4-12 and Figure 4-13 shows 20 of the leading BESS products with pictures of the product. This provides a sampling of some of the leading integrated storage products as of June 2024 but is not definitive. Given that there has been a trend in the industry toward a 6.1 x 2.4 m (20 x 8 ft) enclosure, the products are shown with that as the most common form. For uniformity, products that have a footprint with a different area have been shown with a capacity normalized to the equivalent capacity in a 14.9 m^2 (160 ft^2) area. Below the graphic, footnotes indicate any notable or distinct aspects of each product.

AESI Terastor: 3.4 MWh

BYD MC Cube T:
6.4 MWh

Canadian Solar
Solbank 3: 5.0 MWh

EnergyVault
BVault: 3.2 MWh

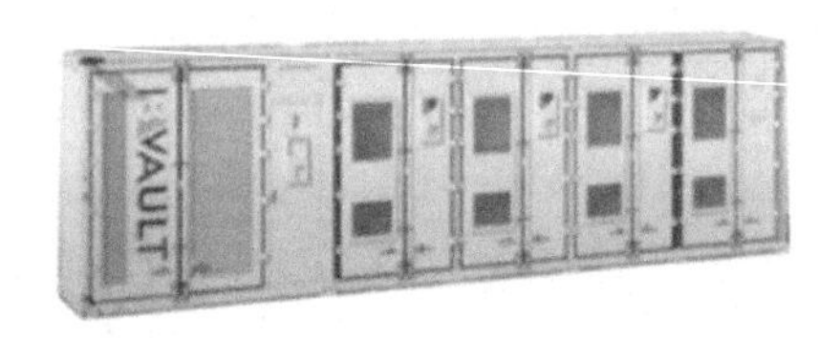

Fluence
Gridstack Pro: 5.6 MWh

Hithium Block: 5.0 MW

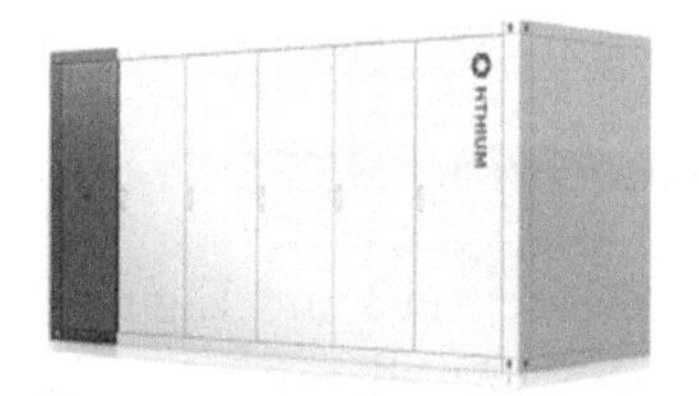

Figure 4-12
Leading BESS enclosures and capacity, normalized to 20-foot enclosure, Part I

Powin Pod: 5.0 MWh

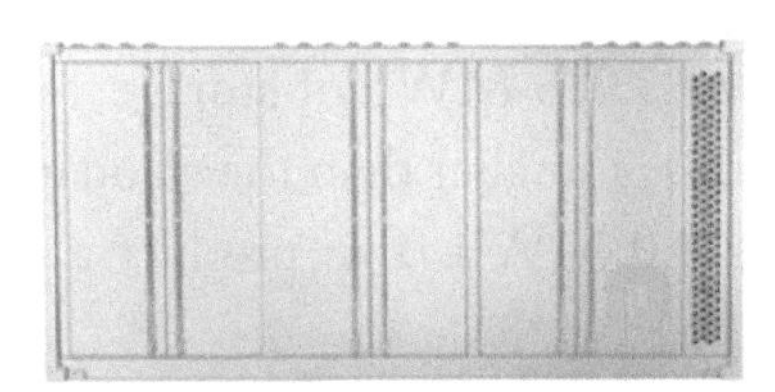

REPT Wending: 6.9 MWh

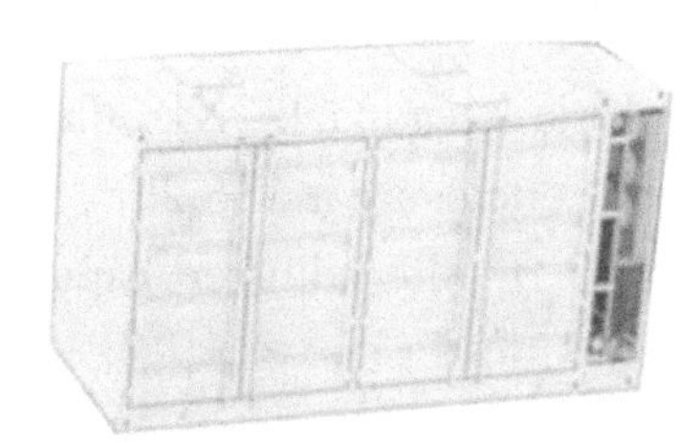

Sungrow PowerTitan 2.0: 5.0 MWh

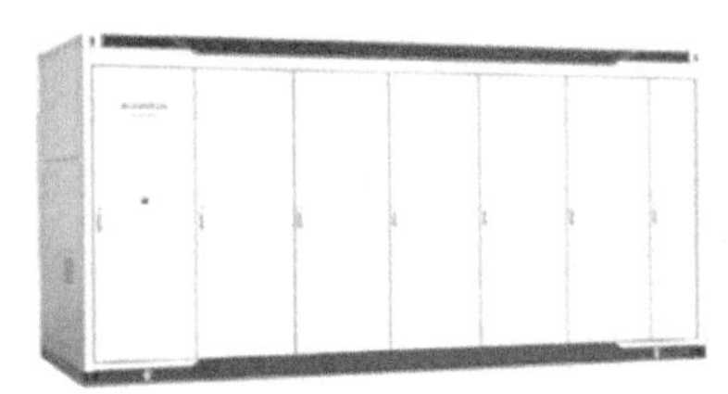

Tesla Megapack 2XL: 4.0 MWh

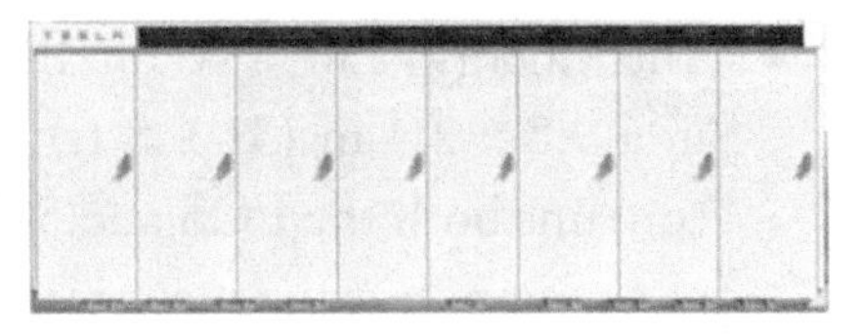

Trina Elementa2: 4.1 MWh

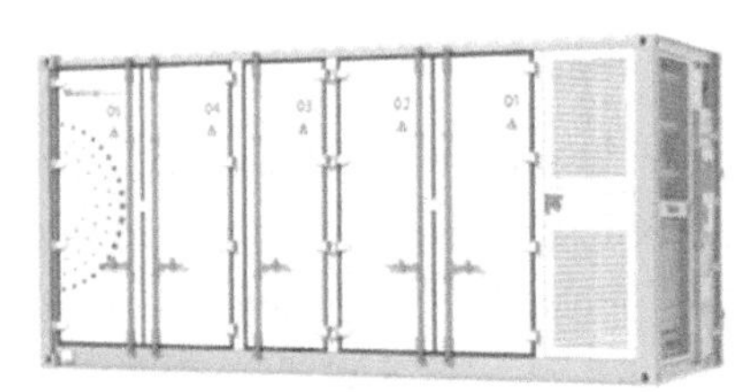

Wartsila Quantum2: 4.1 MWh

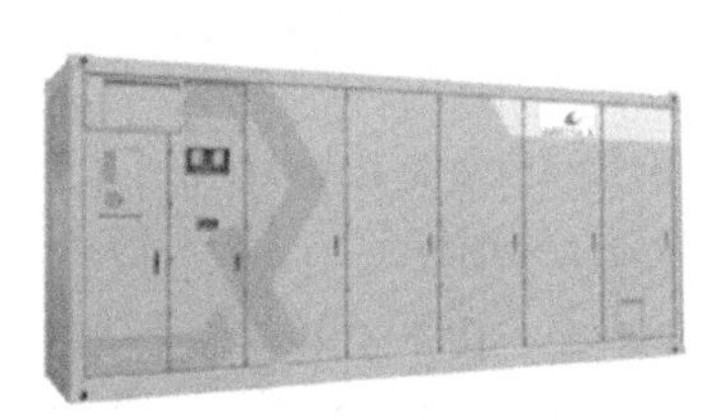

Figure 4-13
Leading BESS enclosures and capacity, normalized to 20-foot enclosure, Part II

Notes on Energy Storage Products:

- The Sungrow PowerTitan 2.0 includes PCSs.
- Tesla Megapack 2XL offers up to 3.916 MWh of storage, but the enclosure is 156.2 ft^2. Since this is smaller than the standard 160 ft^2, the adjusted capacity is 4.0 MWh. This product also includes PCSs.
- The AESI TeraStor offers up to 7.2 MWh AC storage (~7.3 MWh DC) in a single 8.2 x 3.8 m (26.9 x 12.5 ft) enclosure, although it is shipped as four separate units which are joined together on site.
- The EnergyVault B-Vault BV2.2 offers 4.3 MWh of storage in a 9.8 x 2.4 m (31 x 8 ft) container. Note that this product contains both the PCS and 50kVA transformer. Density is corrected to reflect the inclusion of these components.

PCS Providers

ABB: ABB is a global technology leader in electrification and automation. Their eStorage Flex and eStorage Max products are leaders in energy storage PCS technology.

Delta: Delta is a Taiwan-based firm providing power and thermal management solutions. In 2023, Delta Electronics reported consolidated sales revenues of approximately $13.2 billion. Their PCS line includes standalone inverters and products that include skid-mounted MV transformers.

Dynapower: Dynapower is a Vermont-based company that specializes in energy storage and power conversion systems. They market their utility-scale energy storage PCS products under the CPS and PowerSkid product lines, and also sell DC-DC converters. The company has delivered 2 gigawatts of clean energy systems globally. They were acquired by Sensata Technologies in 2022.

EKS: EKS (stylized as eks) is a Spain-based manufacturer of energy storage PCS products. Formerly known as GPTech, in 2023 eks was acquired by Powin, who later sold a controlling interest in the firm to Hitachi. eks has played a significant role in deploying large-scale energy storage systems worldwide.

EPC Power: A US-based manufacturer of energy storage PCS products founded in 2010, EPC Power provides PCS products for utility-scale, commercial, and industrial applications. Goldman Sachs and Cleanhill Power acquired a controlling interest in EPC Power in 2022. In 2023 they sold over 2 GW of PCS products globally. Note: EPC Power is unrelated to the industry term engineering, procurement and construction.

Ingeteam: Founded in 1972 and headquartered in Spain, Ingeteam specializes in electric power conversion products across the residential, C&I, and utility-scale PV and storage industries. In 2023 they supplied 4.2 GW of PV and battery inverters.

Huawei: Huawei is a global Chinese multi-national electronics firm. In 2023 they shipped 42 GW of PCS products globally. Their solutions can be used for solar PV, energy storage, or hybrid PV + Storage projects. According to PV Tech magazine, Huawei exited the US market in 2019. [31]

SMA: SMA is a German company founded in 1981 that specializes in PV system technology. It plans to double its production capacity at its headquarters in Niestetal, Germany, from 21 GW to nearly 40 GW by 2024.

Power Electronics: Power Electronics is a prominent Spanish manufacturer of solar inverters and energy storage solutions. Founded in 1987, the company has provided inverters to over 1,600 renewable energy projects. As of 2023, they have installed over 91 GW of PCS products.

Tesla: Founded in 2003, Tesla is headquartered in Palo Alto, California, and is known primarily for its electric vehicles (EVs). In the utility-scale energy storage sector, Tesla offers an integrated BESS product known as the Megapack. Tesla does not sell their PCSs, but their AC-block BESS product includes integral PCSs as part of their technology.

TMEIC: Established in 2003 and headquartered in Tokyo, Japan, TMEIC's PCS technology is employed in various applications, including renewable energy integration and grid stabilization. Their inverters are used in major solar power installations in India, as well as the 350MW Victoria Big Battery.

WSTech: WSTECH manufactures a range of PCS products with power ratings from 600 kW to 7,500 MW. Their products are available for PV, storage, and hybrid applications. They have traditionally been more popular in Europe than the US.

Leading PCS Products

Table 4-4 lists 10 of the leading PCS products and their power capacities. This provides a sampling of some of the leading PCS as of June 2024. An example of a PCS, along with adjacent MV transformer, is shown in Figure 4-14.

Table 4-4
PCS Products

OEM	Product	Power Capacity [MVA]
Dynapower	CPS Series	3.0
eks	PCS-3MS	3.8
EPC Power	CAB1000	1.5
Ingeteam	Sun Storage	3.7

Power Electronics	FP4390K (PCSK)	4.4
Sinexcel	PWS1-1725KTL	1.7
SMA	Sunny Central Storage SCS UP-XT	4.6
Sungrow	SC3450	3.5
TMEIC	BESS Inverter	2.5
WSTech	APS5000	5.0

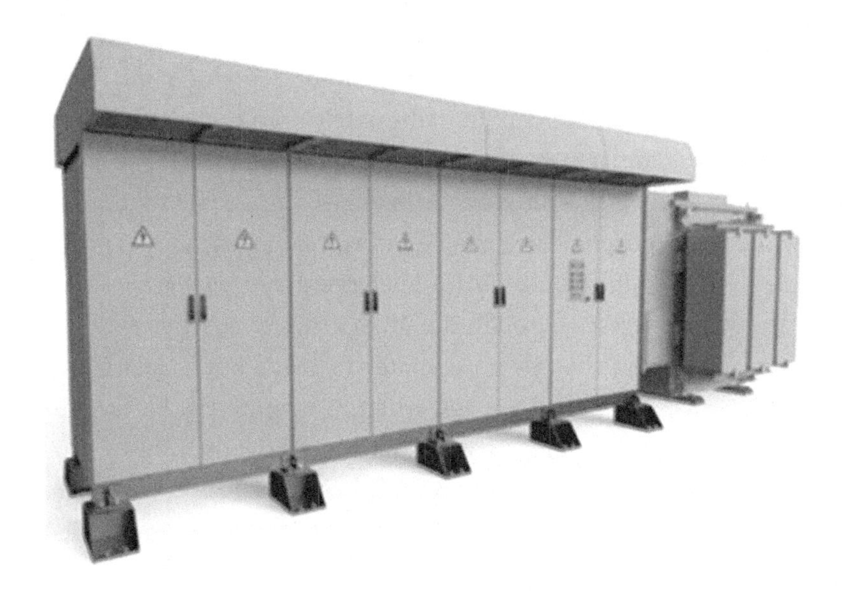

Figure 4-14
Example of PCS: the eks PCS-3MS

5. BESS CONTROLS

Any operational battery system involves controlled flows of both power and control signals. As described in Chapter 4, power flows into and out of the battery through the conductors which link the battery to the PCS, transformers, and ultimately onto the power grid. The control signals are transmitted by fiber-optic, ethernet, and wireless connections, which link the various battery components together and allow them to communicate, passing information back and forth about the state, status, and condition of all battery system parts. The signals are then processed through on-site hardware together with cloud platforms, allowing the system operator to make both real-time and strategic decisions about how and when to operate the battery. Collectively the hardware and software systems that collect this data and make these decisions are known as BESS **controls**.

Although they may seem trivial compared to power components, controls are critical to the performance of a battery system. Without them the system cannot operate, even if all power components are in good working order. This section explores the functions of battery controls, key components, and how the controls system works in concert to ensure that a system is operated safely, efficiently, and profitably.

The heart of battery controls is decision-making. A battery system has thousands of decisions to consider every second the battery is operating. For example:

- Is there any abnormal condition that needs to be evaluated?

- Should the system be idle, charge, or discharge energy?
- Are the cells too hot or too cold?
- Is the fire alarm system working properly?
- Can the power grid accept the requested power discharge?
- Has the system come close to violating any terms of the warranty?

Most utility-scale systems have multi-tiered levels of hardware and software that make these decisions. They are run via a combination of microcomputers, local servers, and cloud-connected servers. Some control systems operate independently of the utility or grid operator, while others are wired into the grid controls and may receive instructions from the grid operator's servers several times per second.

Battery control systems are designed to require minimal inputs from human operators. This is done by designing algorithms which tell the system what motives to prioritize and giving the system rules to follow for making decisions based on the inputs received. In a properly functioning system, the operator may have to sign on only every few weeks or months. However, there are some critical decisions, especially those related to safety, in which a human operator is notified. In general, if there is ever a question of a system operating safely, the systems are programmed to sacrifice performance (and profits) in favor of investigating and clearing any safety red flags before returning to safe operations.

Poorly designed controls systems may result in increased battery capacity degradation or lost revenues, and in some cases may contribute to critical safety incidents.

The key functions of battery controls are:

- Ensuring the BESS always runs safely using proactive and reactive methods:
    - Proactive: System monitoring conditions indicate a potential safety event is imminent

  - Reactive: In the event of a safety incident, warning the necessary parties, limiting the effects of the event, and managing the appropriate response

- Coordinating the BESS operations to optimize charging and discharging based response to utility requests or a grid condition, which determines when the BESS will charge or discharge
- Ensuring the system is ready to deploy for unexpected conditions, such as power outages or service calls
- Coordinating the operations of the BESS with other related or nearby systems, such as with a coupled solar PV or wind plant that is operated in conjunction with the BESS
- Maintaining a historical record of the battery's operations, known as a data historian

Since the BESS controls system consists of many different components, it typically is divided into multiple controls "layers," which are linked together in a hierarchy. Together this hierarchy operates like a military command structure, allowing the individual components to pass information upward, and for commands to flow down to the individual components according to the strategy chosen.

Each layer makes decisions based upon the status of the control layer below it, and on the sensors that are locally fed into this layer. This provides a level of protection that allows quick and independent responses, ensuring that control is coordinated across all layers.

In some cases, there may be conflicting commands – for example, if the controller sends a signal for the BESS to discharge, but an individual battery module senses an abnormally high operating temperature. In ideal operation, the central command would not make this request, but if for some reason it does, the module can resist this command, protecting itself from unsafe conditions. This illustrates the fundamental purpose of battery control – multi-layered decision-making that includes both centralized and autonomous elements to balance performance and safety.

This chapter reviews the six key components that compose a BESS controls system. While battery control systems vary widely from system to system, most battery systems include these elements:

- **Battery management system (BMS)**: The BMS, the most basic level of controls, is at the cell and module level, collecting and making sense of inputs (such as voltage, current and temperature). The BMS is responsible for the safety and operation of each individual battery cell and makes control decisions based upon a grouping of these cells. BMS may also encompass a Battery Rack or Bank Controller (BRC) which aggregates data at the rack level. The BMS is also responsible for balancing the voltage of the cells.
- **Energy management system (EMS)**: An EMS aggregates information collected from the BMS and/or BRCs to make system-level operational, safety, and communication decisions. The EMS is also responsible for determining the available power and capacity of the battery system. In addition to information from the BMS, the EMS may take into account other pieces of information such as:
    - Ambient site weather data
    - External controls to first responders
    - Fire safety signals
    - Utility or grid operator controls
- External data reporting from components such as PCS, transformer, switchgear, meters, or auxiliary transformers
- **Power Plant Controller (PPC)**: The PPC directs the EMS to coordinate its decisions with an energy control system (ECS)

such as an external transmission network or a Supervisory Control and Data Acquisition (SCADA) system. These systems each consist of hardware components, software systems that operate on top of the components, and a user interface that allows the operator to verify the system's operations. In a site that has both battery and other distributed energy technologies, like PV, there will be other technology specific plant controllers that may be aggregated together in a single overarching SCADA system.

- **Utility EMS:** This final layer of control is responsible for controlling the safety and operation of the electrical grid connected to the site. It is called the Utility Energy Management System (EMS). This control layer will command the site to provide one of more energy products (see Section 6.2 on Use cases) to the grid.
- **Fleet-level controls**: These are used for systems that include several different battery assets in different locations, such as Virtual Power Plants (VPPs). These controls ensure that all the systems function in concert to achieve the overall fleet-level goals.

Although not a separate layer, each of the above five components has critical elements which contribute to the overall security and access control of the site. Collectively, these systems are known as the project's **cybersecurity controls**. These systems consist of hardware and software that protect the battery system from malicious threats from outside actors. Given that nearly all utility-scale battery systems are part of the nation's power grid, a well-designed and properly maintained cybersecurity program is essential for the individual system and the broader power grid.

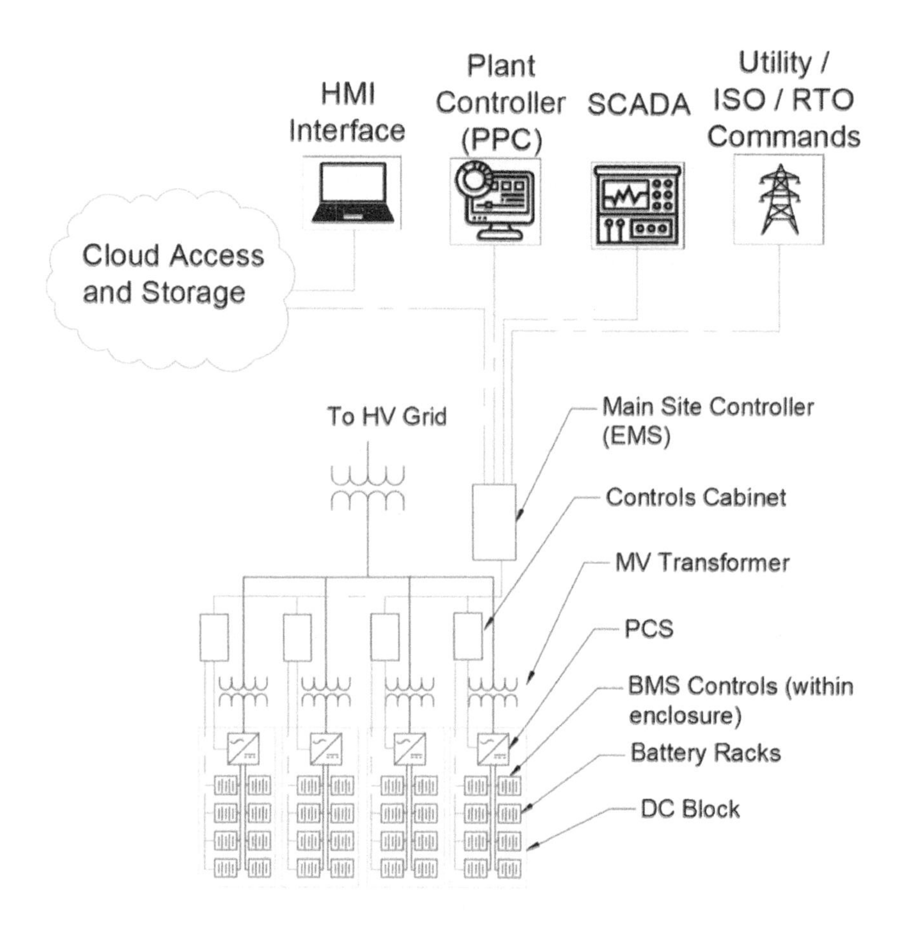

Figure 5-1
Schematic of typical BESS communications diagram

5.1 Battery Management System

The most fundamental level of battery controls is the Battery Management System (BMS), which operates at the battery cell or module level. The BMS collects the most basic battery cell measurements: temperature, voltage, and current.

The BMS is responsible for:

- Collecting battery warranty data such as cell temperature, voltage, and current of the cells it manages.
- Determining the State of Charge (SOC) and State of Health (SOH) of the cells it manages.
- Balancing the voltages of the cells of the cells it manages to prevent cross currents.
- Monitoring the battery support equipment (such as the HVAC system and Fire Detection system) to ensure cell safety.
- Providing DC Contactor control to prevent the cells it manages from being used in unsafe conditions (for example, if the cell SoC is too low or too high).
- Determining Power Limits of the cells it manages for charging and discharging the system to prevent damage to the cells.
- Providing alarms to the EMS to warn against unsafe conditions.

The BMS is often referred to as the last line of defense since it is closest to the battery cell and is responsible for measuring critical indicators such as cell temperature and voltage, even when other systems fail. The job of the BMS is to ensure the safety, longevity, and overall optimal operation of the battery cells. One way to think of the BMS is as the BESS' involuntary nervous system - it ensures that the BESS stays safe and healthy for the expected life of the system without requiring real-time decision making from a human. It is typically hardware fitted to the cells, though it may also correspond to cell groups composed of paralleled individual cells, or modules of cells or cell groups in series. Different manufacturers refer to their BMS in different ways, with some firms using a **Battery Control System (BCS)** which serves as a collector of various cell-level BMS data.

Once the data is gathered, the BMS communicates the information in real time up to a hardened industrial computer. This aggregated data is monitored and fed through algorithms to make decisions on how to run the system in real time. Typical decisions made by the master system controller include:

- Send a signal to the energy management system (EMS) to stop the BESS from discharging or charging if cells are overheating or have reached upper or lower voltage thresholds
- Turn on localized fans for directed cell cooling
- Send temperature data to an HVAC controller to ensure temperature thresholds are respected
- Log cell level data to a data historian for long-term storage and analysis
- Report voltage levels to a balancing circuit microcontroller which will make determinations on whether to balance a cell's voltage up (charge) or down (bleed) in order to move the voltages of specific cells toward a targeted average voltage

Battery management systems come in many flavors, all of which are proprietary and have unique ways of balancing cells and monitoring cell parameters.

Typically, a BMS is not a design option as it comes standard with any battery product. For those systems that do include an integral BESS, the designer should research the different BMSs on the market and get an idea of the idiosyncrasies of different manufacturers' offerings. Be careful when replacing an OEM BMS with a non-OEM BMS as this may affect the battery warranty.

BMS Protocols

The BMS communicates to an EMS using standardized protocols, which are pre-defined methods for sending data. Although each BMS is a proprietary design feature of a battery, they tend to use similar methods of communication. The most common protocol used by a

BMS is the *CANBUS*. CAN stands for controller area network (CAN), which is a standard used for electronic communications in a variety of industrial and automotive applications. CAN is a message-based protocol which connects various electronic control units (ECUs), also known as nodes. CAN uses a system of voltages to relay binary signals between the various nodes to transmit data. Given that it can be designed to facilitate high-speed communications between many different components, CANBUS is an ideal platform to allow the many thousands of individual cells, modules, and controllers relay signals.

While there are other protocols for data transmission, CANBUS is the most common, and the protocol used by most commercially available BMS systems on the market.

BMS Failures

The BMS is designed to be fail-safe, in that it provides a last line of controls which can isolate the battery system. However, there are ways that a BMS can fail, resulting in either a loss of connectivity or ultimately in a potential safety incident.

The most common method of BMS failure is in communications. If a BMS is not able to accurately read the temperature, voltage, or current of any given battery module, it cannot provide the appropriate signals to the EMS to take appropriate decisions. A communications failure could happen via a mechanical event in which the battery cells were physically severed, or from electronic interference which doesn't allow for clear messages to be transmitted. Regardless of the cause, a properly-designed controls system should interpret any loss of communications as a serious hazard – in other words, if there is no news from any part of the system, the controller should assume there is a safety hazard present, and the system should take the necessary precautions until communications can be restored. In some cases, despite many layers of controls, a loss of communications has allowed an unsafe condition to occur.

As an example of this, the Victorian Big Battery (VBB) fire which occurred on July 30, 2021 in Geelong, Australia. According to the independent report of this incident conducted by Fisher Engineering, Inc. and Energy Resources Safety Group (ESRG) [32] a key factor in the fire was that the telemetry systems, fault monitoring, and electrical fault safety devices were disabled or operating with limited functionality, due to the system being commissioned. The lack of this connectivity ended up resulting in a fire which destroyed two battery enclosures before being brought under control. The effects of the fire are shown in Figure 5-2.

Figure 5-2
The Victorian Big Battery fire July 2021 [17]

It is important to note that there are some safety incidents which even a fully functional BMS cannot prevent. One case of this would be the dendritic growth of crystals within a battery cell, which can cause an internal short-circuit within the cell, in turn triggering a thermal runaway event. Opening all circuits in the offending cell will not stop the chain reaction, meaning that no element of the controls system can stop this. However, the BMS can provide data leading up to and during an event of this type which may allow the issue to be caught prior to becoming a more serious safety hazard.

Interfacing with a BMS

For the most part, the owners and operators of a battery plant do not need to know about the inner workings of the BMS. This is because it is part of the under-the-hood communications and control of the battery system, similar to the electronic equipment used in a vehicle to ensure its safety. As most battery manufacturers provide their systems with fully functional BMS and EMS controls platforms, the end user does not have to be involved in the day-to-day BMS events.

Some battery manufacturers do not allow the information collected by the BMS to be made public to the owners of the system, while others allow full access of this data to the higher tiers of control. Although the data is out of sight, it is important to know that such data exists, and in the case of safety incidents or near-misses, it may be helpful to request this data from the battery provider.

Balancing

One additional function of the BMS is to fine-tune the voltage of cells. It is a common issue in BESS operation that the voltage of some cells may be higher than others in a module or rack. This results in sub-optimal performance of the system, since a cell higher or lower than others doesn't allow the system to charge to the very maximum (100%) or minimum (0%) SOC. This problem is typically solved using two methods:

- **Passive balancing**, in which a resistor is used to bleed off some charge, bringing down the voltage of the cell or module

- **Active balancing,** in which a trickle charge is used to bring up the voltage of individual cells or modules

Passive balancing can be achieved relatively easily by using resistors on each cell or module that are used to bleed off charge.

While most BMSs have passive balancing, only certain products have opted to have active balancing, since it requires having an additional circuit running throughout the rack.

5.2 Energy Management System

If the BMS is the involuntary nervous system, the EMS is the voluntary system. The EMS allows the various battery system components to communicate with each other, and to interface with the outside world. Together with the BMS, the EMS works to optimize and operate the system according to algorithmic or hard-wired run settings. It is typically accessible via a human machine interface (HMI). In practice, this is either a secure web-based controls page or a physical terminal located in the battery controls cabinet on site. The HMI gives the authorized user direct control of the system if desired, but can also be set to respond to various situations automatically.

Typical decisions and control capabilities of an EMS include:

- Aggregating the SoC data from each BMS to calculate a site-wide SoC and energy capacity
- Aggregating the alarm information from each BMS and PCS to calculate a site-wide available energy and power
- Allowing the entire system to be either manually controlled on site or automatically controlled by a utility or market trader
- Aggregating battery warranty data such as cell temperature, voltage, and current of all the BMSs and storing it in a local or remote data historian

- Coordinating output control between multiple technologies on a single site (BESS, PV, gH2, Wind)
- Controlling the PCS, which sets power levels and moves energy in and out of the battery from or to the grid
- Setting reactive power levels on the PCS as desired by the operator or required by the grid operator
- Allowing the user to determine the operational status of the entire BESS
- Receiving commands from the grid operator based on market bidding results
- Monitoring grid and BESS state parameters and responding according to a predetermined algorithm
- Communicating status of the BESS to outside parties
- Allowing outside parties to remote into the system to monitor and control the BESS

Coordinating Safe Operations

The EMS is often referred to as the brain of the system, balancing the commercial objectives of the battery with the safety requirements of a system. A properly designed EMS always prioritizes safe operations over profits. This means that any portion of the battery system with abnormal signals can be disconnected temporarily, until its signals return to normal. Alternatively, the portion of the battery system can be flagged for human intervention, conducted either remotely or in person. Once the signals are brought back into normal ranges, the portion of the system which was taken offline can be re-energized and continue operating as normal.

As mentioned earlier, a command may be issued from the EMS which a lower-down layer of controls recognizes as incorrect. For example, if the SoC of a certain battery cell was at a voltage indicating it was fully drained, the BMS could open its contactors and set off an alarm to avoid further discharge of the rack containing this cell. This is an important aspect of controls design – while the system is engineered to optimize for performance with a top-down approach, it also gives individual units the autonomy to make decisions on their own behalf.

In some cases, this approach may result in **nuisance trips**, or cases in which safety systems are too sensitive, going offline when not necessary. This is a common issue which must be addressed as part of the testing and commissioning of a system. Although it may cause frustration, nuisance tripping is based on the design principle of erring on the side of caution – once an operator identifies a nuisance trip, the appropriate settings can be adjusted to ensure that the controls system trips when necessary.

As artificial intelligence (AI) becomes more advanced, it can be used in novel ways to provide additional layers of safety to battery systems. For example, if an AI algorithm can identify a pattern of voltages which has preceded fires in other systems, a red-flag can be raised even if the observed voltages have not strayed beyond the module's operational limits. As AI continues to evolve, there will undoubtedly be more ways in which it can use analysis of large data sets to improve both the safety and performance of battery systems.

Native versus Third-party EMS

Some integrated battery systems on the marketplace today include some type of EMS as part of their service offering, which is known as a "native" controls system. For some systems, the use of this native controller may be required, while others offer it as an optional feature.

If the owner does not wish to use the EMS offered by the battery provider, there are several third-party EMS providers who can furnish and integrate the hardware and software to control the battery system as a whole. In most cases this is not recommended, as doing so may mean that the warranty or performance guarantees offered by the battery manufacturer on the system are inferior, or in some cases unavailable.

Rather than replacing the EMS offered by the manufacturer, some owners instead opt to integrate a third-party controls system on top of the controls system offered by the battery provider. In practice, this would mean that the battery provider's EMS is configured as a "follower" and the third-party EMS is configured as a "leader." In other words, instead of being the command center, the EMS looks to the third-party controller to receive its marching orders, although it still would be responsible for controlling the communications with the BMS and other inputs internal to the battery enclosure.

Using a third-party controller introduces an added layer of complexity, as well as added expense to the project. However, it offers some key advantages as opposed to using the native controller:

- A native controller is recording all data about its own operation, including faults which may indicate it has operational problems. Having a third-party monitor and store this data can avoid having a "fox-watching the henhouse" scenario.
- In the case of a dispute with the battery manufacturer, it is preferable to have the operational data held by the third party for comparison to the same values measured and stored by the battery manufacturer.
- Most native EMS solutions do not include the EPC of the on-site communications wiring or controls cabinets; some third-party controller companies can include this in their scope, allowing for a single firm to handle this complete scope for both construction and operations. Section 5.7 includes some of the prominent companies operating in this space.

Data Historian / Analytics

The BMS hardware has sensors which are continuously monitoring cell temperature, voltage, and current of each cell. Since cells typically range from 100 to 900 Wh, larger systems sometimes consist of over one million individual battery cells. The BMS may be polling each cell to report its values as often as every 2 seconds. This results in a torrent of data, in some cases well over 1 GB per hour. One of the key functions of a BESS control system is to store this data in a secure place, which is commonly known as a data historian. The information stored in the historian may be kept on the local server temporarily, though typically this data is uploaded to a cloud-based location for access by remote users.

The storage of operational data typically involves some compression due to the large and unmanageable amount of information. Typically, all data is stored locally for a certain period, such as two weeks. This allows the full data set to be stored when looking back to a critical event, such as a cell going into thermal runaway. After the two weeks have elapsed, the data is condensed – for example, rather than recording the temperature of every cell on the 2-second interval, the data may be compressed to reflect the maximum, minimum, and average cell temperatures of a group of cells, and further may be compressed by reporting these values over 15 minutes rather than two seconds. This enables systems to offer a comprehensive record of recent operating history for the prior two weeks. If there is a safety incident, this allows the detailed data set to be captured in full to examine what occurred. During normal operation, after 2 weeks the record will include the data summary, but the data gathered at seconds-intervals, such as BMS data, will be discarded and the data summary will be available for future use.

Parsing the massive amounts of data can be a herculean task, but there are now many software solutions designed to assist in providing these services. By running algorithms on the available data, a third party can gather important insights on everything from mismatched voltages to potential safety issues. This is often done using artificial intelligence-enabled analytics, which allow the system to process the gigabytes of data and output key insights. Firms that perform these services vary from full-scale controls firms to specialized analysis-only platforms. Some of the more common providers are listed below in the Usual Suspects section.

Interaction with Outside Controller

Battery sites that have a straightforward use case, like daily load shifting, can be controlled entirely by the EMS. In other words, the EMS can be programmed to charge or discharge at certain times on certain days, weeks, or months of the year. However, given that batteries can respond to control signals within fractions of a second, more sophisticated use cases can be employed which require more sophisticated control algorithms that cannot be implemented in the EMS. In more sophisticated battery markets the utilities or grid operators may require the battery to respond to a signal that changes based on the conditions of the grid.

For example, a system might be configured to perform frequency regulation (more on this under Value Streams in Section 6.1), in which the system may be switching from charging or discharging every few seconds, depending on whether a grid is under- or over-loaded. The required behavior of the grid cannot be predicted, so instead the grid operator may require the battery controller to connect with a secure portal which will issue the control signal telling the system what to do. This requires the grid operator to provide instructions for the EMS controller to perform a secure integration with the controls system, which takes place at the time of system commissioning. Once this system is in place, the EMS can receive and act on the signal automatically, eliminating the need for human control of the system, allowing the battery to respond within milliseconds to control signals. Typically, systems that are performing this type of functionality will have a minimum required response time which must be guaranteed by the grid operator.

Figure 5-3 shows the time lapse of a typical system response to a discharge signal. In this diagram:

- The x=0 value indicates the time at which the signal is sent by the outside controller
- t_d is the time it takes for the PCS to begin to respond to the signal
- t_r is the response time
- t_s is the time for the system to respond and to provide a stabilized output

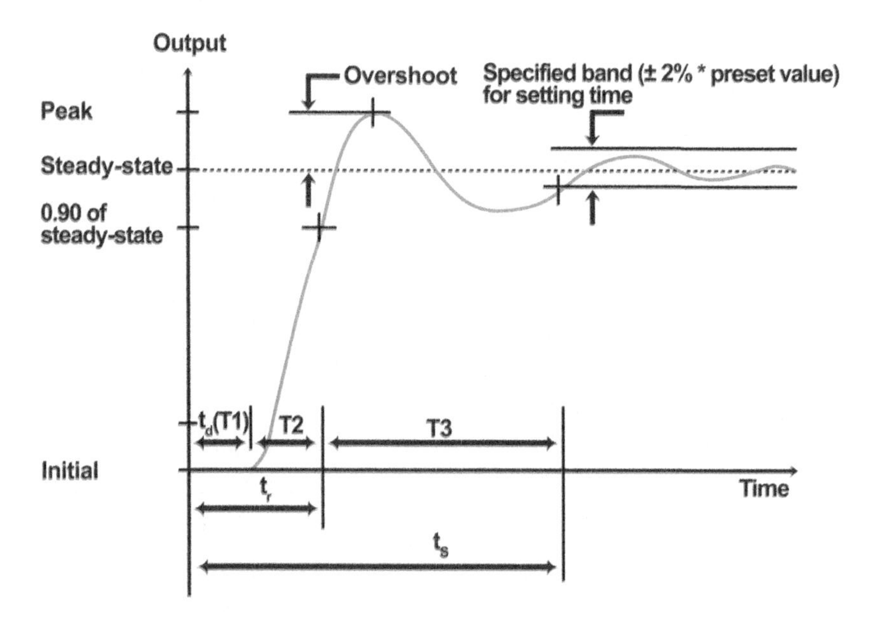

Figure 5-3
Typical response to an outside control signal

For systems that respond to this type of signal, the system is not able to respond perfectly. This may be either because the system has times where it is offline or unable to respond to the signal, or because the system operator has determined that it is more financially viable to deliberately avoid responding. Most systems of this type are given a score for the settlement period which quantifies how effectively it was able to respond to the signal. Depending on the contract there may be a minimum score required to stay in the market, LDs assessed for low performance, or a "grace score" in which the system is compensated in full despite not responding perfectly.

5.3 Power Plant Controller and SCADA Systems

The larger the battery system, the more likely it will need to integrate with a larger utility system or a Remote Operations Center (ROC). A SCADA (supervisory control and data acquisition) system is used to monitor and control other systems either locally or remotely. It is typical for larger installations and provides data visualization and data analysis features. The lines between the SCADA system and EMS may blur, but in general they will both use a networked switch connected to on-site hardware such as a modem, remote terminal units (RTUs), and other devices that enable multiple parties to be involved in monitoring and controlling the system. Expect to see a SCADA system integration required if connecting the BESS to the transmission system, a high voltage substation, or perhaps even at certain medium voltage connection points, as the utilities and independent system operators will expect to be able to interface with the BESS using protocols that are common to equipment in those systems.

Some examples of integration of SCADA with existing controls might be a BESS SCADA designed to interact with a PV SCADA at a single control point, or a BESS SCADA that had a terminal at a local utility substation for the utility to be able to monitor the status of the battery.

5.4 Fleet-level Controls

The connectivity of BESS has led to the deployment of many batteries in the same facility, state, or region. While most battery systems operate in isolation, battery systems can also be operated together to achieve maximum benefits. Operating in concert, the batteries can be considered a fleet and require another level of controls to ensure a coordinated response.

5.5 Virtual Power Plants

While this book is focused on utility-scale battery projects, it is important to note a growing asset type that behaves like a utility-scale battery, even though it does not look like one.

Smaller battery installations, such as those in homes or at commercial or industrial locations, are normally too small to participate in the ancillary services or capacity markets – while each region is different, typically projects must be 10 MW or larger to provide this type of service. While smaller batteries can technically perform these services on a smaller scale, these minimum requirements make it impossible for these assets to participate in larger markets.

One solution to this problem is to operate many smaller energy storage installations as if they were a single large plant, in what is known as a **virtual power plant**, or VPP. In a VPP, individual batteries, whether they are residential, commercial/industrial, or utility-scale, use advanced controls to dispatch the batteries in unison, as if they were a single large energy storage asset. Figure 5-4 shows a sample connectivity arrangement for a VPP.

In many instances the batteries are not owned for the sole purpose of discharging in unison, but rather they consist of batteries owned by homeowners or businesses whose installation costs have already been paid; the VPP operates by having these small battery owners enter into an agreement whereby they commit some portion of their battery

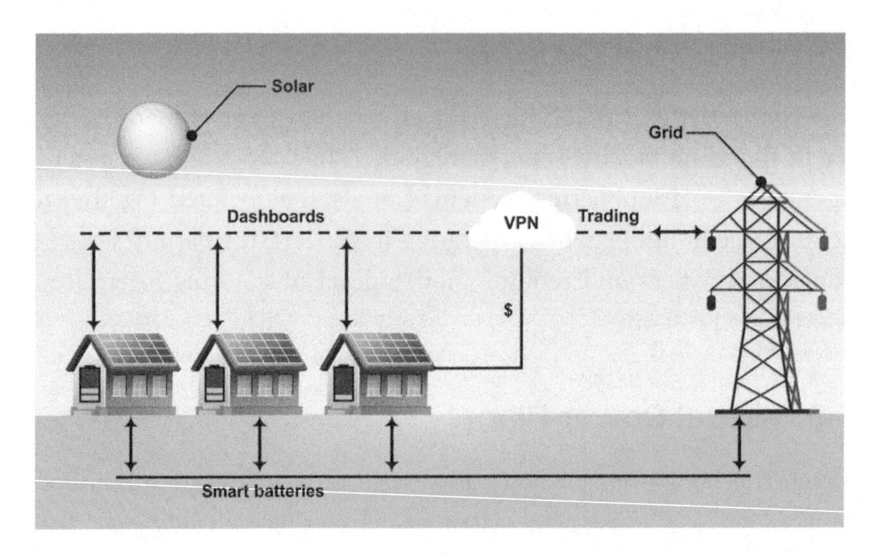

Figure 5-4
Sample VPP connectivity [33]

participate in the VPP, when called upon by the centralized control system. In return, the battery owners are compensated based on either a flat rate or based on how much their batteries are used.

Dispatch of a VPP requires synchronized control systems to ensure that the batteries are discharged in concert, as well as advanced contracting with hundreds or even thousands of participants, to ensure they agree with the compensation that can come from discharging their assets. Since VPPs are often dispatched when the need is the greatest, such as in large seasonal demand peaks, participants may be reluctant to participate with their entire battery, since these are the periods that the systems are most in demand. Most systems of this type will allow users to commit to using a portion of their storage, while not leaving the systems completely discharged.

As an example, if 100 LGChem RESU 16H batteries (a residential product with a nameplate capacity of 7 kW / 16 kWh) sign up to a VPP program in which they allow the VPP to use 50% of their energy. In this case, if all the batteries were fully charged ahead of a discharge event, the VPP would be able to discharge 700 kW for 1 hour and 9 minutes (or alternatively, 350 kW for 2 hours and 18 minutes). While this amount of energy seems small (a single Fluence Gridstack Cube has 724 kWh of energy, for example), these programs could be scaled up dramatically to produce megawatts of dispatchable capacity.

A large trial of a VPP was run by Tesla with Powerwall owners in the summer of 2022 [34]. The California grid was experiencing a peak load event, with the aggregate state loading exceeding 50 GW. The grid operators, and even the governor, Gavin Newsom, urged consumers to limit their power consumption, especially around air-conditioning and electric vehicle (EV) charging, to avoid increasing the load in peak hours. Tesla conducted a voluntary discharge event for Powerwall owners in which over 4000 Powerwalls discharged in the peak hours to provide over 30 MW of power into the California grid. While this was a relatively small amount of power (around 0.06% of total demand), the program demonstrated that VPPs are feasible and can help to provide substantial amounts of power, without requiring the construction of new power plants.

There are various downsides to a VPP: it is quite complex, requiring synchronized communications between hundreds or thousands of individual batteries, it requires contractual commitments from the owners of those units. For the utility or grid operator that is receiving the battery services, measuring the overall contribution of the VPP can be difficult to verify, as it depends on measurements taken by the manufacturer coordinating the delivery of the services, requiring audits to verify the accuracy of the output. Additionally, there is a fear that since the VPP depends on the contributions of many individuals, they can be unreliable - the same drastic events which cause the grid to need emergency services, such as a hurricane or snowstorm, may make the battery owners opt to use their batteries for their own backups. VPPs hold vast promise to transform the industry, as in Texas alone regulators estimate there are approximately 3,000 MW (3 GW) of distributed energy resources (DER) that could contribute power to the grid [35].

Overall VPPs are a relatively new phenomenon, but their use is expected to grow quickly. While the controls are more complex than traditional power plants, VPPs have the benefit of requiring much less infrastructure, since they are built on the backs of smaller units connected at low- or medium-voltage. As of this writing many firms are growing that are pursuing VPP-based technologies.

One massive source of energy storage that could be tapped as a VPP are EVs. If spare capacity from vehicles could be brought online, a process known as Vehicle-to-Grid, or V2G, this could soon dwarf both stationary energy storage resources. For scale, according to the US Department of Energy, the EV manufacturing capability is projected to increase from 55 GWh / year in 2021 to nearly 1,000 GWh / year by 2030. If even a fraction of these batteries are able to be connected with V2G, the results could revolutionize the power generation industry. V2G presents some safety concerns, since it would involve back-feeding energy from a vehicle to the grid, whereas most currently available EV chargers are designed to feed energy exclusively from the grid into the vehicle. Additionally, the program would require drivers to be willing to sacrifice part of their range to participate. Lastly, utilities would have to be willing to allow V2G; to date many have been slow to adapt to this new technology. Despite the roadblocks, V2G holds great promise as an up-and-coming technology. This technology is covered in greater detail in Section 11.3.

On the whole, VPPs are still in their infancy as of 2024. However, given the rising prominence of distributed energy storage, and rapidly evolving communications systems, they promise to be a fast-growing creative solution that can help accelerate the transition to renewable energy.

Subscription Control Services

Many energy storage technologies and products are driven by regulatory policy. Some of these policies create strong incentives for private firms to help customers take advantage of the policies to lower their electric bills. Such was the case in Ontario, Canada where the Independent Electrical System Operator (IESO), the local grid operator, implemented a system to avoid large peaks on the grid, known as the Global Adjustment, or GA. The system is designed cover the cost of building new electricity infrastructure, maintaining, and refurbishing existing generation, and funding conservation efforts [36]. The GA imposes a large demand fee on any large energy consumers, based on their share of consumption during Ontario's top five peak demand hours of the year. The resulting fee can result in hundreds of thousands of dollars in additional electricity bills for large consumers. The users do not know when these hours will occur, so the net effect is for customers to avoid using electricity in times likely to be in the five peak hours (usually in the middle of the day on hot days). The GA has had the successful result of reducing the peak loads on the IESO grid. It has also created an ideal market for behind-the-meter (BTM) energy storage, since by having batteries on site, large providers can reduce their consumption on peak days, avoiding the large GA fees. As a result, many energy storage firms have begun offering a "share-of-savings" service, in which they install batteries for small fees (or in some cases with no initial payment). In return for installing the systems, the storage providers receive a share of savings that the C&I customers reaped by avoided GA fees.

This process is an example of a subscription control service (SCS), in which the services of energy storage are offered as a subscription. As markets evolve, financial products evolve along with them to serve these markets. Many of the energy storage markets are based on reducing peaks, and result in large fees to consumers, providing strong incentives for energy storage SCS products.

Most SCS products are highly dependent on accurate controls systems, along with algorithms to predict when they are needed. Since they are often monitoring markets which require instantaneous decision-making and dispatch. In the case of the GA market described above, the ability to predict the peak hours could use an AI-powered algorithm which takes in past performance data as well as real-time data such as meteorology.

As energy storage markets continue to evolve, SCS markets will expand and continue offering energy storage benefits to a broadening base of customers.

5.6 Cybersecurity

The controls and data associated with a BESS should be considered highly confidential and sensitive information. Since nearly all battery systems require internet connectivity, battery controls systems should be considered vulnerable to cybersecurity threats. Recent ransomware attacks have focused on various points of the power grid, including energy storage assets. For utility-scale systems the **Federal Energy Regulatory Commission (FERC)** and the **National Energy Reliability Commission (NERC)** have guidelines and best practices that should be followed. NERC compliance is mandatory for any sites over 300 MW. Depending on where the BESS is built, the **Independent System Operator (ISO)** or **Regional Transmission Operator (RTO)** may have additional cybersecurity requirements as a condition of interconnecting with the power grid. For example, the California ISO (CAISO) must approve certifications from the controls system's hardware and software. Given the size and growing importance of energy storage systems on our power grids, some installations may be considered to be of strategic national security and subject to further federal regulations.

Even in the event of a breach, a properly designed system would not typically be subject to a safety event, since the controls system cannot normally induce unsafe conditions without triggering safety events such as a fuse blowing or breaker tripping. However, a breach may induce some unwanted effects.

Wherever a BMS or EMS is integrated with a broader controls system like a PPC or SCADA, the cyber-security protocols must be coordinated across the organization to ensure adequate coverage.

Third-party Security Firms

One proven method of avoiding cybersecurity threats is for the project developer to hire experts to design, evaluate, or improve the cybersecurity project. There are many security consultants in the industry who specialize in energy infrastructure and can perform a security audit, probing the project and its platforms for potential weaknesses. This type of audit should be conducted periodically, and preferably without prior notice to the operators.

In the case of a successful breach of the system, a reactive audit may also involve an outside third party to assess the damage and ensure that the system has been restored. At a minimum, the method used by the attackers should be mitigated to avoid any repeat attacks.

A number of controllers in the market may include some level of cybersecurity as part of their service offering. For example, the Fractal EMS provides network firewalls and other security protocols as part of their standard deployments. However it should be noted that any security system is as strong as its weakest link – if a control equipment developer's internal control systems have weaknesses, these may allow bad actors to enter into a project's entire control system.

A growing trend across many industries is the use of ransom-ware attacks, in which a system's sensitive data may be held as collateral in a ransom negotiation. There are now insurance products available which compensate the policy holder in the event of this type of attack. Several preventative strategies exist to back up, mirror, and silo data in ways that decrease the threat that could be caused by hackers attempting, or succeeding in, a ransomware attack.

Physical Security

While the concerns around batteries are traditionally focused on cybersecurity threats, BESS installations should also be protected from physical intrusions as well. Most battery installations do not have full-time staffing, although some have personnel that may be working on nearby facilities. Because of this, most battery plants are monitored by cameras placed throughout the site, which are usually web-accessible and motion activated. The majority of installations are protected by chain-link fences, although in some locations walls may be required (which can serve as both security and a noise barrier).

There have been some reports of batteries being shot at with firearms, particularly in rural areas. While most of the battery enclosures have steel plating that provides some protection against this type of attack, larger caliber rounds may be able to penetrate the enclosure. In the worst-case scenario, if a battery cell is punctured it could cause a short and possibly send the battery into thermal runaway. While it is difficult to prevent this type of attack, some potential remedies include opting for thicker enclosures, having floodlights to illuminate the area around battery installations, and having staffing at installations that may be prone to this type of attack.

5.7 Usual Suspects

Controls providers encompass many various hardware and software services and components. This section is broadly divided into EMS, Dispatch optimizers, Scheduling / Market Operators, and data analytics providers, but there are several firms which provide multiple services across these boundaries.

EMS Solutions

Fractal EMS: Fractal is an energy storage controls firm offering full-service storage controls design, installation, and monitoring for operational systems. Their proprietary solution includes both the hardware required for controls as well as the software to monitor the system, as well as to optimize revenue. It can be integrated either on top of the controllers provided by the manufacturer, or as a standalone EMS. Fractal also provides O&M services, and has a sister company, Fractal Energy Storage Consultants, who provide energy storage advisory services.

InAccess Networks, SA: InAccess is a UK-based global designer, developer, and integrator of renewable energy controls systems. They focus on plant-monitoring, SCADA, plant control, and grid integration, and have become one of the global leaders in BESS controls.

Indie Energy: A provider of energy storage EMS, Indie offers fleet management, site control, SCADA integration, energy market controls and performance analysis. Based outside of Austin Texas, as of April 2024 Indie stated they had completed 110 projects with 5.7 GWh of energy storage.

PXiSE Energy Solutions: Pronounced "Pice" (which rhymes with mice), PXiSE is a member of the Yokogawa Group, a Japanese multinational electrical services company. PXiSE is based in California and offers controls technology for distributed generation, system balance, and power quality. They are one of the leading global controls providers in the BESS industry.

Schweitzer Electronic Laboratories: SEL is the leading manufacturer of controls hardware for protection, control, and automation of power systems. They focus on providing the products often found monitoring BESS substations or controlling key components. SEL gear is found on nearly every utility-scale BESS plant in North America, and is often called out by name on plan sets and specifications. Beyond hardware, SEL offers software and advisory services for integration of their hardware.

TriMark Associates, Inc.: Trimark is a California-based provider of controls solutions specializing in BESS applications. Their hardware and software can conduct advanced metering, data telemetry, SCADA, or ISO communications. They are one of the leading US-based providers of energy storage controls infrastructure. TriMark manages SCADA control for over 9.5 GW of PV and BESS in 37 states.

Swell Energy: Headquartered in Los Angeles, California, specializing in home energy solutions, smart grid technology, and virtual power plants (VPPs). Swell was founded in 2014 and provides energy storage systems and controls platforms in addition to VPPs. Many of their projects involve partnering with utilities, some of which involve tens of thousands of homes. In 2023 Swell stated they aggregated 350 MWh of VPPs through 16,000 battery systems installed in homes and businesses.

Stem: Stem has significantly impacted the renewable energy sector by enhancing the integration and efficiency of energy storage systems. They have deployed over 1.2 gigawatts of energy storage capacity globally, managing more than 950 energy storage systems across various sectors including commercial, industrial, and utility-scale projects.

Dispatch optimizers

Dispatch optimization is a service that can be provided either as a part of the integrator service offering, or by a third-party company. Some of the leading providers are:

Ascend Analytics is a firm that sells data sets related to energy markets. They also offer a software platform called SmartBidder that is a unified platform for custom bid optimization combined with scheduling services to manage asset performance and operations for storage, renewable, and hybrid assets.

Fluence IQ Digital Platform: That has multiple products covering intelligent bidding and asset management. According to the company's website, they have over 10 GW of assets being managed by their software products.

Tesla Autobidder: Autobidder is Tesla's native real-time trading and dispatch optimization software that comes with their utility-scale storage product, the Megapack. Tesla has stated they intend to operate on over 7 GWh of BESS by the end of 2024 [37].

Wartsila IntelliBidder: IntelliBidder is a branded product in Wartsila's GEMS Digital Platform suite that offers market price forecasting, renewable forecasting, schedule commitment, and bid optimization functions.

Scheduling / Market Operations

BETM: Boston Energy Trading & Marketing is a subsidiary of Mitsubishi established in 1989. They offer energy trading, battery optimization, and scheduling/dispatch services to manage energy assets, including BESS projects.

Customized Energy Solutions: CES specializes in 24-hour market operations, offering monitoring, scheduling, market offer generation and curtailment management. They are one of the leading scheduling coordinators.

OATI WebTrader: OATI is a software platform for financial and physical trading, scheduling, risk, and settlements in energy markets.

PCI Energy Solutions: PCI, also known as Power Costs, is based in Norman, Oklahoma. Founded in 1992 PCI provides software solutions to the energy industry, and they have stated that over 70% of Fortune 500 energy and utility companies in North America use their platform. In energy storage, they offer a BatteryTrader platform, which uses machine learning algorithms to monitor SoC, predict prices, and develop bid strategies.

Data analytics providers

Three leading firms in this space are Zitara, ACCURE, and TWAICE, who are listed under Section 9, the Usual Suspects under Chapter 9: Operations and Maintenance.

6. PROJECT DEVELOPMENT

When Nikola Tesla and Thomas Edison built the world's first power stations in the 1880s, their fledgling utility companies had to build out both the generation, distribution, and metering of the grid in full. In the early years of power grids, these functions were all managed by a single entity – the utility, with line workers to build new grids, meter readers to bill for the power, and engineers who managed the network of power plants, transformers, and other equipment that comprise the grid. As grids became larger and more complex, including new transmission lines interconnecting cities, utilities and governments began to hire separate service providers to perform some of these tasks. This initially began with energy generation – **independent power producers (IPP)** began signing contracts with utilities to provide power at set rates over the long term. The IPPs financed and built generation plants, signing agreements with the utility to provide energy at a certain price. Like a real estate developer, the IPP was responsible for running the project economics to account for all costs over a project's lifetime: initial project development costs, design and construction, debt service, O&M costs, and eventual decommissioning. Given the large costs of planning and developing a power plant, the firms doing this work needed to ensure that the accumulated system revenues were sufficient to result in a positive return for the investors in the plant.

Today, BESS project developers are a form of IPP: they interconnect to the grid at high voltage substations through transmission lines, they require high capital costs to build, and they sign offtake agreements with utilities or grid operators to establish their revenue streams over the life of the assets. However, unlike a power plant, BESS plants do not produce energy. Instead of selling megawatt-hours into the grid, they instead sell their services to the utility, grid operator, or the open market. Batteries can provide a wide range of different functions which are valuable to grids, grid-operators, or corporate buyers, as discussed in depth in Section 6.2.

Given the high capital requirement of these projects and their steady income, battery plants make attractive infrastructure investments. Firms that can successfully bring these plants online are able to make a handsome profit from either selling the assets to other developers, or by holding on to the asset and earning the dividends that come with their operation. However, there are many risks to navigate, and many development projects have been abandoned awaiting finance, have been plagued by problems during construction, or been built but cannot be operated profitably.

The business of planning, financing, building, and operating renewable energy projects is known as **project development**. This chapter focuses on the development process for BESS projects, and how the firms building these plants have made this novel technology into a profitable infrastructure asset. Development is a combination of technical expertise, commercial savvy, and legal knowledge that bring projects to life. Under the tutelage of a developer, projects move from the board room to the drawing board, to the grid, morphing from an abstract idea to a functional plant that provides critical services to the power grid. Along the way, it becomes a functional asset that earns healthy returns for their investors and lenders.

This chapter outlines the basic economic drivers of the project development process, all major phases of development, and the key players throughout each step along the way. We begin with a discussion of the major business models used by project developers. The following section discusses various services a BESS can provide to the grid, and how these services are turned into revenue streams for the project. Next, the early-stage development process covers site selection, permitting, and key equipment selection. Section 6.6 on Contracting reviews all major agreements which a battery project developer enters into. The section on Project Execution covers the developer's role in the steps of managing construction, testing and commissioning, and eventual operation of the BESS plant. These topics are covered here from the developer's perspective, and a more in-depth discussion of these processes can be found in Chapter 8 on Engineering, Procurement, and Construction (EPC) and Chapter 9 on Operations and Maintenance (O&M). Although it is a critical topic for developers, all topics related to the finance of battery projects have been omitted here since they are covered in Chapter 6 on BESS Project Finance.

6.1 Project Development Business Models

From a business point of view, a battery is an asset like any other – it costs a certain amount of capital to build, it operates over a defined lifetime, and it earns revenue that produces a return for the investors of the project over that time. In this sense battery projects are not unlike other infrastructure projects such as natural gas plants, toll roads, and water systems.

Figure 6-1 shows costs and benefits of a typical project with a 20-year lifetime (which is typical of current large-scale battery construction on the power grid), based on current costs of around $400/kWh. As depicted, the initial project construction is expensive but justified by the net profits, which account for operating expenses (OpEx) and system replacements. After considering all expected cashflows over the project lifetime, including development costs, a successful project will show a positive return – otherwise it is not worth building. In the project development world, it is not uncommon to study a project that appears profitable, but on closer inspection has costs that do not justify the profits that make it an attractive investment. This is a very simplistic model - Section 7.3 addresses the process of building a complete financial model of the project.

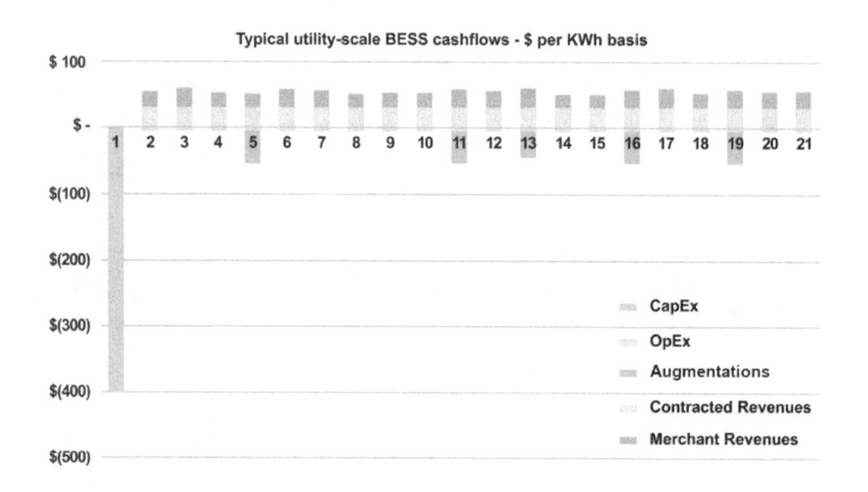

Figure 6-1
BESS lifetime cash flows

Selecting projects that make financial sense is a key priority for all developers. It is common for a developer to simultaneously develop many projects – this ensures that if one project runs into difficulties and cannot be built, there are other projects in the portfolio to cover the development costs spent on the cancelled project. Developers must consider both the economics of an individual project, as well as all their whole portfolio, anticipating a certain failure rate.

A developer shepherds each project through various stages, which are discussed in depth in the subsections of this chapter and shown visually in Figure 6-2.

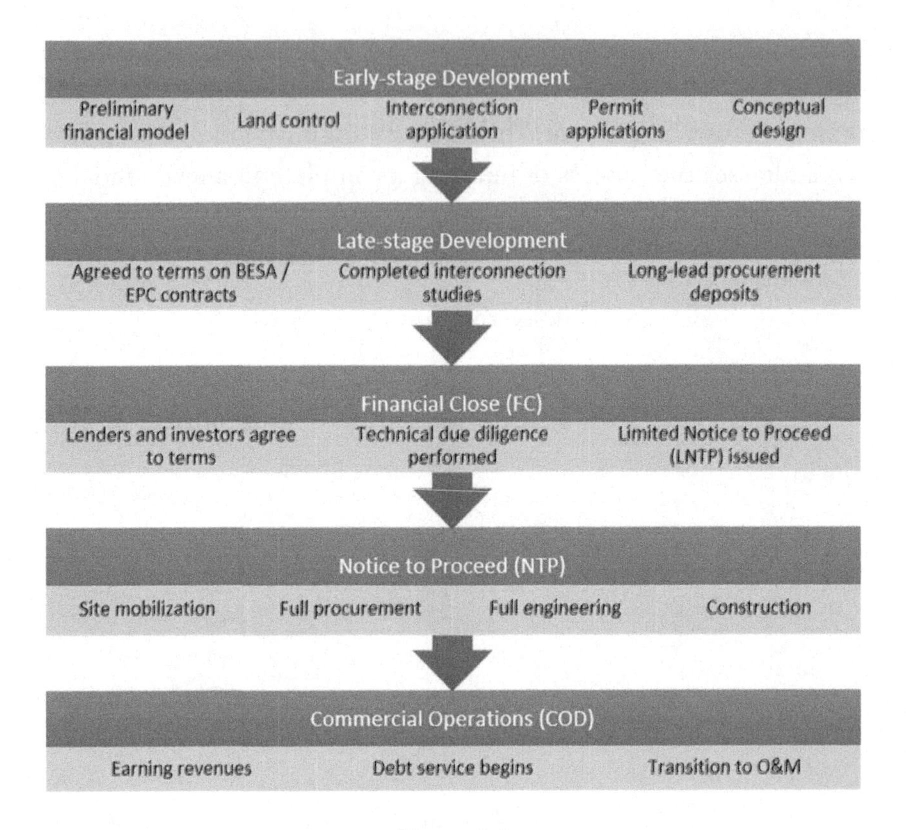

Figure 6-2
Project development stages

At the earliest stage, the project's preliminary financial model is determined, which describes its approximate revenue strategy, either contracted or merchant, and expected costs. The individual project site is assessed for its feasibility and secured through a land lease or purchase option agreement, which is known as **land control**. An interconnection with the utility and/or grid operator is applied for, and the permitting process is begun. The early-stage design begins with limited detail, in a process known as **conceptual design**. If the project progresses to the later stages of development, the contractual relationships are fleshed out in greater detail: suppliers of the BESS and other major components, EPC contractors, and the agreement with the utility and major offtakers. The next milestone is the true point of no return – the project is presented to investors and lenders, who will decide whether to commit to the project or not. This milestone is known as **Financial Close**, or FC. If they decide to fund the project, it will move on to the construction phase, commonly referred to as **Notice to Proceed** or NTP. Eventually the project will make its way toward beginning commercial operations, which is commonly referred to as **Commercial Operations Date**, or COD.

As a project matures, risks become fewer and value increases. Very early-stage projects have less value and large risks, as they must overcome numerous potential challenges such as permitting delays, public opposition, interconnection difficulties, or changing interest rates. Figure 6-2 shows the typical stages involved in project development as it moves from early stages to eventual operation.

As a project gets farther along in its life cycle, there is less risk, as the project shows that it is able to overcome the potential issues and inches closer to earing revenues. A project achieves the lowest level of risk when it is operational since it has already demonstrated the ability to earn revenue. At each successive milestone, the project achieves greater value and lower risks. For large BESS projects in the US, this development process can take anywhere from 3 to 7 years, driven by long interconnection queues and HV equipment lead times.

Throughout the development process, there are a wide variety of expenses such as:

- **Staffing**: The personnel required to perform project development tasks
- **Procurement deposits**: In later stages of development the project may have to put down deposits to mitigate the risk of long lead times

- **Interconnection deposits:** Most utilities and grid operators require non-refundable payments as part of the interconnection process
- **Soft costs:** Legal expenses, travel, overhead, and other administrative costs of running a development firm

Normally the largest project costs, such as the deposits on batteries or the down payment on EPC contracts, are not made until the project has reached FC.

It is notable that projects do not earn a single dollar until they come online, meaning developers must have sufficient capital to endure several years of development in the hope that future revenues of the project justify these expenses. Because of this, it is rare for a developer to single-handedly take a project from early-stage development through to COD. Many developers prefer to move the project from one stage to another, and then sell the project to a developer focused on adding value at a different portion of the project life cycle. Others may sell a portion of the project but retain the project throughout its life. Common methods of BESS project finance are covered in Chapter 7.

Developers range from small operations with fewer than 10 people and a handful of projects, to massive publicly traded firms with hundreds of projects at various stages of development. Regardless of the size, every developer is committed to increasing the project value throughout its development, culminating in the project becoming operational and earning its anticipated revenues.

The following sub-sections describe some of the more common development business models followed by BESS project developers.

Greenfield Opportunities

The earliest stage in developing an energy project is known as **greenfield development**. This work involves evaluating attractive markets for projects, scouting land parcels, and obtaining preliminary permitting and interconnection approvals. Greenfield developers often start with many potential project sites, and these are gradually weeded out if any major issues arise. Many greenfield developers sell their projects at an early stage, prior to securing the equipment to be used, obtaining interconnection approval, or having any committed investors. Especially in hot markets, they can often earn large returns by marketing their projects to developers who are better capitalized.

The term "greenfield" comes from the business world, where it is used to refer to expansion of large, sometimes multi-national corporations. When the parent company chooses to enter a new market, they can do so by acquiring an existing company, or by building a new subsidiary from the ground up. Where the latter approach is taken, it is being built in a green (undeveloped) field. The same logic has been extended to renewable project development to refer to building projects from scratch, rather than acquiring an existing project.

A point of potential confusion: The term "greenfield development" is unrelated to **brownfield development**, which involves building on abandoned or unutilized property, typically large industrial sites, which may also have some form of environmental contamination. This has become a source of interest in the US as there are tax incentives around brownfield development, as discussed in Section 7.4.

Pure-play Development

Developers that develop projects with the intention of selling them to buyers, rather than seeing them through to the operations phase, are known as **pure-play** developers. These firms often pursue projects in markets that are expected to become hot in the future, similar to land speculators in real estate markets. Since pure-play developers do not build the projects themselves or manage any operational assets, their staff is typically leaner than firms that perform see their projects through to later stages.

One key component of pure-play development is the submittal of a preliminary interconnection application. Since many crowded BESS markets can become inundated with these requests, firms that have secured spots in interconnection queues are able to command a high price for their projects. As interconnection queues in the US have become saturated, some grid operators have imposed larger fees to discourage placeholder projects which may not be built.

In contrast to pure-play developers, IPPs intend to operate renewable energy assets in the long term. Some IPPs develop their own greenfield projects, while others prefer to acquire projects from pure-play developers. IPPs are discussed in greater depth below.

BESS Project Acquisition

Many firms that enter BESS markets choose to acquire projects, rather than to develop projects from scratch. The acquisition can happen at different project stages, ranging from early development through to projects that are close to reaching COD. When acquiring a project, the purchasing firm typically conducts extensive due diligence of the project to identify, and where possible mitigate, technical, commercial, or legal risks. This may be done in-house or using outside consultants specialized in evaluation of BESS projects.

For developers with large capital reserves, acquisition is often a preferred option for entering new markets, since it can dramatically accelerate their timeline for entering a new market. However, buyers must be cautious since high demand and commercial pressures can often drive up prices for assets in hot markets. As of mid-2024, the hot market for project resale has led many project sellers to execute competitive processes to court buyers, soliciting high bids for those seeking to develop in the competitive marketplace. Larger developers typically have internal review committees perform a commercial review prior to executing any acquisitions. Depending on their ownership structure, many developers must pass a vote to arrive at

Final Investment Decision (FID), a commitment to move ahead with the project.

Given that many developers are working on several projects in parallel, it is common to see several projects bought or sold together as a portfolio, sometimes called a pipeline. As BESS projects become widespread, the transactions around their sale and purchase have become ever larger and more frequent.

Exit Strategies

Many developers have planned exits, or partial divestments, built into their business model. While pure-play developers aim to sell projects at an early stage, other developers may aim to sell a portion of their interest as the project moves toward FC. Even after COD, many investors have a specific timeline for exiting a project – for tax equity investors this is typically from 5 to 10 years after COD. Later stage projects that are operational may be acquired by institutional investors such as pension funds, since at that stage projects represent a lower risk and a stable return on investment. It is rare for any one developer to hold a stake in a project from the initial greenfield development to the project's end-of-life, but some larger developers can maintain this position throughout. Instead, many developers have a clearly defined exit strategy which allows them to focus on whatever part of the project life cycle they are specialized in.

Independent Power Producers

IPPs, in the context of renewables, are companies that own and operate renewable energy projects. Some IPPs perform development of projects in-house, while others acquire projects via acquisition transactions and continue to develop and build them through to operation. They frequently have in-house teams to perform or manage engineering and construction activities. Many large developers that have resources, both human and/or capital, prefer to operate the projects to be able to retain revenues in the form of contracts or merchant revenues. IPPs may also "build and transfer" to other entities who wish to acquire the assets, such as utilities.

IPPs need to be able to consume significant risks in all phases of the project. Large firms in this space benefit from having extensive development pipelines (in terms of number of projects) that lend credibility for obtaining land, project funding and securing key equipment at competitive prices. They tend to have a sizeable number of employees in specialized departments carrying out functions for project development and operations, or have the financial backing to be able to manage these activities using external entities. The functions include land acquisition, permitting, market evaluation, contracting, grid interconnection, engineering, procurement, legal and operations management – these are further explored in the upcoming sections.

Prior to the year 2000, most IPPs operated portfolios of primarily fossil fuel thermal generator assets, alongside some hydroelectric, nuclear, or geothermal plants. As wind farms began to become larger and more economically feasible following the turn of the century, they began to become a feature of IPP portfolios. 2010 brought the emergence of renewable IPPS, when solar PV plants began to form large assets in these portfolios and batteries began being included in IPP portfolios in the late 2010s. Some developers emerged that focused exclusively on renewable energy and began bringing larger and larger power grids online. The growth of new additions to the power grid can be seen in Figure 6-3.

The business of building renewable assets, including BESS projects, is a long game that requires vision, planning, financial capital and talented human capital to be successful. Many new firms are entering this space with financial backing from banking and investment institutions because of the lucrative nature of these projects. But picking the right projects to develop in strategic locations and managing

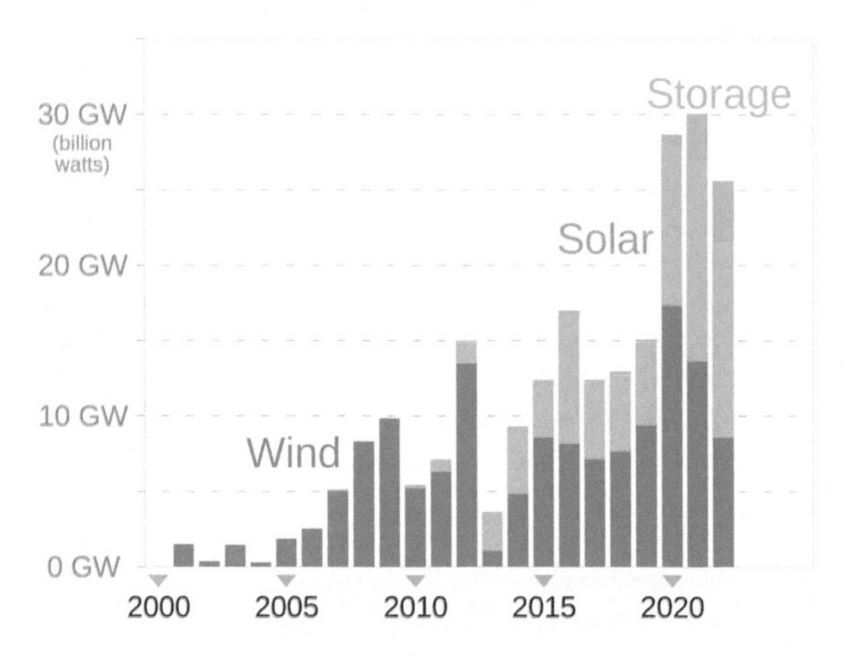

Figure 6-3
New installations of power in the US [38]

activities in a timely manner are critical to developing financially sound projects that are sustainable.

6.2 Revenue Streams

The project must be shown to provide enough economic value to justify the expense of building and operating the project. A source of revenue that a BESS can monetize is known as a value stream, which will be dependent on the project's use-cases (more on use-case in the next section). This section discusses the most common value streams identified and used to finance Li-ion batteries today.

Value streams are quite dependent on the market in which batteries are being built – for example, the PJM market in the US was one of the first to set up a frequency regulation market and attracted large batteries starting in 2014. To use another example, remote locations that have frequent power outages make resiliency a strong value stream, since customers are willing to pay more to ensure that they will have backup power.

In energy, an **offtaker** is any party that purchases the services of an energy project. Offtakers for utility-scale projects are most often large entities such as utility companies, grid operators, or corporate entities. As we will review, having a strong offtaker is the cornerstone of an energy project, since the project's viability depends on the revenues that come from that offtaker.

Broadly, value streams can be divided between **contracted revenue** and **merchant markets**, which are covered in the following two sections.

Contracted Revenue

Contracted revenue is, as the name implies, revenue that comes from a formal agreement, or contract, between the Project and the offtaker. The battery project commits to provide a service continually to the offtaker, and the offtaker commits to purchase the services the battery is producing for a set period, typically from 10-20 years for large projects. This agreement means that if the battery is up and running, it will be available to generate revenue. This arrangement is advantageous as it is a stream of income that the project can count on, provided the offtaker is creditworthy. Most utility-scale BESS projects have some form of contracted revenue stream.

Merchant Market

A merchant market differs from a contracted revenue stream in that the earnings are dependent on market fluctuations, rather than a fixed contract. While this brings some added risk to the project due to the uncertainty of value, it also grants large sources of revenue, which can become the majority of the revenue for a project over its lifetime. Often a project may be simultaneously earning contract revenue and merchant revenue, which may change based on the time of day, time of year, or at different stages in the life of the project.

Whoever is running the market (most often an ISO, RTO, or utility), has criteria that must be met to participate in the market. Typically, this includes having an interconnection agreement, meeting certain safety standards, or minimum participation rules. Provided the project satisfies those requirements, they are then free to bid into the market. Depending on the market, the battery may bid into the market on a daily, hourly, or weekly basis, or whatever the terms of the market are. Some markets operate with bids being issued every few seconds. Given that most battery controls are automated, so too are most of the bidding systems. Many battery projects hire a third party to manage the bidding, which is discussed in greater depth in Section 9.6.

Bidding into a marketplace means that the future amount of compensation is unknown and may fluctuate over time. For example, the fast ramping frequency regulation market in PJM, known as RegD, pays batteries for frequency regulation services on a $ / MW basis. The weighted average clearing price for regulation was $17.46 per MW of regulation in the first six months of 2022 [39], an increase of 49% over the same price in the first 6 months of 2021. While this presents some difficulties in estimating the expected lifetime revenues, it means that the BESS can devise strategies that change over time on how to monetize the services that can be provided.

Offtaker

The offtaker buys services produced by a battery project. In traditional gas, wind, or solar projects, these were typically buyers of energy that agreed to buy power at a certain price, through a power purchase agreement (PPA), measured in dollars per megawatt-hour ($ / MWh). Since batteries can offer many different services, as covered next, a project can have one to many different offtakers, some of which change throughout the years of project operation.

Offtakers take many different forms based on the type of project and market in which it is built. They may be a utility company that buys energy from a PV+Storage plant, an ISO looking to purchase grid services, or a corporate buyer looking to store energy and release it at another time of the day (a use-case known as energy arbitrage). Since many projects provide several different services to the grid, they may have several different offtakers at different points of the project.

Since the offtake agreement is ultimately the guarantee of a project's revenue, it is a critical aspect of the project at the planning and finance stage. Section 8.5 discusses contracts in more detail.

BESS Use Cases

Now let's consider **use-cases**, the potential services that a battery can offer. Gas plants, solar farms, and wind turbines are relatively straightforward in the service they offer to a power grid: they generate power (MW) which feeds energy (MWh) onto the grid. Typically, these systems are compensated by the amount of energy that is fed onto a grid at an agreed-upon tariff ($/MWh). Although there are some additional services that these plants may provide, the vast majority of project revenues stem from this energy payment.

In contrast, batteries do not produce energy on their own – they must first be charged, and in fact always consume some energy between charging and discharging (see round-trip efficiency in Section 4.4). On top of that, BESS plants are quite expensive: on a power basis they typically cost between $1,000-$1,600/kW. For comparison, wind plants cost around $1,300/kW, solar plants cost around $1,100/kW, gas plants are $450-$650/kW. Lastly, at 20 years they have one of the shorter useful lifespans when compared to wind turbines (up to 25 years), PV plants (up to 35 years), or combined-cycle gas turbine plants (up to 35 years). Given these drawbacks, what explains the vast interest in battery systems?

The answer lies in the diverse range of use-cases that batteries can provide, some of which are worth much more to utilities and consumers than energy production. Some batteries can sell their services onto a grid at a certain price, known as a **tolling agreement**. Others are used to purchase energy at times when it is cheap and sell it back onto the grid when it is expensive, which is called **energy arbitrage**. And others may provide services such as support if the grid goes down, known as **resiliency**. Each of these different use-cases provide value to the end user and typically there are three categories of energy users:

- Individual electricity customer (ranging from residential to commercial to industrial)
- Distribution utility
- Independent Service Operator (ISO) / Regional Transmission Organization (RTO) – regional organizations that handle operation of the grid, market participation, and bulk electric system planning

The services may be provided at one of three levels:

- Customer level (also known as Behind-the-meter or BTM)
- Distribution level
- Transmission level

The various intersections of these categories are shown in Figure 6-4, highlight 13 different use-cases, in a graphic from the Rocky Mountain Institute (RMI) [40].

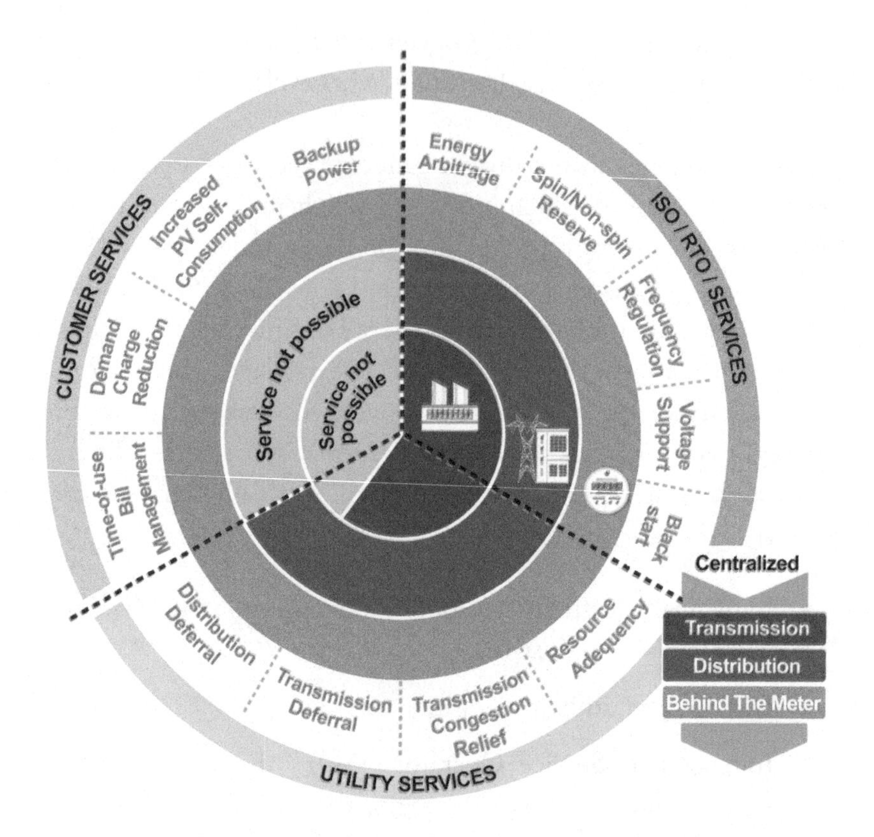

Figure 6-4
Battery use-cases from Rocky Mountain Institute [40]

One of the key differentiators for battery systems is the ability and ease to be able to perform more than one use-case with the same capacity, meaning at the same time. Many BESS projects engage in this practice, called **stacking**. While the BESS can only produce a single charge or discharge signal at each point in time, it may be providing multiple services to different offtakers, allowing the project to earn additional revenues. Certain use-cases may require the full technical capacity of the battery, meaning that it cannot be used for other uses at that time, but other revenue streams may only require the battery to act sporadically, making it easy to balance the competing responsibilities.

In addition to performing services simultaneously, other systems may vary their use-cases around time of the day or season of the year. Stacking is very favorable for BESS project economics. And note that an efficient and effective EMS is critical to the enablement of stacking use cases.

Typical Contracted Battery Value Streams

Resource adequacy

One of thc fundamental planning aspects of load-serving utilities and/or balancing authorities is to ensure they have access to sufficient resource capacity to meet their estimated loads. Thus, plants are paid to have their capacity available and online – formally called a Resource Adequacy, or RA, contract – and it is one of the most reliable revenue streams.

Capacity planning is especially critical for managing the highest peaks that the power system faces. In most markets this corresponds to the hottest time of the year, cooling loads drive demand. These peaks may coincide with other problems caused by hot weather, such as low water levels in hydroelectric reservoirs, lower efficiencies in thermal plants, or solar plants dipping in output as the sun goes down. In extreme cases, grids may be forced into **load-shedding**, or suffer rolling blackouts in portions of the grid. In California in September 2022, the grid hit a peak of 52,061 MW [41], and the major utilities and even the governor put out calls requesting the public to reduce load. Since many of these loads can be predicted based on weather, and since they tend to be only for the hottest time of the day, batteries are a good candidate serving these loads. Further, some states like California have embraced BESS as standard resources to provide capacity, akin to conventional peaker plants, typically using batteries to meet evening loads that increase as solar resources ramp down.

Equipment / Distribution deferral

Similar to the requirement that grids have a certain level of excess capacity, transmission lines are often subject to requirements that they do not exceed a portion of their capacity, to ensure their reliability. Given that transmission lines are very expensive and take many years to plan and build, grid operators must plan many years into the future. They must anticipate the grid's load growth, plan on construction of new generators, and account for existing and future distribution or transmission lines. Often grids may have the issue of a line getting overloaded or having excessive current during peak usage times. This presents a risk to the grid that a sudden event could overload the line, causing power quality problems, or even tripping the breakers in place to protect the line. The traditional solution to this issue was building new lines or expanding the capacity of the existing ones. However, with the advent of batteries, a new solution emerged. If a BESS can be built at the end of a long transmission or distribution line, it can be on standby to inject power into the grid in times of high congestion, reducing the load on the grid and hence reducing the likelihood that one of the lines becomes overloaded. While batteries are certainly not cheap, they may be a more economical solution than building a new line, especially considering some of the constraints of land and permitting. From battery projects' perspective, this means that the system must be extremely reliable, often accompanied by high liquidated damages if the project cannot discharge when needed. However, like capacity markets, the BESS is often able to participate in other markets, especially in periods where there is unlikely to be a transmission or distribution bottleneck on the grid.

Time-shifting / Peak-shaving

Most grids in industrialized countries have some system to incentivize power generation and demand in certain periods. While some simpler offtake and supply agreements may still be compensated by the MW or

MWh, regardless of the time of day, many have moved toward a system where different rates are paid for the generation and consumption of power depending on the time of the day. This has long been common for larger consumers, who pay demand charges which may vary based on time of day, or month of the year, and is increasingly common for residential consumers with the implementation of smart metering. Based on these rates, it may be common for both a consumer and generator of power to have a strong incentive to consume or produce power at a certain time of the day. On the demand side, this can be enabled by installing a BESS plant behind the meter at a commercial or industrial power user. This allows the battery to charge up in times when energy is cheaper, discharge during times when the demand is at its peak, thus reducing the demand charges owed to the utility. This behavior is known as **peak shaving** and can be an excellent use case for batteries.

Similarly, on the generation side, a power plant may be paid a higher rate at peak demand times than at lower demand times. This gives the plant a similar incentive to save power produced during low-demand times and shift it to discharge during peak times, which may not be possible without including batteries at the plant. This is common with solar PV plants, which have high production in the daytime, during peak sun hours, but are paid higher rates at peak demand time, which is often after the sun goes down. Pairing BESS with PV is covered in greater depth in Section 6.1.

Tolling Agreement

Up to this point, most of the operational decisions of the battery have been made by the owner. While there may be a contract to fulfill for an offtaker, the owner is often able to decide which markets they would like the BESS to participate in. However, there are some scenarios where a utility would like to retain full operational control of the BESS but have someone else build and maintain the plant. In this case the offtake may take the structure of a tolling agreement, which means that the user pays a flat fee, typically in dollars per MW per month, and the owner guarantees that the system will be available for their use. This type of arrangement is akin to a lease agreement, rather than purchase, of a car. As with a vehicle, the owner is responsible for the maintenance and upkeep of the plant, while the user is allowed certain criteria in which they must operate. In the case of the car this would be kilometers driven, while in the base of the BESS this is normally MWh of discharge, or cycles, that the BESS can be used for.

This type of arrangement provides for flexibility to the utility since they may not know until the BESS is operational which use-cases they want. While this may end up being more expensive, it relieves them of the burdens required to finance, build, and operate the battery, or to address problems with the system should they arise. From the perspective of the battery owner, this arrangement is beneficial since it gives a contracted revenue stream; however, it is less flexible since the BESS owner cannot choose to operate in more lucrative markets given the capacity is committed to the provisions of the tolling agreement.

Tolling agreements typically have stringent requirements for availability and corresponding liquidated damages to ensure that the service cost comes with minimal downtime and maximum benefit to the buyer.

Typical Merchant Market Value Streams

There are a wide variety of merchant markets available depending on the geographic region and local utility / grid operator. This section reviews some of the more common merchant markets served by battery projects.

Frequency Regulation

A critical role of grid operators is to balance the generation and load on the grid. This is vital because any deviation in frequency can damage consumer equipment, or worse, lead to a collapse of the grid. This balancing is done by system operators constantly predicting what the load and generation will be. Load is predicted using complex data analytics, which involves historical grid consumption, meteorological analysis, artificial intelligence, and other tools.

Based on the load prediction, utilities then do their best to schedule generation of plants to match the target as closely as possible. Given that modern grids have hundreds of generators and millions of consumers within each region, generation and loads are never perfectly aligned. Fortunately, a slight mismatch does not spell disaster, but can cause the grid frequency to deviate from its normal value (60 Hz in the US, 50 Hz in most other markets around the world). If generation exceeds load, the frequency will increase slightly above target value, while if generation lags load, the frequency will decrease slightly below its target value. To correct this imbalance, most utilities run a special type of marketplace for generators with rapid response to inject power into the grid (in the case of an under-frequency event) or absorb power from the grid (in the case of an over-frequency event). The result is alignment of demand and generation. Figure 6-5 shows the net result of frequency regulation, for which the resulting difference between generation and demand is lessened, although not completely eliminated.

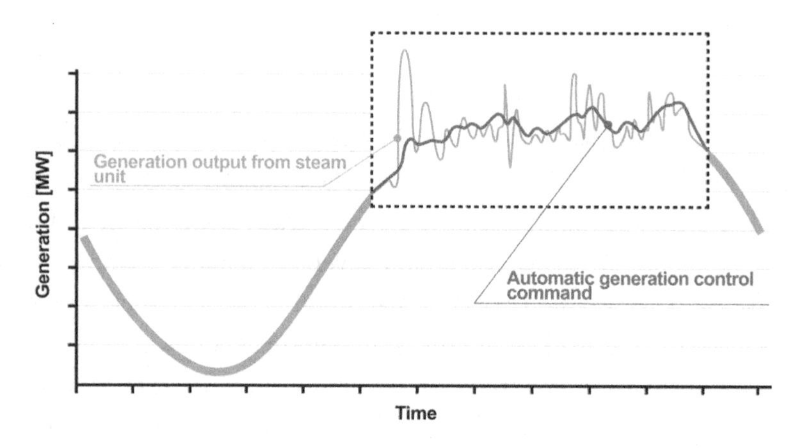

Figure 6-5
Typical frequency regulation correction

Prior to batteries coming online, frequency regulation service was often provided by natural gas turbines, which can respond in under 10 minutes to this signal, when properly configured. However, given that battery PCSs can respond in less than a second or faster, and can quickly swing from injecting to absorbing power, frequency regulation is a perfect job for a grid-connected BESS.

Utilities that have a market for frequency regulation typically require generators to connect to a secure portal which issues a command indicating whether the given plant should inject power, absorb power, or stay idle. This command is issued on intervals which are often just a few seconds apart. These signals are transmitted to the battery system's EMS, and eventually the PCS's operation is modified to absorb or inject the required power. Battery systems bid into the marketplace ahead of time based on the amount of power output (in kW or MW) they can commit to frequency regulation. This may be the full power of the battery, or it may be a fraction of the battery's power capacity. Based on whether the bid was accepted or not, the system will participate in the market for the given bid period.

Most battery projects use a third-party scheduling service to interact with the frequency response service – this is discussed in greater depth in Section 9.3. The Project provides the operating and bid strategy to the scheduler, and the scheduler then transmits those instructions to the grid operator and ultimately collects payment from the grid operator, which is then passed to the Project.

Frequency regulation can be a very lucrative operation, and often forms a large part of the merchant revenue of many BESS projects. However, it is notable that the total amount of frequency regulation required for any grid only constitutes a small amount of the total generation of that grid, typically under 10% of the total generation capacity. Therefore, while demand is likely to continue growing, the grid may have an excess of generators providing frequency regulation, resulting in a decrease in prices.

The term **frequency response** may be mentioned alongside frequency regulation. These are similar functions with an important distinction. While both services either inject or absorb power from the grid, only frequency regulation receives a signal from outside the system, typically provided by the ISO / RTO or the utility. In contrast, frequency response responds to the frequency as measured at the point of interconnection (POI). Based on an algorithm contained within the EMS or power plant controller (PPC), the system translates the measured frequency into instructions about whether to inject power, absorb power, or to idle. Frequency regulation is common to larger, more sophisticated utilities, since it requires a secure communication link to the grid operator's server. Frequency response is required of generators in the US and is considered an ancillary service. It differs from frequency regulation in that it does not require receipt of an outside signal.

Spinning Reserve

A grid operator that governs grid reliability typically has strict requirements for a grid to be able to handle a range of contingencies while still offering service. For example, the grid may be required to lose some of its generators, or to have a transmission line go down, without affecting the grid. One of the common ways this happens is by mandating that the grid has sufficient extra power generating capacity that can turn on, in the case of one of these contingency events. This is often a natural gas plant which can turn on quickly, or a hydroelectric plant that can start generating when required. These services are often divided into spinning and non-spinning reserve, with spinning reserve being defined as a generator that can come online within 10 minutes or less, while non-spinning reserve may take 10 minutes or longer to begin feeding power into the grid. This is because spinning reserve, as the name indicates, is a plant that is already synchronized with the grid. Since most power generation involves a spinning rotor to generate power, this refers to the fact that it is already spinning when its services are required. In contrast, non-spinning reserve requires startup and synchronization, which takes longer, depending on the type of generation.

Typical behavior of spinning or non-spinning reserve is shown graphically in Figure 6-6.

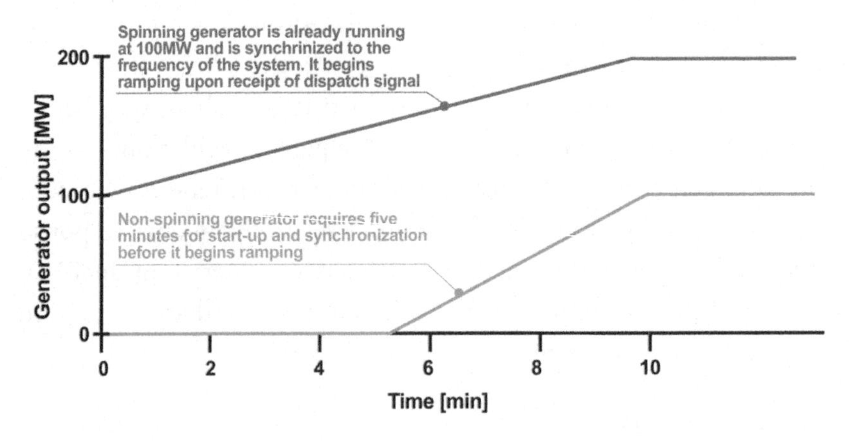

Figure 6-6
Spinning vs. non-spinning reserve

Since batteries use power electronics in the form of a PCS, they can normally ramp up to full power in the range of milliseconds rather than minutes. This makes them even more "spinning" than spinning reserve, despite not having a rotor like most power generators! The ability to produce power so quickly is highly sought after by utility grids, since it provides a form of backup to ensure reliability of the grid. This reliability is guaranteed through **capacity markets**, in which a power plant guarantees the ability to start up when needed, most often in times of peak demand. The capacity payments are made based on the market price for capacity, which varies over time. When there is expected large demand on a power grid, such as during a hot spell in summer, the capacity markets will increase the price, since there is greater demand in the form of greater need for power plants that could come on if the grid were to become overloaded. In contrast, the price will drop in periods of less demand, since there are more power plants available to cover the potential shortfall in generation capacity.

Since spinning reserve revenues are based on a marketplace, it is difficult to know how much the system will make. The most common methods are by evaluating historical performance of the markets, and by obtaining data to attempt to predict the future performance in the marketplace. Although these predictions are uncertain, merchant revenues often make up a large portion of the revenues of the project, so a best estimate of these values is critical to the project being financed. Predictions of revenue are one of the key inputs into a financial model of a project, which will be discussed further in Section 6.

Energy Arbitrage

Power plants on an electrical grid make money by selling their power and energy into the grid. Most large solar and wind plants do this in the form of a PPA, in which the utility agrees to purchase energy at a set rate over the life of the facility. Simple arrangements often have a set price per MWh, often with an annual escalator to cover inflation. More complex arrangements have provisions which pay different rates for the power or energy based on when they are provided to the grid. The price paid for energy is lower prices in times of low demand or plentiful supply, and higher during peak periods, when demand is high but supply may be scarce. The resulting effect is to encourage more power plants to be built and to operate during the periods when most load is needed. Some markets, such as ERCOT, the grid operator in Texas, have real-time pricing, which operates like a type of stock market, determining the pricing for different parts of the grid constantly throughout every hour of every day. Figure 6-7 shows a sample of the pricing map, color-coded to reflect different prices at different parts of the grid. This map can be found at https://www.ercot.com/content/cdr/contours/rtmLmp.html.

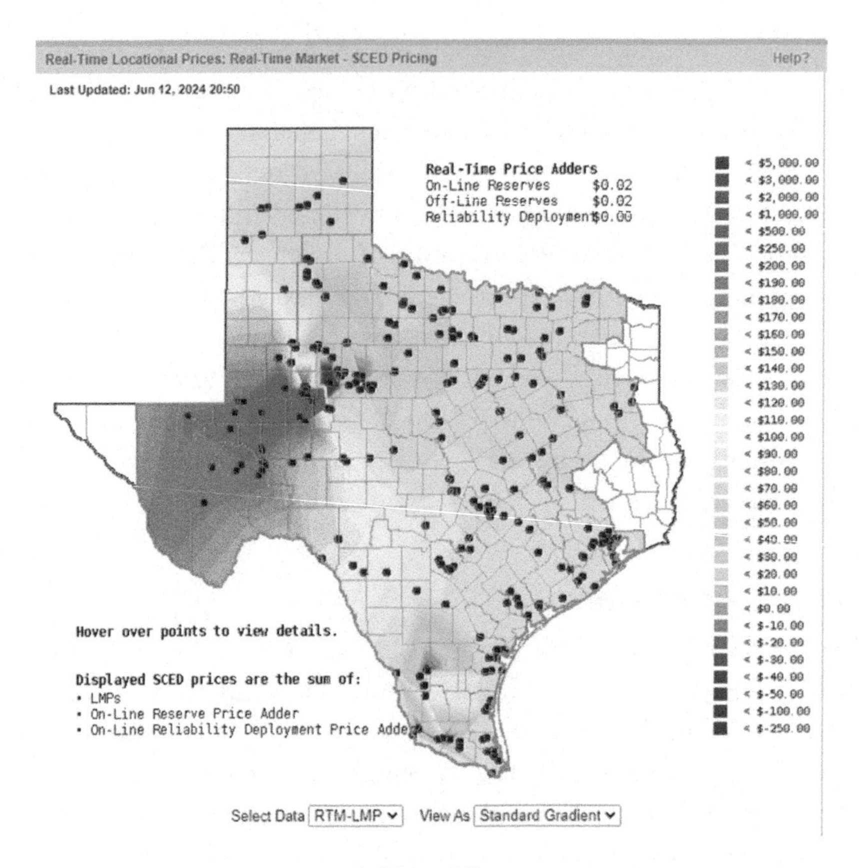

Figure 6-7
ERCOT real-time pricing [42]

In the last 30 years, the addition of large-scale wind and solar power plants have introduced more variability in the market, since these generators can only feed power into the grid when the wind is blowing or sun is shining. In fact, too much generation can drive down energy prices for the plants and eventually go negative, meaning that the grid will pay someone to accept the power. Examples of grids with this phenomenon are California, which has a high penetration of solar PV, and Texas, which has a large portion of wind on the grid.

For batteries, the differences in pricing at different times of the day represent an opportunity for arbitrage. Similar to the way a stockbroker wants to buy low and sell high, the battery can charge up when generation prices are cheap and discharge when the generation prices are higher. Batteries are especially well-suited to capture real-time price arbitrage opportunities, which can be substantial, due to their rapid response time. This requires sophisticated dispatch strategies enabled by predictive analytics. Since the pricing swings are dependent on market conditions, the exact amount of revenue from arbitrage is hard to know up front, but in some BESS projects it represents one of the larger revenue streams on the project.

Solar+Storage

One of the most common use-cases for utility-scale batteries is for them to be co-located with solar PV plants. This arrangement is known as solar plus storage, or simply PV+Storage. This is primarily because, unlike wind, solar produces all of its energy during a certain portion of the day and produces no energy at night. Most grids have their peak demand in the evening, which is when power prices are highest. As described in the above section on arbitrage, this creates an opportunity for batteries to absorb power in the daytime and release it after the sun has gone down when prices are high or at other times as needed. This use is known as energy time-shifting. This may allow a PV+Storage project to produce much higher revenues than PV alone.

One additional use of batteries is capacity smoothing or firming, in which the batteries are used to compensate for the intermittent nature of power production from renewable sources. Since these plants are being powered by the wind and the sun, they are subject to large dips in power production due to cloud cover or sudden drops in wind speed. These drops may be only for a minute, in the case of a passing cloud, or may remain for several hours or longer, in the case of a larger weather event such as a storm. In a large PV system without energy storage, large drops in PV may require the natural gas plants to ramp up their production to compensate for the drop in power. If PV is coupled with energy storage, the net effect can be a much steadier output of power to the grid with fewer drops in power, easing the utility's burden of compensating for these events. A typical PV+Storage system performing capacity firming is shown graphically in Figure 6-8.

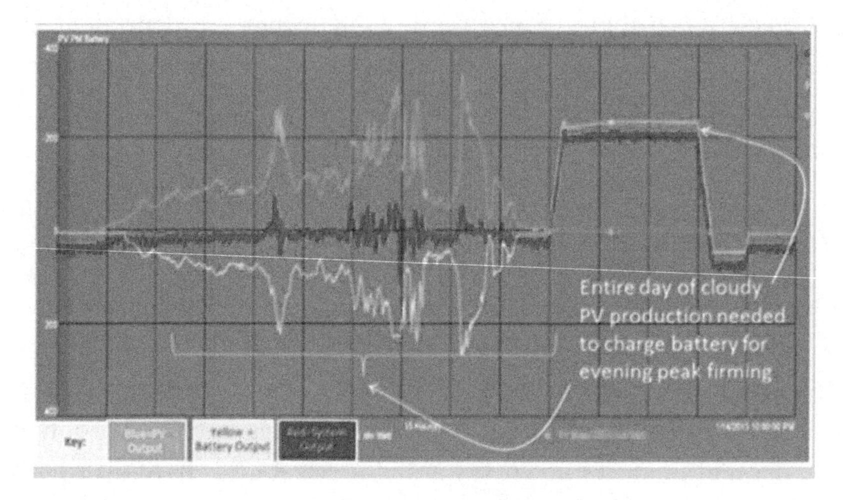

Figure 6-8
Firming function of PV+Storage [43]

Yet another benefit of pairing PV with a BESS is to capture POI clipping losses (as opposed to inverter clipping, which is not covered in this book). In many cases, PV plants are forced to curtail their production because of congestion on the grid. In other words, the grid may not be able to accommodate all the energy produced at certain times of the day. This results in valuable renewable energy getting clipped especially during sunny times of the day. When a battery is paired, especially coupled with the PV on the DC-side (behind the same PCSs) the battery can charge up from this energy that would otherwise be wasted. This phenomenon is shown graphically in Figure 6-9.

Although no project likes to waste available power, the economics show that this will optimize revenue for a PV-only project. However, with the decreasing cost of batteries, it has become more profitable to co-locate batteries with the PV system. This configuration allows for the solar to route the PV to the batteries in the times when it would have been clipped and exporting it later in the day when the PV output has decreased, which may also allow for that energy to be sold at a higher price.

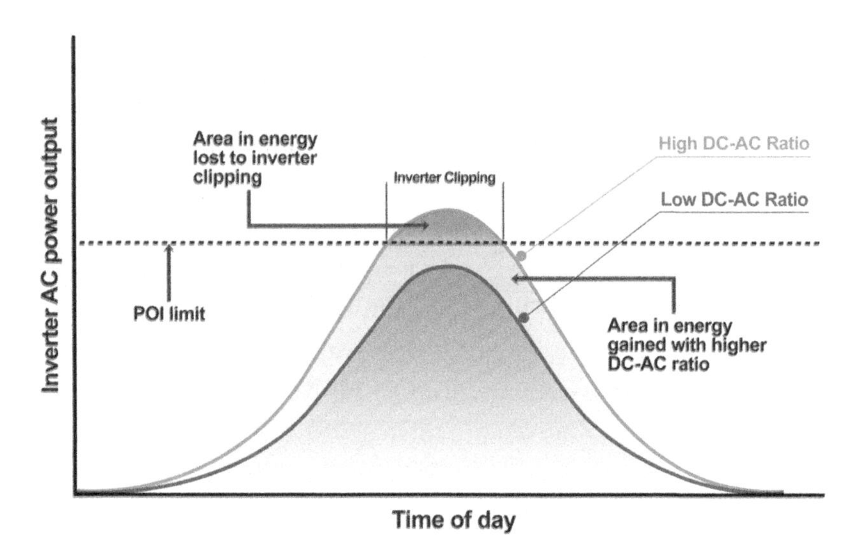

Figure 6-9
DC clipping losses

In 2022, 95% of all PV projects in the California interconnection queue had some type of battery [44].

Solar and storage plants may be one of two different configurations: AC-coupled or DC-coupled. This topic is covered in full in Section 8.1, but briefly: in AC-coupled systems each battery block and solar block has its own PCS feeding on the AC bus, while in DC-coupled systems, as touched on above, each block consists of a single PCS fed by both the solar and battery. AC-coupled systems are more common, and allow the system to either work together or to function independently, while DC-coupled systems typically work to shift energy from the solar energy. DC-coupled systems are the only architecture which allows for capturing of clipping losses.

Microgrid Applications

Microgrids may take many forms across many different applications. A hospital with a backup generator is a form of microgrid, although it may be connected to a power grid 99.9% of the time. A copper mine located hundreds of miles from a utility grid, running on heavy-fuel oil with solar PV is also a microgrid. In its own way, a nuclear submarine is also a microgrid. Since it is common to have multiple sources of power for a microgrid, batteries have been included in many recent microgrids. Their fast response time allows microgrids to cover outages with low outage time, and in some cases even no outage at all (what is known as a "bumpless" or "blinkless" transition.

This book is focused on utility-scale systems, while most microgrid applications are small-scale (1 MW or less). However, large mining applications require microgrids, and these may be quite large – sometimes exceeding 50 MW in capacity. Any utility-scale BESS product can be configured in a microgrid, but some products, such as the Tesla Megapack 2XL, have marketed themselves specifically for this application.

Whenever a BESS project is used to avoid an outage, this is referred to as a **resiliency application**, since it is based on maintaining a facility with power, rather than participating in some market for power, energy, or grid services. While battery-based residential and commercial/industrial microgrids are growing in popularity, utility-scale systems rarely provide resiliency-based services. This is because large power systems require the grid to operate and sell their services. Without the grid, they would have little application for microgrids.

The closest thing to a utility-scale resiliency application is a service known as transmission deferral. This application is based around utilities' contingency planning requirements. Most utilities are required to have sufficient transmission lines to be able to keep the grid up and running even if major lines are forced to shut down. To do this, many lines are required to operate at a limited portion of their capacity. If a certain capacity level is exceeded, the utility must expand existing lines or build new ones to be prepared for an outage. Since building or expanding transmission is very expensive, some utilities have turned to large-scale batteries as an alternative solution. The battery is required to be charged and standing by during peak periods.

Notably this is one of the use cases of the Waratah Super Battery, a 850 MW / 1.6 GWh battery deployment in Australia, which as of this writing is one of the world's largest planned utility-scale battery installations. This project is being developed by Akaysha Energy, an Australian energy storage developer, with the services being purchased by EnergyCo, part of the New South Wales (NSW) government. This project is covered in greater detail in the Case Study in Section 13.1.

Offtake Agreements

The offtake agreement is the contract that governs the purchase of the contracted battery revenue stream. Traditionally most independent utility-scale power plants nearly always sold their energy production to a utility company via a PPA. Under the terms of the PPA, the generator would be paid a set amount per unit energy, usually expressed in cents per kWh or dollars per MWh. Unlike traditional generators, batteries do not produce energy, but rather have the ability to charge and discharge, or to provide other valuable services to the grid. Since they are not simply producing power, batteries do not have PPAs – instead, the agreements under which BESS plants are compensated are more generally known as **offtake agreements**. Although some still refer to offtake agreements as a PPA, this is a holdover from prior days, similar to how many refer to a PCS as an inverter.

Offtake agreements vary widely, but they are typically based on available power, often expressed on a $ / kW or $ / MW basis. In some cases they may be based on providing a certain power level for an amount of time – for example a project might be required to sustain a 40 MW discharge of power for a period of two hours.

Some offtake agreements require the system to be ready to discharge in the case of an outage. When paired with a solar PV plant, batteries may operate under terms of the solar PPA, which may define a special $ / MWh price after the sun has set – the plant can only achieve this price by discharging solar energy from the battery. Offtake agreements sometimes comprise hundreds of pages in a vast mix of technical, legal, and commercial requirements. Some utilities will issue a competitive process, inviting developers to bid on the agreement, while others may be negotiated bilaterally between developer and utility. Since most utilities have to comply with existing regulations, their offtake agreements may sometimes require approval from state, provincial, or federal government, as well as regulatory bodies such as a public energy commission.

Offtake Term

The length of an offtake agreement, often called the "term" or "tenor" of the project, for utility scale battery systems is typically from 10 to 20 years, although some offtakers may prefer a shorter or longer time based on their needs. Typically the offtake agreement is aligned with the useful life of the asset, which in most cases for batteries is 20 years. Most industry standards consider either 20 years or 60% state-of-health (SOH) to be the useful life of a battery. When the offtake agreement reaches the end of its term, some agreements allow for their extension, while others have a fixed length.

Compensation

To a developer, the compensation is the most important part of any offtake agreement – after all, this is the reason they are building the project! The compensation structure depends on the use-case for the battery, but the payments are typically made monthly corresponding to the battery's performance in a given month. In some cases it may be a fixed payment, such as in a tolling agreement, while in others it may be variable depending on the services provided in that month.

In some cases where payment is a risk, there may be a requirement for payment guarantees to ensure that the project is able to collect. Where the offtaker is a large utility or TSO/ISO, this is typically not required since they are very creditworthy, but for offtakers such as corporate entities, bank guarantees may be needed, depending on the project. In the case of projects in developing countries, it is standard practice to obtain a "sovereign guarantee," which requires the host-country government to backstop the payment via their treasury.

Metering

Most offtake agreements include provisions about where and how the energy and power are metered, since this is generally how the project demonstrates its energy, power, or other performance criteria. Utility-scale meters are composed of a set of current transformers (CT) and voltage transformers (VT), which feed information into a physical meter, which then feed to a server or online location, depending on the utility's procedures. It is common to have several meters required on a project, and typically the contract stipulates who must install the metering. More detail on the technical aspects of metering is covered in Section 4.6.

The most common location for metering is the POI, which is often on the high-voltage side of the main power transformer (MPT) of the project. It is common for other contracts, such as the BESA and the BoP contract to also reference this point, so there is no confusion about where it is located. Alternatives to measuring at the POI are measuring at another point, such as the DC or AC PCS terminals, the battery terminals, or at the medium-voltage level.

Testing/Commissioning

For any battery to connect to a grid, the offtaker is typically very concerned with its ability to perform its duties safely and to the established requirements. The provisions of an offtake agreement which discuss testing are some of the most important. They often may include passing specific tests required in a given jurisdiction or showing that the battery can safely disconnect from the power grid. Where batteries are being connected to a controls system operated by a grid or ISO/RTO, they must demonstrate the ability to successfully receive and respond to those signals.

Typical offtake agreements include the following provisions:

- Submittal and acceptance of a test plan, which is the responsibility of the developer, though it may be done collaboratively with the offtaker.
- Scheduling provisions for the test (often 30-60 days prior to the test date)
- Passing criteria for the test
- Re-test provisions in the case of a failed test

Once the test is performed, the data is provided to the owner, or in some cases the independent engineer (IE) to demonstrate its compliance. If the project passes its commissioning tests, it is permitted to begin operating on the grid. In the case that the project does not meet its performance requirements, the offtake contract outlines how the contractor must address this. The most common next step is for the project to repeat the test. If the re-test is successful, the project comes online or stays online. If not, the contract may impose one of two conditions:

- Delay the operation of the project until test conditions are met. Although this may be very costly to the developer and suppliers, this provision provides a strong incentive to oversize the project to ensure there is an ample buffer in testing to avoid this condition. This is more common for provisions of tests at commissioning.

- Impose performance liquidated damages that penalize the developer for having missed the performance criteria. This is more common for annual testing than it is for initial commissioning tests.

It is important to note that the testing provisions and liquidated damages required by the offtake agreement often roll down to the battery supplier agreement and EPC contracts. This is because the project does everything it can to minimize its exposure to risk – wherever possible it passes the offtake requirements on to others. More can be found on this and the actual procedures of testing and commissioning in Section 8.7.

Liquidated Damages

Liquidated damages, commonly referred to as LDs, vary widely but are a common feature in most offtake agreements. LDs provide an incentive for the project to hit its required timeline, level of performance, or other metrics, by imposing a financial penalty if the project misses that metric. When imposed, they are normally deducted from the planned payments to each contractor. It is important to note that most times the owner can choose to waive liquidated damages if they wish to do so, depending on the situation.

Delay LDs are imposed where a project misses specific milestones, such as the commercial operations date, substantial completion, or notice to proceed. They are most commonly imposed on a dollars-per-calendar-day basis by reducing an anticipated payment to the contractor. Delay LDs are usually not imposed if the delay is out of the contractor's control, such as if a force majeure event occurs, or if the delay was due to actions of the owner.

Performance LDs are imposed when a project misses a specific performance requirement, most commonly in the form of missed test criteria. The most common performance LD provision in battery agreements is in the case of an energy capacity shortfall. As an example, suppose a capacity test is required to show 40 MW discharge over 4 hours (160 MWh), but has an LD provision of $10 / kWh of shortfall. If that test shows only 157 MWh, the Project would be liable for the missing 3 MWh (3000 kWh), which would mean they owe $30,000. This penalty is often deducted from the monthly or annual invoice, but this varies from project to project.

Similarly, RTE and availability often have provisions that are pegged to the percentage, or fraction of a percentage, by which the project underperforms. LDs typically have a cap imposed, which is the maximum amount of penalties that the project can pay over a month, year, or its lifetime. LD caps are often a required provision to the lenders of the project, since excessive exposure to LDs means the project carries more risk.

Typical contracts governing BESS projects and their key terms are outlined in Section 6.6.

Hedge Agreements

Since many battery projects are dependent on merchant revenues, the expected returns of these projects are hard to project exactly. They are dependent on the performance of existing markets, the possible change in market rules, or the introduction of new markets. For example, in 2016 the PJM Interconnection, a grid operator in the Northeast US, had a thriving market for some of the first utility-scale batteries deployed in the US Many projects were developed and began operating lucratively in the marketplace. However in January 2017 the market rules changed [45], which resulted in accelerated degradation of the batteries, and ultimately significantly decreasing project revenues.

Lenders that have provided debt to utility-scale batteries are wary of the risks caused by a variable revenue stream. While such projects can be lucrative, they have sought ways to limit the potential downside of such projects, in case the market revenues decrease, or the market changes. One tool they have used to do this is a hedge agreement. There are many types of hedge agreement, but the most common version of this mechanism is a financial instrument that acts as an insurance policy. The project typically pays a fixed premium, and in return, the firm issuing the policy agrees to provide a "floor" to the revenue that the project earns. In other words, the project is guaranteed a certain level of revenue, regardless of whether the prices have fallen. Using this type of agreement the project is still subject to variability in the amount of revenue it may earn, but it has protection to avoid very low amounts of revenue that would endanger the financial viability of the project.

As of this writing hedge agreements are still relatively novel in the finance of utility-scale BESS projects, but they are continually become more accepted by major lenders for projects that are highly dependent on merchant revenues.

6.3 Site Selection

Selecting an appropriate location may be the most critical decision for BESS project success. There are a variety of considerations that impact the decision of where to site the project, called siting. Many developers have a certain niche they occupy, whether siting is determined geographically or based on a certain use case, while others cast a broader net to identify potential sites. The developer may perform the siting in-house or may contractor a third-party firm that specializes in GIS analysis.

The first criteria for siting is the project size; this determination depends on available CapEx funds, development capabilities, expected market growth, and risk appetite. Size is often a result of the project's financial modeling, which is covered in greater depth in Section 7.3. This section assumes that a size has been selected for the project in question, and reviews some of the most common criteria used to identify potential sites for battery projects.

Market Identification

Since BESS revenues vary widely depending on the policy, the target market is often one of the driving factors that determines where to site a project. For utilities that issue RFPs or tenders for storage projects, there will typically be a specific geography that is eligible for the project.

For projects that intend participate in merchant markets, the project can typically be located anywhere in market – in the US this is usually within the territory of an ISO or RTO.

For contract revenues, projects may need to be physically close to the offtaker, although in some cases the battery can sell its services at a distance by trading on the transmission grid. Although there are always losses associated with transmission, they are low and may provide a valuable avenue for a project to monetize its returns.

Whatever the geographical bounds of the market at hand, this is often the starting point in narrowing the search for a project site. For example, if a project is bidding into the ERCOT market, it will be limited to the boundaries of that ISO's service territory. For projects seeking revenue from sources such as energy arbitrage, it may make an important difference where the project is located, since the markets may fluctuate widely at difference nodes of the system.

Anticipating lucrative merchant markets may be a complex exercise that involves studying the historical data to find when prices have been favorable. Many consulting firms conduct studies to predict how markets will behave in the long run, which is an exercise that considers a variety of inputs, such as policy, macroeconomic considerations, and load growth on the grid.

Interconnection

After deciding which market to operate in, a developer hunts for a suitable place to interconnect the project. As discussed in Section utility-scale BESS projects must be connected at high voltage (usually 69 kV or higher). This may be at existing utility substations, or, less commonly, a place to connect the project can be created on a transmission line – known as a 'line-break' or 'in-and-out' substation. And the interconnection requirements are determined by the project size.

The interconnection point must have the technical capacity, or **thermal capacity**, to both charge and discharge the project's power capacity. Power flow models are used to determine if the point of interconnection (POI) is capable or if network upgrades are needed. Although it is common for developers to make minor upgrades, this work is very expensive, often costing millions in construction costs. Having a feasible interconnection is a key criterion of a BESS project. The interconnection often requires a conceptual, or preliminary design, which gives the utility or grid operator a general idea of the project to be built. These designs are frequently updated with later iterations of the application and may change at further stages of the development cycle.

The developer works to identify suitable substations within the chosen market: substations voltage levels and thermal capacity must be sufficient for the size of the project. It is preferable to build projects adjacent to substations, though this is becoming more difficult as competition buys available land and builds projects. And this is just the start of the interconnection process: the interconnection agreement is discussed in Section o and the engineering of the interconnection is pursued in Section 8.1

Land Acquisition

Acquiring the land on which the project will be built is one of the most important aspects of any project. Although other aspects of the project may be changed, the site is one that defines the project and locks it in on a given location. Most developers lease the land for the projects, negotiating a per-hectare (or per-acre) price while some prefer to buy the land. In either case, the landowner must agree with the effects of the construction of the project, the long-term use of their land, and any operational effects such as truck traffic or noise from the project. Developers are in fierce competition over sites with preferable qualities, which may result in landowners entertaining multiple competing offers, increasing the value of the land.

Land lease agreements are usually executed as options, in which a payment is made for the developer to have the right to lease the land in the future. This allows the developer to secure the other aspects of the project, such as interconnection, supply agreements, and offtake agreements, prior to committing to the long-term lease of the project.

Neighbors of the land used for the project, while not directly involved in the transaction, are also important stakeholders to consider. Most projects hold comment sessions and meetings for the local community to learn about what the project is and its impact on their land. While batteries occupy a smaller footprint than most wind or solar plants, there are many places where the communities may be concerned about safety risks, displacement of farmland, or impacts of the construction of a project.

Every community and land project is different, but it is always wise to maintain a clear and positive relationship with the landowners on a project, as they can cause large delays or even cancellations of projects.

Many jurisdictions require public comment meetings, which are held to describe the project to the local population and allow them to voice concerns. Although many jurisdictions have approved battery storage projects without incident, others have had meetings where the local population is hostile to battery projects. These concerns may be based around safety or publicity of battery fires or may be simply due to dislike of a novel technology that is perceived as hazardous. Every community is different, and different developers have different strategies, but wherever possible it is best to hold these comment sessions early in the development process to identify and address concerns before they can threaten the project.

6.4 Permitting

Numerous permits are needed to build a BESS project, ranging from environmental to safety, local to federal, and interconnection to zoning. Given the nuances across a range of permits, it's wise to enlist permitting experts. Let's explore the common permitting requirements across the US regions.

Given the enormous size and impact of a BESS plant, there is a large amount of permitting and compliance work that must be performed to ensure that the project meets all municipal, state/provincial, and federal requirements. Many projects have been abandoned at a late stage due to unexpected permitting delays, or insurmountable roadblocks. Given the high stakes of permitting, it is one of the avenues that should be pursued early and often.

Federal Environmental Restrictions

Most countries have rules at a national level intended to protect the environment. In the US this legislation is the 1970 National Environmental Policy Act (NEPA), which requires federal agencies to consider environmental impacts of projects the government conducts. In addition, the rules require that private infrastructure investments undergo study to understand their effects on the environment. As infrastructure projects, BESS plants are typically required to study these effects.

NEPA typically requires any project to complete an **Environmental Impact Assessment, or EIA**, which evaluates the potential environmental consequences of the project. The EIA covers all possible effects of both building and operating a battery plant. Its key components are:

- **A baseline study**, which establishes the current environmental conditions at the project site
- **An Impact assessment**, which assess how the project will impact the air, water, flora and fauna on the site, as well as impacts to the local population such as socio-economic effects
- **Mitigation measures**, evaluating how the impacts can be minimized or abated
- **Public consultation**, which provides a forum for the community, interest groups, and governmental agencies to express concerns and promote local knowledge of the project
- **An environmental management plan (EMP)**, which outlines how the project will comply with all required regulations

Compared to fossil fuel generators or even wind turbines, batteries are relatively innocuous. However, like any construction project, there are effects associated with the construction of the plant and from the presence of people, vehicles, and equipment that remains once it has gone into operation. The goal of the EIA is to study all these effects.

A more comprehensive evaluation of environmental effects under US law is the **Environmental Impact Statement (EIS)**. The results of the EIA determine whether or not an EIS will be required. The EIS includes a study of potential alternatives to the proposed project, a description of the affected environment, and an analysis of the environmental impacts of each of the possible alternatives.

Notably, most environmental studies do not describe specific actions which are required or not – instead they are designed to provide more objective information that allows parties involved to make decisions about the effects of the project, and its proposed alternatives.

Most BESS projects are found to have limited environmental effects, but they still must undergo the rigorous permitting process required in each jurisdiction.

In addition to the NEPA requirements, there are several other federal agencies which may have requirements for each project. These include the following:

- Department of Transportation (DOT)
- US Fish and Wildlife Service, part of the Department of the Interior
- Environmental Protection Agency (EPA)
- Bureau of Land Management (BLM)
- Department of Energy
- Department of Agriculture
- National Oceanic and Atmospheric Administration (NOAA)

State/Province Compliance

In addition to the federal regulations, each individual state or province in which a project is constructed may have their own set of environmental regulations. Most states typically have regulations that are at least as stringent as federal environmental regulations, and they may include special provisions related to the conditions or priorities of each state. For example, the state of California has its own environmental protection agency, CalEPA, which must be complied with in addition to federal requirements.

Where possible, state, and federal agencies communicate to try to avoid duplicating inspections for the same facility, but the federal environmental agencies will typically defer to the state as to how their rules are enforced. However, in certain scenarios, such as emergencies, significant violations, or projects with interstate impacts, the EPA or other relevant agency may choose to act directly to enforce rules on the project.

Local Authorities Having Jurisdiction

Regardless of where the system is built, and what codes apply, the approval process is typically performed at the local level. In the US, this is done at the county level. Local permits may include the electrical inspection, fire departments, environmental regulators, and other municipal, county, state, or federal agencies which may require approval. The local approving agency that must provide final sign-off on the project is known as the **Authority Having Jurisdiction, or AHJ**. In the US this is typically the local electrical inspector, but this may vary depending on the project location.

In most jurisdictions, plans must be submitted to the AHJ to obtain a building permit, and their approval may include several rounds of comments and questions from local authorities, including site visits and/or public comment sessions. Following these approvals, the project may proceed, and normally requires a final inspection when construction is complete, a step referred to in many areas as "signing a job card." This closeout of permits gives the AHJ a final look at the facility as built, and is often a requirement for the facility to reach substantial completion (SC).

While the AHJ is bound by their local codes, there is wide latitude among AHJs about what is required for the project to be completed successfully. For example, some AHJs may push to have a water tank installed on site for fire protection, while others may be less stringent on this provision. Others may require that all projects receive UL 9540 field certification, while some are content for the BESS to meet UL 1973 and UL 1741 certification.

The best way to ensure a positive relationship with an AHJ is to meet early and often with them to understand the needs and requirements for each product. Many projects have been delayed in a late stage due to late stage concerns raised by an AHJ. Nearly every project will have some issues that require resolution, many of which will require buy-in from the AHJ. Wherever possible, projects should strive to maintain a positive relationship with open lines of communication among all of the individuals and organizations required to permit the project.

The AHJ may commonly involve the local fire department, who may have to provide a separate approval of the system. In addition to approval, fire departments are involved in many other aspects of the project such as training, interconnection of fire-alarm systems, and approval of fire hydrants or on-site tanks. Given their importance in the safety of a battery project, it is always best to proactively involve the AHJ and fire department early and often in the development and construction process. Even after commissioning, many projects continue to check-in with fire departments to conduct drills, update them on any notable events on site, or provide continuing education.

ISO/RTOs

Most large grids consist of several different utilities responsible for generating, transmitting, and distributing power. As grids grew larger, the need arose to have an independent organization to coordinate among the hundreds or thousands of entities that make up the power grid.

Independent system operators (ISO) and Regional Transmission Operators (RTO) were the entities formed in the US to reduce government oversight, foster competition, and to make grids run efficiently and reliably.

Specifically, ISOs and RTOs are responsible for the following:

- Dispatching power from generators planning for grid expansion and introduction of new generators
- Helping the grid avoid congestion, and manage it when it occurs
- Making sure that transmission lines operate safely and efficiently
- Continuously balancing the load and generation on the grid

- Forecasting and scheduling power generation to limit the risk of grid blackouts or failures
- Designing and operating a market for suppliers of energy, power, and ancillary services

Most ISOs/RTOs have extensive protocols, forms, and documents that must be followed for new generators to apply to and eventually connect to the grid. This includes performing studies, anticipating new equipment, allocating costs, and eventually performing the testing that will allow power plants to connect to the grid.

Given the high demand for new projects connecting to the grid, this process typically takes years to complete in most ISO regions and may involve outside consultants or contractors. Therefore, it is one of the critical processes to start early, and to maintain open lines of communication with the ISO/RTO of the project. The Federal Electric Reliability Council (FERC) issued a landmark set of reforms in their Order 2023 aimed to make interconnection processes more efficient and swifter across the US Some of these include grouping of projects into clusters to enable concurrent studies, more firm requirements for projects to enter queues, and higher deposit commitments from developers. It is expected that these reforms will significantly reduce processing times and speed up interconnection milestones.

IFC / Equator Principles

For international projects, and especially those that may receive concessional finance from the World Bank, the International Finance Corporation (IFC), or aid agencies, it is common for the projects to require compliance with international norms. The most common of these are the Performance Standards on Environmental and Social Sustainability, commonly referred to as the **IFC Performance Standards**. These are a broad set of requirements governing the environmental, social, and labor conditions on a job site. The standards are quite broad and typically require an internationally recognized third-party inspector to prepare a report assessing the project's compliance with these norms.

Closely related are the **Equator Principles**, a framework for risk management that is adopted by funding agencies, and based on the IFC Performance Standards. These standards are listed as criteria by over 100 financial institutions and 37 countries.

As of this writing the current version of the IFC standards is 2012, with guidance notes issued in 2021. The Equator Principles were updated in October 2020.

6.5 Key Equipment Selection

After selecting a site and moving permitting along, another key activity is selecting the equipment that will be used by the project. As described in the following section, there are a variety of contractual arrangements for battery construction, but the most common is for the owner to purchase the key equipment directly from suppliers, while contracting a separate firm to design and procure all the remaining equipment, and to eventually build the plant.

The key equipment on a project are the battery, the PCS, and the transformers. Together these pieces of equipment comprise the majority of the CapEx on any given BESS project. Successful selection of these components is critical to the project's technical success. The firms that produce these components, often referred to as the **OEMs (Original Equipment Manufacturers)**, are key stakeholders in the project. The battery OEM often also supplies a long-term service agreement (or LTSA, covered in depth in Section 9.1) – making them a partner who will remain with the project through its entire life cycle, which may be 20 years or more. This section reviews some of the considerations for key components.

Other equipment decisions, such as the controller and transformers, come later in the development timeline and are secondary to the battery and PCS selection.

BESS Selection

A variety of considerations are used when selecting the BESS hardware used for a project. Leading criteria are:

- Price
- Reputation and commercial backing of the manufacturer
- Technical capabilities of the product
- Product delivery timeline
- Country of origin
- Experience in a certain market or type of project

As discussed in Section 4.2, there is an option to use an AC-block or a DC-block. This will be driven by product tier, C-rate, and design flexibility versus simplification. AC-blocks are much simpler: the batteries and PCS are one unit, which simplifies engineering, procurement, integration, and construction. However, DC-blocks are more versatile as they can pair with different PCSs (options for contracting, pricing, and integration design), allowing for C-rate flexibility, and all for DC augmentation (see Section 8.1).

Price is, for obvious reasons, one of the key drivers in selection of a product. The marketplace currently includes a wide variety of price-points, from publicly-traded firms with hundreds of years of experience to new firms who have arisen in the past few years with recently deployed products. Projects with high revenues and margins tend to use more expensive products backed by larger corporations. Given the long commercial relationship that a BESS project entails, it is common for large consideration to be given to the commercial organization and reputation, in addition to the technology.

Some media firms, such as Bloomberg New Energy Finance (BNEF) have defined specific tiers of battery supply, although there are no industry-wide objective classifications of these tiers.

From a technical point of view, there are several key considerations. To summarize them here:

- **C-rate** is a key factor, with typical values for utility-scale projects ranging from 0.25 C to 1.0 C. Most LFP suppliers offer products that meet both 0.25 and 0.5 C-rate requirements, so this decision does not reduce the selection pool much but is critical for selecting a product.
- **Chemistry** is a consideration – although the marketplace is currently dominated by LFP, some rare projects may be a better fit for an NMC or LTO chemistry. Some more innovative projects may employ newer chemistries such as sodium-ion or iron-air.
- **Cycling** defines how much usage a battery is designed to use, and over what lifetime.
- **Useful Life** is a definition of the expected lifetime of the battery in years, which effectively determines the length of the project. Typical values range from 10 to 25 years, with 20 years currently the standard for utility-scale projects.
- **Degradation**, or the expectation for the project to lose energy capacity over its useful life. As described in depth in Section 3.9, there are wide variations in the degradation of each battery project, which are typically backed up by a performance guarantee for the life of the project.

Beyond these technical criteria, another key consideration is the BESS integrator's ability to meet the project schedule. Most supply agreements will contain liquidated damages if batteries cannot be received or be brought into service at the milestones defined in the contract. Although typically the developer is pushing the battery supplier to stay on track for delivery, the supplier may also push the project to ensure that they are ready to accept the batteries.

PCS Selection

For AC-block BESS products, there is no need to select a PCS, since it is already integral to the battery enclosure. However, for the more common DC-block products, a PCS must be selected. Most battery manufacturers will provide a list of PCS suppliers they have successfully integrated on past projects. If possible a project should avoid using a combination of battery and PCS that do not have extensive experience working together on successful projects.

As with batteries, PCS suppliers come in many nationalities, price-points, and technical capabilities. Some of the key considerations in selection of a PCS are:

- **Grid-forming** vs. **grid-following**, which defines the PCSs ability to form its own grid (forming), or to synchronize with an existing grid (following). Grid-following PCSs are the norm for utility-scale projects; however, there are some cases where grid-forming capabilities may be required, such as microgrid applications and systems that are required to black-start a grid.

- **Reactive power capabilities**, which define the product's ability to supply power at various power factors. Some utilities or grid operators may have special requirements for reactive power that cannot be met by all PCS manufacturers. The most common is a four-quadrant PCS, which means it can operate at any power factor.

- **Power rating** for utility-scale systems typically ranges from 1.0 to 4.5 MVA. This value can have a large effect on the layout of the project, since it is common to have a single PCS connected to multiple battery enclosures. Although larger power ratings exist, they are rare on utility-scale projects.

- Ability to integrate with the BESS or controls system. Since the PCS gates the charging or discharging of the BESS, it is critical for its communications to be robust. The easiest way to guarantee this is to use a PCS that has a long track record of being used on successful projects with the same battery product and controller.

As with batteries, there are many non-technical considerations, such as the PCS supplier's corporate backing, project experience, or experience with a specific market or type of project. The selection of PCS is typically an iterative process that involves the battery supplier, designer, and technical teams on both sides. Given its large impact on the project's success, it is not a decision to be taken lightly.

Main Power Transformer

Also known as the Generator Step-up Transformer, or GSU, the MPT is one of the most expensive single pieces of equipment used on a utility-scale project. For large projects this may be a massive piece of equipment that weighs hundreds of tons and requires specialized transport and unloaded equipment. Many large projects have a single MPT, although some may have two or three to limit the risk if this piece of equipment were to ever be lost.

These pieces of equipment are custom-built for each project, and often are designed to be oil-cooled or air-cooled. The frequency, high-side and low-side voltages, power rating, impedance, type of cooling, and dimensions are all aspects that must be engineered for the specific application.

Given the outsize cost and importance of the MPT, it is frequently procured by the project developer through a separate agreement as owner-supplied equipment. Figure 6-10 shows a 160 MVA transformer that steps voltage up from 34.5 KV on the low side to 238.5 kV on the high side.

Figure 6-10
Main power transformer – 160 MVA at 238.5 kV

6.6 Contracting

Every major purchase or sale of equipment, goods, or services on a large BESS project is governed by a contract. On the purchasing side, the project must acquire land, hire firms to build the BESS, and have intricate contracts around the terms of the financing of the product. On the sell side, the project must negotiate an offtake agreement with whomever is purchasing the services provided by the BESS.

Minor procurements may be governed by simple agreements of only a few pages, while the major purchases are governed by contracts that have appendices running over 300 pages. While this may seem like dry subject matter, having appropriately structured contracts for all aspects of a project helps make it **bankable**, meaning that it meets the risk requirements of commercial lenders. In practice, this means that contracts are designed in ways to limit the risk to the project as much as possible. All major risks on the project are described, and wherever possible they are put off onto an insurer, separate contracted party, or contractor. While this may sound overly legalistic, it is a key aspect that allows projects to be financed by large corporate lenders and to receive tax equity investments, as will be covered in Chapter 6.

Given the critical importance of negotiating the interconnected agreements which govern a renewable energy project, the counsel representing the developer is a key partner of the development team. Some firms use in-house counsel to negotiate these agreements, and often the lead counsel may be a shareholder in the development firm.

Contracts governing renewable energy projects are complex and are the subject of several books of their own. This chapter summarizes some of the key agreements which govern most BESS projects, their role in the development process, and elements which should be considered in a successful project.

Key Agreements

The following is a list and description of the key agreements which are used on nearly every BESS project.

Table 6-1
BESS project key agreements

Agreement	Acronym	Contracting Parties	Description
Financing Agreement	FA	Project, lenders, equity investors	Overarching agreement that governs the capital contributions, disbursements, dividends, and commercial relationships between the Project, its lenders, equity investors, and independent engineer (IE). Includes provisions for what will occur in the case of default and the hierarchy of parties' abilities to be reimbursed.

Equity Capital Contribution Agreement	ECCA	Project, Tax equity investor	Agreement under which a tax equity investor agrees to provide funding to the project once it has been built and met certain conditions.
Land Lease or Purchase Option Agreement	n/a	Project, Landowner	Agreement under which the Project commits to lease or purchase a piece of land from the landowner. When the project moves toward financial close, this option will be executed and a new purchase or lease agreement will be entered into by the same parties.
Battery Energy Supply Agreement	BESA	Project, Battery integrator	Contract which covers the technical requirements, supply, delivery, testing, and commissioning of the battery system. May also cover the PCS, and may include a workmanship warranty or performance guarantee.
Long Term Service Agreement	LTSA	Project, Battery integrator	Contract which binds the battery integrator to provide preventative and corrective maintenance services throughout the useful life of the product or a fraction thereof. Typically includes performance guarantee for degradation, round-trip efficiency (RTE) and availability

Operations and Maintenance Agreement	OMA	Project, O&M Provider	Agreement between a third-party firm responsible for operations and maintenance of the BESS plant, typically covering all materials outside of the battery itself.
Engineering, Procurement, and Construction Agreement	EPC Contract, BoP contract	Project, EPC / BoP Contractor	Agreement which governs the commercial, technical and legal terms for the design, purchase, and construction of the BESS plant. Covers from limited notice to proceed through to construction itself, substantial completion (SC), and commercial operations date (COD). May exclude certain owner-supplied equipment. Often has provisions for bonding, retention, and termination.
Interconnection Agreement	IA	Project, Utility, ISO/ RTO	An agreement with the utility and/ or ISO/RTO to allow the battery to be interconnected with the power grid. Includes both technical and commercial requirements for interconnection. May include several levels of study to demonstrate the technical repercussions of incorporating the plant on the grid. May describe required upgrades to the grid to allow the project to come online, and who is financially responsible for this work and equipment. Often includes schedule with mandatory milestones for both parties.

Offtake Agreement	n/a	Offtaker, Project	The agreement which covers the purchase of services provided by the battery over the course of its useful life. Critical agreement that is the cornerstone of the financing of the project, includes provisions for performance requirements, compensation, delivery schedule. Typically has a set term, may have provisions allowing for extension.
Hedge Agreement	n/a	Project, financial services firm	Financial arrangement which allows the project to limit commercial risk through a contractual mechanism, similar to an insurance policy against the offtake agreement or merchant market performance.

6.7 Project Execution

The final step of the project development process begins when the project reaches FC. This is when all contracts are in place, all risks have been evaluated and mitigated, and the banks are ready to commit their capital to the project. When all parties have signed on the dotted lines, the funds begin to flow from the investors and lenders, the EPC contract is executed, and the contractor can be granted **notice to proceed (NTP)**. This allows both the contractor and owner to place purchase orders (PO) for key equipment if they have not done so already.

In this stage, the developer is playing an oversight role to keep the project on track. They typically must keep the lenders and investors apprised of the progress of the construction, often with independent verification from the IE. As the work proceeds through the various milestones, the contract requirements must be met for the project to draw its funds and for the EPC contractor to be paid. The flow of these funds is governed by the terms of the financing and EPC agreements.

The EPC of the project proceeds through key milestones such as:

- **Mechanical completion (MC)**, when all physical equipment has been installed on site (some projects have this in the requirements, while others do not).
- **Commercial Operations Date (COD)**, when the project has met all utility and grid operator requirements for being energized and synchronized with the grid.
- **Substantial completion (SC)**, when the EPC contractor has completed all technical requirements of its contract, can demobilize, and must only complete the remaining punch list items.

At this point, the project enters full operation. Typically, this means that the project passes from the portion of the development team responsible for its construction to a separate team responsible for its **Operation and Maintenance (O&M).** As discussed in Chapter 9, this may be handled internally by the developer, or by a third party.

Once the project has come into operation, the contractual relationship between battery supplier and the project shifts from BESA to the LTSA. Although the BESA may have provisions for workmanship warranties, the LTSA is the primary agreement outlining the performance requirements of the battery.

From a financial perspective, once the plant reaches COD it has finally begun to generate revenue after many long years of development and capital expenditures. At this point it passes to the developer's **Asset Management** team, or to a third party if this is not performed in-house. It is often combined with other assets to form an operational portfolio. As detailed in Section 7.5, the financial model continues to govern the system, with the asset manager tracking whether it has his its revenue targets and kept expenses below the anticipated costs in the model.

As apparent, the development process of a battery plant is a long road, but a required journey for making battery projects a reality.

6.8 Usual Suspects

This section includes a brief overview of some of the leading BESS project developers around the world. Since this book is focused primarily on North America, this section is the most extensive. Although not definitive, the lists attempt to cover some of the larger players in each of these regions.

North America

AES: Applied Energy Services was founded in 1981, and today they are a publicly traded utility and power generation firm with over 30 GW of installed capacity, including coal, gas, and renewables. Within AES, AES Renewables has been a market leader in deployment of renewable energy projects, inaugurating 3.5 GW of renewable projects in 2023. Including 600 MW of BESS capacity. Together with Siemens, AES was one of the founders of Fluence, a leading BESS integrator, although AES builds projects with a variety of integrators.

Akaysha: Australian storage developer Akaysha Energy was founded in 2021 and was acquired by Blackrock in 2022. As of mid-2024, Akaysha has over 2 GW of energy storage projects in their Australian pipeline, including the world's most powerful battery at 850 MW (see Section 13.1). Beyond Australia, Akaysha is actively developing BESS projects in Japan, the US, and is seeking to expand into additional global markets. The case study in section 13.1 describes Akaysha's Waratah battery, a massive 850 MW / 1680 MWh project.

Amp: Toronto-based multinational renewable developer with 7 GW pipeline. Amp is primarily focused on solar PV projects, but more recently has been involved in energy storage projects.

Avangrid Renewables: Large PV developer that is part of the publicly traded Spanish energy generation and distribution company Iberdrola. Headquartered in Portland, Oregon, Avangrid has built more than 7 GW of renewable energy projects including wind, solar and batteries.

Aypa: Backed by Blackstone Private Equity Group, Aypa has 22 GW of utility-scale energy storage and hybrid renewable energy projects in development, and 33 projects in operation or construction across North America. Formerly an arm of Canadian developer NRStor, Aypa is currently headquartered in Austin, TX.

Brookfield Renewables US: Based in New York City, Brookfield is a leading owner, operator and developer of renewable power. They have a diverse portfolio of renewables on five continents worth about 33 GW of operating assets making them one of the largest renewable energy companies

Canadian Solar: A vertically integrated PV module manufacturer that has arms that develop, design, build, and maintain projects. Originally in PV but have branched into PV+Storage and Standalone storage. They have their own brand of BESS product sold under the name SolBank.

GridStor: Goldman Sachs-backed energy storage developer based in Portland, OR, founded in 2022. They have a pipeline of 3.2 GW of energy storage projects across the US.

Hecate: Founded in 2012, Hecate is a top-ten developer of solar PV and energy storage plants in the US. They have 3.6 GW of contracted projects and a development pipeline of over 40 GW, including over 10 GW of storage projects, according to their website.

Invenergy: Based in Chicago, Invenergy is a multinational developer, primarily in wind energy, some solar PV, and more recently storage. Invenergy has developed over 30 GW of total projects, including over 19 GW of wind projects.

KeyCapture: A New York-based energy storage developer with 600 MW of storage in operations or under construction, and a development portfolio of over 9 GW, according to their website. In 2021 they sold a controlling stake in their company to SK E&S, part of Korean multi-national holding company SK Inc.

NextEra: One of the largest renewables developers in the world, and the leading owner of energy storage assets in the US with over 2.8 GW of operational storage assets. NextEra is a publicly traded firm that owns Florida Power & Light (FPL), a leading utility in the US South.

NRStor: A leading Canadian developer of energy projects, NRStor is based in Toronto. They are in the process of building Canada's largest battery, the 1 GWh Oneida project in Ontario.

Plus Power: Developer founded in 2018 focused exclusively on energy storage. According to their website they state having a portfolio of 1.7 GW and 4.2 GWh of BESS in construction or operation, with over 10 GW in interconnection queues.

Primergy: Small-scale and utility solar PV, and more recently energy storage developer, 19.5 GW, including 8.6 GW of BESS projects under development.

RWE Clean Energy: The US arm of RWE AG of Germany, RWECE is one of the leading renewables developers in the US with over 24 GW of onshore wind, solar and BESS projects. RWECE has an operating portfolio of 8 GW of BESS in the US, and as of 2024 recently completed three new BESS projects in Texas and Arizona, totaling 190 MW with an additional 770 MW (2,280 MWh) under construction. The Case Study in Section 13.2 describes RWE's Fifth Standard project as an example of an AC-coupled BESS as part of a PV+Storage plant.

SBEnergy: A subsidiary of the Japanese conglomerate SoftBank. In 2021 SBEnergy signed a supply agreement for 2 GWh of energy storage projects with iron-flow battery supplier ESS.

Strata Clean Energy: California-based PV and BESS developer with 4GW Developed, and a 14GW pipeline of projects. Primarily a solar developer, Strata has pivoted into storage and has many standalone and PV+Storage projects in development and construction.

Vistra: Has over 7.8 GW of renewable assets operational. One of their notable ESS projects is Moss Landing in California which was the largest project at the time of COD.

Europe / UK

Acciona: Spanish energy storage developer, focused on PV projects, more recently energy storage. According to their website, they have 190 MW of operational projects, and a pipeline of over 1 GW.

EDF Renewables: An early entrant into the PJM market, EDF Renewables, a subsidiary of the French utility EDF, has over 12 GW of projects between wind, solar, and energy storage. They are also operational in the US market.

Engie: In 2023 French energy giant Engie acquired Broad Reach Power, with approximately 1 GW of operating and pipeline BESS assets, and they have continued to develop projects with the Broad Reach infrastructure. They are also active in the US market.

Lightsource BP: is the largest solar developer in Europe with 9.5 GW of solar projects developed worldwide.

New Energy Partnership: NEP is a leading BESS developer in the UK, who has stated they seek to deploy over 2GW of BESS projects by 2025.

Penso Power: Focused on large-scale BESS projects, Penso is developing the Hams Hall BESS with a 350 MW / 1.75 GWh capacity.

Zenobe: Zenobe Energy is the largest independent owner and operator of battery storage in the UK.

Latin America

Enel Green Power: EGP is a subsidiary of the Italian energy company Enel. EGP specializes in generation from wind, solar, geothermal, and hydroelectric power. The firm manages over 1,300 power plants globally with more than 63 GW of installed renewable capacity, including in Latin America and especially in Chile.

Solarpack: A Spanish company which operates in 14 countries, Solarpack has over 2.5 GW of projects constructed and 1 GW currently in operation or construction. They have been actively growing their BESS practice in Latin America and globally.

Africa:

ACWA Power: ACWA was founded in 2004 and is headquartered in Riyadh, Saudi Arabia. Their portfolio includes renewable energy projects, as well as fossil generation and desalination plants, with a total capacity of over 53 GW.

BioTherm Energy: BTE has developed renewables both wind and solar photovoltaic (PV) projects. They are active in Kenya and several other African markets. BTE's biggest project in Africa is the

Golden Valley Wind Energy Facility, which is a 123 MW wind energy project. It reached COD in May of 2021.

Enel Green Power: See Latin America section.

Engie: See Europe Section

Lekela Power is a UK-based developer of renewable energy projects focused on renewable energy development in Africa. They have over 1.3 GW of installed wind capacity and a pipeline of 2.8 GW. They signed an offtake agreement in 2023 for one of the largest standalone BESS projects in sub-Saharan Africa, the 40 MW / 160 MWh Taiba N'diaye project in Senegal. In 2023 Lekela was acquired by a JV of Egyptian firm Infinity and UAE's Masdar.

Scatec Solar: Headquartered in Oslo, Scatec operates in numerous emerging markets, such as South Africa, Egypt, Brazil, Pakistan, and the Philippines. One of their largest construction programs include Kenhardt has a 225 MW, 4-hour battery.

Middle East:

ACWA Power: See Africa Section.

Emirates Water & Electricity Company (EWEC): The purported largest BESS developer in the Middle East, EWEC is a state-owned company in UAE. They have built two 150 MW BESS projects, some of the largest in the region. Beyond UAE they have been active in Saudi Arabia and Egypt.

Enerwhere Sustainable Energy (ESCE): ESE is headquartered in Dubai, UAE, and founded in 2012. EnerWhere is one of the leading operators of solar PV plants in both Africa and the Middle East.

Masdar (Abu Dhabi Future Energy Company): Masdar is headquartered in Abu Dhabi, UAE and is focused on the development of renewable energy projects across Africa.

Asia

Adani Green Energy (India): As of the end of 2023, Adani's operational renewable energy capacity exceeds 8 GW. This includes nearly 5 GW of solar PV projects and nearly 1 GW of wind power.

China Energy Engineering Group: A state-run conglomerate in China, they are involved in developing large-scale energy storage projects across China, including BESS projects and several other technologies.

Invenergy (Japan): Invenergy has a robust pipeline of renewable energy projects in Japan, including 220 MW of late-stage development onshore wind projects. Their ongoing projects also include expansions in offshore wind and advanced energy storage solutions (Invenergy).

Jinko Solar (China): Together with Sunrev, Jinko Solar a leading developer of BESS projects in China. They have over 1 GWh of BESS projects under construction in the Jiangsu province of China as of 2024.

Ming Yang Smart Energy Group (China): MYSE is a leading Li-ion battery project developer in China. They have a 320 MW project in Inner Mongolia region of China which is anticipated to reach COD in 2024.

ReNew Power Ventures (India): ReNew Power has commissioned over 7.7 GW of energy projects in India. The company aims to double its capacity to 10 GW by 2025.

Sindicatum Renewable Energy (Singapore): **Sindicatum's** is a developer operational in key markets across Asia. They are active in regions with high demand and focused on renewables. They have developed and operate projects with a total capacity of 250 MW.

State Grid Corporation of China: The SGCC is China's state-owned grid company, the largest utility in the world serving over 1 billion customers. SGCC has made many strategic investments in energy storage, including 3 GW as of 2022 with a target of 100 GW by 2030.

7. PROJECT FINANCE

Battery projects are expensive. Even a modestly sized utility-scale system, such as 5 MW / 20 MWh requires a capital cost of approximately $5.5 million at the current price of approximately $275 per kWh for an installed system. For the largest systems, investments may even exceed $500 million. Given the very capital-intensive nature of this technology, the ability to successfully finance the project is one of the key drivers of whether it will be able to be built.

BESS projects are seen by many as a burgeoning technology that is enabling decarbonization of our power grid, one of the key steps necessary to address climate change. Others see battery projects as a technological marvel which allows our grids a flexibility that was never possible before the arrival of Li-ion cells. Both these things are true – yet batteries also play an important financial role: they are an attractive infrastructure investment for equity / tax equity investors and lenders. In many ways, batteries make ideal infrastructure assets: they require large capital costs to deploy, they have relatively long project lifetimes, and their operation is critical to economic growth of a society – making them likely to be supported by governments. A recent example of this is the inclusion of **investment tax credits (ITC)** under the IRA for standalone BESS projects. When properly financed and executed, battery projects can become attractive equity investments as well as sources of large loans to be made by banks. This chapter is focused on the art of procuring and deploying the necessary capital to build a BESS project, which is known as **project finance**.

When viewed as an infrastructure asset, an energy storage project has one primary goal – to make returns for its owners and investors over its lifetime. While there are numerous financial incentives available to BESS, if there are not competitive returns to be had, new projects will not be built. Given that most energy projects take years and millions of dollars in development work before earning a single dollar in revenue, it is critical that the project's economics be studied thoroughly as part of the planning process.

The chapter first reviews the basics of project finance as it applies to utility-scale battery projects. While project finance is an in-depth topic that could fill volumes, the goal of this section is to familiarize the reader with some of the key finance concepts that support BESS projects as they are built today.

The first part of the chapter reviews the basic principles of project finance of renewable energy projects, with an emphasis on utility-scale BESS projects. After a brief history of this type of finance, we turn to the key elements of finance for project development and construction: the financing agreement and financial model. Next, we review how BESS projects perform as financial assets, and how they are managed. Lastly, the Usual Suspects section reviews some of key players in BESS project finance for the North American market.

7.1 Typical Financing Structures

As of the writing of this book, the largest wind plants in the US exceed 1 GW, while solar PV plants of 500 MW and larger have been announced. The largest BESS projects are now exceeding 500 MW, such as the 850 MW Waratah Super Battery (featured in a Case Study in Chapter 13). The capital required for these projects has risen to match some of the world's large infrastructure projects such as highways, bridges, and ports.

The financing of an infrastructure project of this size is a significant undertaking requiring many months, or even years, of planning, and dozens of stakeholders with various concerns. The goal of a successful infrastructure investment is to have all the stakeholders come to a consensus around the project's key elements, execute the project, and convert that capital into an asset producing consistent cashflows. The work of gaining this consensus and converting capital into cashflows is the primary goal of a successful developer.

Developers range in size from small outfits with only a handful of individuals working on the project to large publicly traded firms with thousands of employees. Regardless of size, every developer has access to a limited amount of equity to invest in their projects. Smaller developers may use their own capital for a transaction, but most larger projects involve the developer soliciting an **equity investor**. Typical equity investors in renewable projects include infrastructure funds, IPPs, corporations, or even pension plans.

From the developer's point of view, equity is the most expensive capital they have access to, since an equity investor captures a portion of the profits of the operational plant. Therefore, developers seek to leverage this equity with debt, allowing them to build as large a portfolio as possible with the minimal investment. In practice, this leads developers to seek lenders who are willing to contribute the largest possible amount of debt. Most projects receive from 70-80% of their capital costs in the form of debt, typically provided by large corporate banks, or syndicates of several banks lending together. This is similar to any real estate investment, such as a family purchasing a home in a fast-growing real estate market – they often want to minimize their down payment, borrowing as much as they can through a mortgage. This allows the family to capture more of the home's rising value, while paying a steady interest rate to the bank that granted the mortgage.

As with any investment, the lenders are keen to deploy their capital into sound projects with the minimal level of risk. However, even the best of projects carries some potential hazards. For utility-scale battery projects, the risks include events such as:

- Delays and/or cost overruns related to the construction of the project
- Technology malfunction, such as PCS failures or accelerated degradation
- Commercial risks such as lower-than-expected market for the services of the BESS
- LDs imposed by the project offtaker
- Safety risks such as batteries overheating or entering thermal runaway

Before contributing capital, investors and lenders do everything they can do to understand, quantify, and mitigate all risks associated with a battery project. There are many large banks vying for the same investments, so there is a balance between a lender understanding and mitigating risk, and not presenting an undue burden that will delay and complicate a project.

The balance between investors, lenders, and their priorities is at the heart of project finance for renewable energy projects.

7.2 The Mechanics of Project Finance

In any project finance transaction, there are three main key stakeholders: the project sponsor, the equity investors, and the lenders. In markets with investment tax credits, there may also be tax equity investors. Although many subcategories of these exist, these are the main drivers of any project.

While all four typically contribute capital to the Project, each has different motivations, expectations, and requirements for that capital in the transaction.

Sponsors

The **sponsor** is the firm that is spearheading the construction of the Project. This is often a developer of energy storage projects, as discussed in Chapter 6, although it may be an IPP, a bank, or even an individual. The sponsor is the main representative of the project that collaborates with the other stakeholders. The sponsor is ultimately responsible to organize the finance of the project, coordinate the execution of the works, and take responsibility if there are problems. For smaller projects, the equity for the project may be entirely committed by the sponsor. However, for larger projects, it is common for the sponsor to contribute a smaller amount of capital, while partnering with a larger equity investor to build a larger project.

Equity Investors

The **equity investors** of a project own a portion of the project, typically in the form of shares in the **Special Purpose Vehicle (SPV).** Like any investor, they are seeking a return on their capital. Equity ownership of large projects is commonly split between several investors, since any one of them may not be able to finance the entire project. It is common for many projects to change hands throughout their life cycles – many times the original greenfield developer may sell the project to another set of owners who may see it through to construction. Those owners may in turn sell the project to another set of investors who may operate the project in the long term. Equity investment typically has a target return that it is trying to achieve over its life, whether that may be in the early stages of development, the first few years of the project, or over the project's full operating life. Given the relatively large size of the investment required to finance a BESS project, the equity investors are typically asset management firms, infrastructure funds, investment banks, or private equity firms.

As the project moves forward, the risk gradually decreases – projects that are in their early stages have less value because there are many possible events that could derail their completion, such as local permitting, interconnection delays, or supply chain issues. As a project nears commercial operations, its risk decreases, increasing the value to the owner. At an early stage, most projects are typically owned by the sponsor of the project itself. As the project matures, gaining in value and declining in risk, a portion of the project is often sold to larger equity investors. These firms often own portfolios with many assets, targeting returns based on the performance of the asset in the long term.

Some owners may be hands-on with their projects, while others may be strictly contributors of capital, content to allow the sponsor to guide the development of the plant. At any given stage of development/construction, the project's sponsor is the key decision-maker responsible for moving matters forward, but its equity investors often have a large say in the direction and pace of the project.

In the US, a special type of equity investor exists who is focused on tax incentives available to the project – this is known as a **tax-equity investor**. The role of these investors is covered in depth in Section 7.4.

Lenders

In contrast, **lenders** are banks who lend capital at a certain interest rate, seeking a steady return on their investment as the debt is repaid. As described above, since these loans are not backstopped by the equity investors, the loans they make are referred to as **non-recourse debt**. The lenders may, however, have a lien on the project itself.

Debt provides most of the capital to build projects under a project finance framework. Given that these loans may be hundreds of millions of dollars on large projects, they are usually provided by large commercial banks, some of which have special divisions specifically devoted to finance of energy projects.

Banks are risk-averse, and since they contribute a large portion of the funding, they have a large say in how the project is executed, how the capital is disbursed, and what happens in the case of a default on the project. Insurance requirements, escrow transactions, and even the structure of supply agreements may all hinge on the bank's risk tolerance.

The terms of the involvement of lenders, as well as that of equity investors, is outlined in the **financing agreement**, which is discussed in depth in Section 7.4.

Project Risks

To guarantee that investors receive their expected return and lenders are paid the service on their debts, each party is keenly focused on ensuring that the project is a success. To this end, both lenders and investors are constantly trying to evaluate any potential pitfalls faced by a project, and how the associated risks can be mitigated. Despite all precautions, as with any investment, there is always a risk that a project may underperform, or in some cases to fail entirely. This could be caused by a variety of sources: natural disasters, unexpected market performance, technical issues with the battery product, or the bankruptcy of one of the parties involved.

The first course of action when a project is in trouble is to attempt to right the ship before the issue becomes serious: this could involve repairing or replacing portions of the physical infrastructure of the plant, hiring a different asset manager, or other adjustments in the operation. If the issue cannot be resolved amicably, the project may result in legal action to attempt to recoup the investment. If there is an insurance policy to cover the issue, a claim may be made to compensate for the loss. In the most serious of cases, and after exhausting all other legal avenues, a project may default on its debt. In this case, the banks who have lent the money to the project will pursue every legal avenue to recoup their money.

If the project encounters serious problems and is forced to default on its debt, the lenders have two options: either repossess the project to attempt to operate it or sell the project to recoup its investment.

If the project is owned by the sponsors directly, the lenders would be able to pursue civil litigation to collect on their loan. This could include going after the other assets owned by those companies, their equipment, or possibly even the shareholders of the company. This would mean that equity investors would be much more reluctant to lend to the project, due to the high risks they would incur.

In the early 2000s, when renewable energy projects began to become more profitable and less risky, developers began to structure their investments in a separate entity to hold all assets of the project, rather than holding them on their balance sheet. These SPVs hold all assets and conduct all business on behalf of the project.

The Special Purpose Vehicle

An SPV has several other names – it may be referred to as "the project," or occasionally "the Project" with a capital P because it is a formally defined term in the project's financing agreement. Many countries allow for entities that play this role, known as limited liability companies (LLCs). Throughout the book these terms are used interchangeably to refer to the entity which holds the project, operating as an SPV. Other common names used are the "project co" or "Separate Purpose Entity".

The SPV purchases the equipment and hires the contractors to build the plant, collects revenues on behalf of the plant, and eventually repays the loans and distributes dividends to the project's investors. When the project takes on debt, the lenders are making a loan to the SPV; the offtake agreements are typically signed with the SPV as are EPC contracts for the facility. The SPV may consist of a single large corporate owner, or it may have many members with differing levels of ownership. Most SPVs have one owner who controls the majority stake in the project, serving as the lead sponsor of the project.

The only collateral provided by the SPV to backstop the loan are the assets of the project itself. In the case that the project defaults on its debt obligations, the lenders can pursue the project as collateral, but are not able to sue the underlying equity investors in the project. Because of this limited recourse for the lenders, project finance is often referred to as **non-recourse finance**.

The arrangement is a balancing of risk, where **lenders** typically:

- Contribute the highest amount of capital to the project
- Bear the lowest risk if the project defaults
- Receive no large upside if the project is more profitable than anticipated
- Are not able to pursue the equity investors in the case of default

While **equity investors**:

- Contribute a smaller amount of capital to the project
- Bear the highest risk if the project defaults
- Stand to earn a significant upside on their investment should the project meet or exceed its expected profitability
- Have their balance sheets shielded from lenders, should the project default

The relationship between the various parties in a project finance transaction is represented graphically in Figure 7-1. In it, the SPV can be seen as the main channel through which all investments are made, revenues are earned, and dividends are paid.

This structure has allowed for massive growth in the renewable energy sector. The main drivers of this growth have been sponsors' gaining experience building bigger and bigger projects, improvements in technology, and the evolution of contracts to protect lenders from project risks. All these elements contributed to a declining default rate for these types of projects, making them attractive to lenders. At the same time, the limited risk has attracted a large flow of capital from

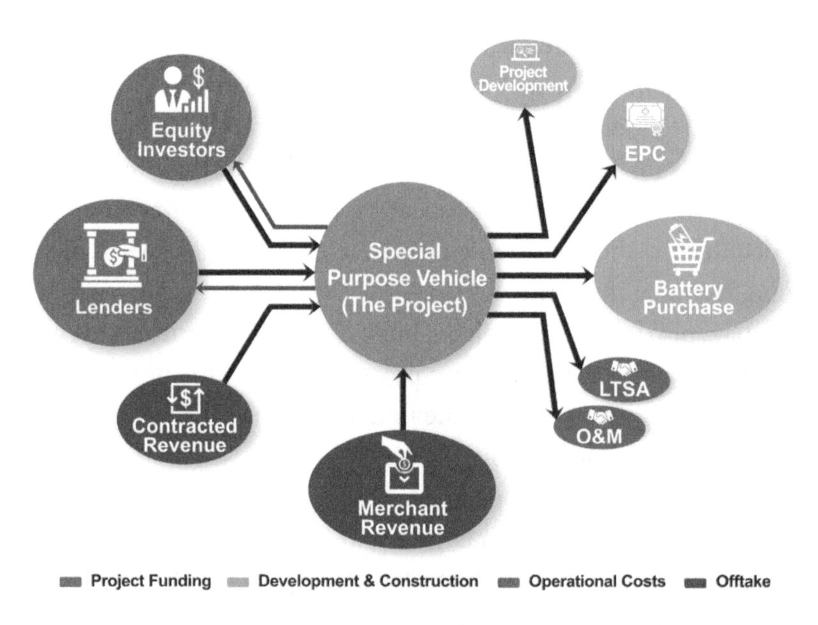

Figure 7-1
Asset cashflow map

equity investors. This growth coincided with the expansion of tax policies in the US that offered large incentives to projects, which is covered in Section 7.4.

Evolution of Project Finance

The first examples of project finance style loans date back to the maritime ventures in Roman times [46]. Throughout the centuries that followed, the model has been refined and modified to suit various industries, such as shipping, port-building, and development of oil and gas extraction. The energy sector began to finance projects using this model in the late 1970s and early 1980s, when it was primarily used to finance natural gas infrastructure in Europe [47]. In the 2000s, the model came into use in the US renewable energy industry, led by developers such as Jigar Shah at his firm SunEdison, and later his fund Generate Capital. While wind farm and solar PV plant investments were originally seen as risky investments on novel technology, lenders gradually became comfortable with the risks and began to make non-recourse debt available to the projects. This structure was possible only because the relative risks of the projects had declined to the point where banks felt comfortable with making an non-recourse investment.

Today, nearly all the projects built in the US, as in many other parts of the world, are built using project finance. For the foreseeable future, this will continue to be the underlying model, as projects continually set new records for size and value.

Lenders provide a large portion of the funding, and they are generally risk-averse, giving them a large say in how any project is executed. This is primarily determined through the **financing agreement,** which is the document that outlines how the money will be contributed and distributed, as well as what happens if the project is delayed, cancelled, or is performing below its revenue targets.

7.3 Financial Modeling

Investors commonly refer to a project's ability to "pencil out" – projects that have excessive costs and make insufficient returns "don't pencil," while those that can access markets that are lucrative enough to offset the costs of developing, building, and operating the BESS do. The mechanism used to evaluate a project's returns is the financial model, a crucial component of any successful project.

Building a Model

The financial model is spreadsheet that takes a full accounting of all expected cash flows on a project from its inception to its final decommissioning. Typically built in Microsoft Excel, the model takes into account all the possible expenses and liabilities of the project, and ultimately produces an internal rate-of-return, or IRR, which indicates if the project is profitable. Most developers have entire teams devoted to building their financial models, while some use outside consultants. Models may range from simple calculations on a year-by-year basis to massive spreadsheets with macros to evaluate differing cases. This section describes the basics of this modeling process to allow the reader to understand how the process is used to evaluate energy storage projects as a financial resource.

Revenues

As discussed in the prior subsection, revenues are the key drivers of value streams which will earn the project money. Once the value streams have been selected, the model must then be used to estimate the amount of earnings that the project will make over its lifetime.

For contracted value streams, such as fixed payments from tolling agreements, this can be a simple prospect. For example, if a project is being paid $10 per kW per month for operating a 1 MW / 2 MWh battery, the monthly revenue can easily be stated as $10,000 / month for the life of the project, assuming there are no reductions based on LDs or other possible discounts to revenue. If there is an escalator, a common feature in offtake agreements, then the revenue may increase by a fixed percentage per year.

For more complex revenue streams, such as those from bidding into merchant markets, estimating revenues can be more complicated. Predictions for merchant markets often require purchase of data sets which estimate the project revenues over its lifetime. Such data sets are available from advisory firms, such as Power Advisory or Ascend Analytics in North America. These data sets provide forecasts of the market value of services that a battery may be participating in, such as capacity, frequency regulation, or day-ahead energy markets. For projects that are targeting energy arbitrage, the spread between the difference in prices must be estimated to estimate the revenue.

Once the value of individual value streams has been established, the project must determine the most cost-effective streams to pursue. Since a BESS can participate in multiple markets simultaneously, there are choices that are required to optimize a project's revenue. For example, a PV+storage project may allow for a project to earn revenue from storing the PV energy for release in the evenings, as well as participating in frequency regulation markets and being paid to participate in a resource adequacy. At various times of the day, one of these three value streams may be more lucrative than others, meaning that the project may be pursuing differing strategies across different times of the day.

Collectively, these decisions are known as the battery's **dispatch strategy**. Determining the optimal dispatch strategy may have a large effect on the project's revenues, and therefore is an area that has been studied deeply. Most developers have algorithms that can predict the optimal strategy. These often employ some sort of artificial intelligence (AI) to analyze past trends and predict future market performance. Some developers perform this optimization in-house, while others use tools that are available to evaluate revenue streams, such as DNV's Hybrid Energy Resource Optimizer (HERO). Many software platforms that come with BESS products have their own market dispatch tools – these are summarized in Section 5.2 on Battery Controls.

The final revenue model for a given project feeds into the financial model, providing an estimate for the overall earnings that a project will generate over its lifetime. While every model is based on some significant assumptions, it is critical to have a baseline forecast of what the project earnings will be. Traditionally, developers tend to take a more optimistic view on what the project's earnings will be, while lenders and investors take a more skeptical view of the expected earnings. Often these assumptions may be the subject of debate, including the IE on the project. For BESS, the most uncertainty with respect to merchant revenue estimation is on the future market conditions, especially volatility in energy prices and real-time energy arbitrage potential, which is governed by more renewables coming on to the grid in most markets. Since renewable energy production from solar and wind are unpredictable, they can cause price swings, which batteries can capitalize on, but it is very hard to forecast these swings out in the future.

Battery finance experts commonly talk about the "upside" and "downside" cases. The upside case is a condition that is favorable to the project – high revenues, for example, or few incidents of corrective maintenance. Conversely, a downside case represents where adverse conditions affect a project – accelerated degradation of the battery, for example, or perhaps a natural disaster which might require the project to pursue an insurance claim and may lead to extended downtime. Savvy developers and financiers consider both the upside and downside cases, and the statistical probabilities of each, in order to assess the risk to their investment as accurately as possible.

Costs

Besides revenues, the expected costs of a battery project are of course one of the most important inputs into the financial model. Battery costs typically include the initial capital expenses of building the BESS, but also must include development costs, operational costs over the life of the system, preventative and corrective maintenance, augmentation (if applicable), and decommissioning costs. When the financial model is developed, many of these numbers may be estimates, particularly those occurring far in the future, but it is important that they be approximated for the purposes of forecasting the expected financial performance of the project.

The cost of a BESS is typically expressed in cost per kWh of storage, since costs generally scale with energy capacity, but for some purposes it may be expressed in cost per unit power (kW). Figure 7-2 shows the 2022 cost estimate for construction of a utility-scale (60 MW) BESS, for various durations (2-hour to 10-hour).

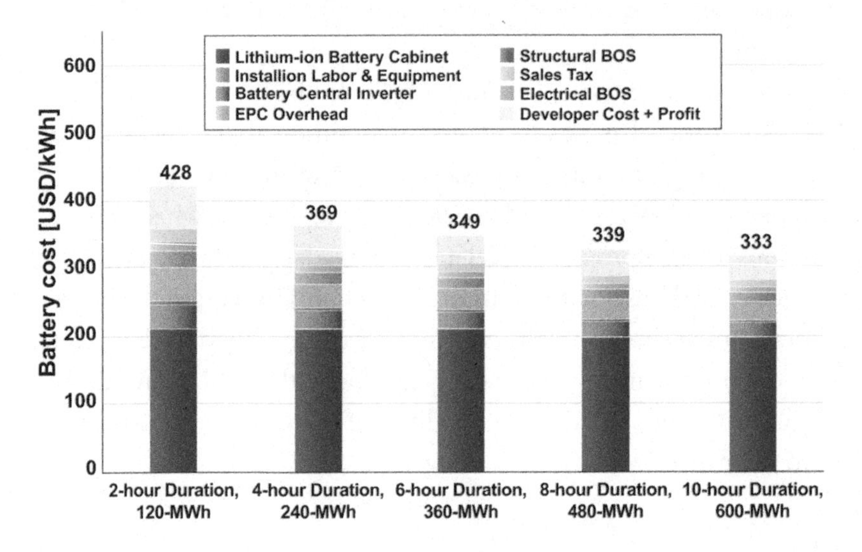

Figure 7-2
Utility-scale battery cost per kWh [48]

Developing a battery project requires a high amount of spending long before the system can collect any revenue. These expenses include, but are not limited to, salaries for the development team, land lease or purchase cost, and preliminary design development. The project also often pays fees or deposits that are required for the interconnection of the project, legal fees required for the various contracts involved in the project, and costs for preliminary permitting of the project. Collectively the expenses prior to the financial close (FC) of a project are referred to as development costs. These costs vary depending on the type of project, location, and how it was acquired, but typically range from 5-10% of overall project CapEx.

The EPC costs typically account for 80% or more of the CapEx. As expected, the major equipment (battery systems, PCS, and transformers) tend to make up the lion's share of the CapEx. The battery and PCS are priced as $/kWh and $/kW respectively – these are converted to total CapEx by multiplying with the number of units or volume (MWh and MW) required. The volume requirement or sizing of the project needs to account for the expected contractual obligations and use-case of the BESS, lifetime, degradation, and compatibility of batteries and PCS.

The financial model also accounts for the incentives that a project earns, such as the federal ITC available as a provision of the 2022 IRA, or other incentives such as bonus depreciation. These incentives often form a crucial aspect of making project economics pencil out, as it may provide a tax credit equivalent to 30% or more of the CapEx. The economics of these incentives are further explained in Section 11.4.

After the project is operational, the financial model must account for all the costs incurred as the project enters operations. This includes all costs to service the debt on the project, disbursements of profits to the investors, and O&M of the BESS plant. The model must also account for the final decommissioning of the project at the end of its operating life.

Quantifying Risk

The financial model developed by the project sponsor often has more aggressive assumptions than lenders may like. For example, they may assume that the battery is 99% available, even though the availability warranty provides for only 98% availability. Alternatively, they may assume that the marketplace in which the battery is operating will provide strong revenues, or that the battery will degrade more slowly than the warranty.

Conversely, the lenders or investors often argue for a more conservative approach – they often prefer that the model's base case assume that the battery will degrade exactly as described in the warranty, rather than some level above that, which may be reasonable. Alternatively, they may base their assumptions on a market performance that is lower than the historical average. After all, the investors and lenders want to know that even if the project does not perform up to the optimistic projections of the sponsor, they still will be able to receive strong dividends (in the case of investors) and to cover their debt service (in the case of lenders).

One common way to quantify the risks of a project is to measure them in a probability value. A P50 value refers to the performance which is expected to be the most likely performance of the project – in other words, if 100 of these facilities were built, 50 of them would perform better and 50 of them would perform worse. P90 refers to a conservative estimate for the facility's performance, indicating that 90 of the hypothetical 100 would perform at this level or better. A P99 case refers to a level of performance that will be achieved by the asset in all but the most extreme cases.

These metrics come from the solar PV and wind finance world, where they are more centered around meteorology and data measurement. For solar plants, the question is around production, which can be estimated with accurate energy assessment methodologies and solar insolation predictions. For wind it is based on measurement data from towers erected to gather data on each individual site, coupled with long-term weather prediction.

Since batteries do not directly generate energy, the P50, P90, and P99 cases involve some more nuance – typically they involve various levels of market performance, battery degradation, or expectations around battery replacement costs. As of this writing, there is no standard financing methodology used to evaluate these risk levels – each lender or investor has its own methodology to make these determinations.

The debt service that the project must pay to lenders is typically one of the largest expenses after the project has become operational, particularly as of 2024 when interest rates are at the highest they have been in several decades. Lenders' primary concern is making sure that a project can pay its debts, even if it performs somewhat worse than anticipated. A common metric used to analyze the debt cost, as well as to quantify risk, is to compare the ratio of the cashflow of the project to the cost of servicing the debt. This value is commonly known as the **debt service coverage ratio,**, or DSCR. This metric changes over the life of the project and is a general indicator of how financially viable the project is throughout its life. If the DSCR ever drops below 1.0, this indicates that the project would be unable to cover its debt obligations, which would mean that the lenders to the project would be unable to be paid and make their anticipated return. From the lender's perspective, this would be a risk of the project's ability to repay the debt. DSCRs of 1.3 or more typically indicate a healthy cashflow relative to the amount of debt.

A common P99 DSCR for a BESS project may be 1.0, indicating that in all but the most extreme cases, the project will still be able to cover its debt obligations – but without any additional revenue to support other project expenses.

To quantify the risk associated with the project, it is common for many financial models to run a series of **stress tests**, which are scenarios of low probability, high-impact events that might be encountered by a BESS Project. For example, a drop in market prices for a certain merchant revenue stream, a higher-than-expected rate of inverter failures, or the failure of a battery warranty. Although each lender or investor has different criteria for these cases, most transactions involve running some form of differing scenarios or tests to understand the financial implications of a given risk playing out in the field.

Operational Cost Modeling

Although the financial model typically places a lot of weight onto the CapEx and expected revenues of a project, the O&M expenses are also critical inputs that will determine the financial health of the asset. The O&M costs of a project are typically divided into the preventative and corrective maintenance of the BESS, which is typically paid to the battery manufacturer via an LTSA, and all other costs required to maintain the BoP and for the system to participate in the markets it operates in. The technical aspects of O&M agreement are covered in depth in Chapter 6. The costs for these services that are reflected in a financial model normally include:

- **Preventative and Corrective Maintenance**: A long-term service agreement (LTSA), as further described in Section 9.1, is an agreement between the battery integrator and the project to ensure that the battery operates to its described specifications throughout its useful life. The LTSA covers the cost of the planned work to be done for maintenance throughout the battery's life, which is typically 15-20 years. This fee varies based on the contract and ranges from $10-20 / kW / year.
- **Insurance**: Like any other asset, batteries can purchase insurance to protect against any unexpected failures. It is particularly important to cover acts that might be excluded from the standard BESA or LTSA, which are often referred to as **force majeure** conditions. These include but are not limited to weather events, acts of god, acts of war, or other things that are outside of the control of the supplier or service provider. Often battery integrators may have some insurance policies which come as part of their products or components, such as backstops to cell degradation. Other insurance policies may cover more conventional risks such as flooding, hurricanes, or earthquakes, depending on the area in which the project is built. With the increase in extreme weather events caused by climate change, many of these policies are costing renewable energy projects a greater share of their revenues than they have in the past.

- **Capacity maintenance agreement (CMA):** Some LTSAs include a provision for a capacity maintenance agreement, or CMA, in which a minimum energy capacity will be maintained throughout the life of the project. LTSAs that include CMA provisions have a significantly higher price than a traditional LTSA, since the cost must cover the necessary replacement of batteries. As of 2024, CMAs have become less common in utility-scale projects because developers have preferred to carry the augmentation / degradation risk themselves.
- **Augmentation or replacement of batteries:** Most utility-scale projects are required to augment their BESS or replacc the battery modules in order to maintain the required capacity of the project (which is often a requirement of the relevant offtake agreement). In the financial model this requires estimating both what augmentations will be required, and what the cost of batteries will be some years in the future. The most common method of generating these estimates is to assume that the batteries will degrade in line with their guaranteed capacity and assuming the cost of batteries will align with publicly available forecasts, such as those offered by Bloomberg New Energy Finance (BNEF) or NREL. Battery topology is also a key concern – particularly whether the system will be AC or DC augmented. Augmentation is covered in depth in Section 8.1.
- **Balance of plant O&M:** The remainder of the plant, including PCSs (if not contained within the BESS), transformers, fire protection, communications and all other systems outside the battery itself, may be maintained by the project directly, or through third-party contractors. This cost is typically in the range of $8-20 / kW / year [48]. This budget often includes a certain amount of money set aside to cover replacements of components in later years. The most common component replaced is PCSs, which normally occurs between years 10 and 15.

- **Plant operational staff and contractors**: A battery project often has some staffing and contracting expenses required to operate whatever market they are in. This may include paying a scheduling coordinator to dispatch the battery, security on the battery site, or fees to an operational center that monitors the operation of a BESS plant.
- **Decommissioning**: Following the end of a battery's useful life, most projects plan on the project being decommissioned. The technical aspects of decommissioning are covered in Section 9.7. The fees for decommissioning may include recycling of a battery, clearing a site, and performing environmental studies to ensure that no contamination has occurred. Decommissioning costs are speculative since the process for recycling batteries is likely to change significantly in 20 years.

7.4 The Financing Agreement

When project finance is used to finance a battery plant, all parties must agree on how the project's cashflows will work. The **Financing Agreement (FA)** is an overarching contract which contains all these provisions. This document may be several hundred pages long with dozens of attachments. It may also be known as the Credit Agreement or have other names. The FA outlines all of the key legal, commercial, and technical terms of the project: how funds will be contributed, what will be held in escrow and where, the structure of sign-offs and certificates needed for the flow of funds, provisions for dispute resolution, requirements of how the project will come online and start generating income. The agreement usually goes into detail on what will happen if there is an incident on the project, or ultimately a default, including the pecking order of who will be repaid first and in what order, should the investment fail.

In addition to the legal and commercial terms, the FA describes the role of the IE, who is an advisor working on behalf of the lenders to assess the technical risks and how they can be mitigated. The role of the IE is discussed in greater detail in Section 7.6.

The entire financing process usually takes several months to complete, from identifying a lender, trading drafts of all the applicable contracts, and negotiating to sign the eventual FA. When all parties have been satisfied as to how the project will operate, typically trading many drafts between lawyers, the project is ready to achieve **Financial Close (FC)**. This is when the project "closes," meaning that all parties (sponsors, lenders, investors, the IE, or their representatives) have signed off on all documents, and the deal moves ahead. Up until this point, any party may back out of the deal, but once FC is reached, the lenders and investors place their funds in escrow, and the project receives money to begin working on building the facility. For most utility-scale project finance transactions, all major supply and EPC agreements are signed concurrently execution of the FA. This is the critical step allowing the project to move forward from development into construction, which is commonly referred to as **Notice to Proceed (NTP)**. The full EPC process is laid out in Chapter 8.

Typical Structures

Although project finance is the most common method of building BESS projects in the US, in other locations there are different models that are used. This section discusses two of those models – pure equity investment, and corporate financing.

Equity investment

This is the simplest of finance structures. In this method the sponsor has a sufficiently large amount of capital that they can invest in a project without having to borrow money. There are few firms that have a sufficiently large balance sheet to make this type of investment, but for those that are able, this structure has a simpler path to financing since there are no lenders involved. Large equity investors still are expected to perform due diligence on the investment, but the burden is typically less than that imposed by lenders. Equity investments may come in the form of a single large investor, or many smaller investors working together.

An equity-financed transaction may be closed more quickly than other projects, since they have fewer or no lenders to deal with; however, they are relatively rare, since the equity investor bears a large amount of the risk if there is no debt involved. Some of these projects involve some form of term debt which is issued after the project has become operational, when there are no longer any construction risks associated with the project.

Corporate Finance

If the developer uses debt to leverage their investment, one method to do that is to seek debt which is backstopped by the balance sheet of the sponsor themselves. In this structure, a lender provides financing, often up to 70% of the value of the project, allowing the developer to leverage their capital and build more projects. However, the financing agreement includes terms that allow the lenders to seek recourse with the corporate assets. If projects succeed, this is not an issue, as the corporation will repay the lenders and reap the dividends produced by the project. However, if the project fails, the corporation's assets may be at risk, imperiling other parts of the business based on the performance of the project.

Due Diligence

Investors and lenders are in the business of understanding and mitigating risks around a project. While they are typically aware of the basic principles involved in technically complex projects, they often prefer to have a third party perform the diligence rather than having that expertise in-house. Most lenders and investors hire (or require the developer to hire) an IE to perform this work. The IE's role is to assess the technical risks of a project, and where possible, advise on ways to mitigate those risks. This process is normally done through an independent engineer's report (IE Report), which may also be referred to as a Technical Due Diligence (TDD) report.

The IE report usually takes several months to prepare and is an iterative process in which details are shared with the IE as they become available. The contents and depth of an IE report vary from project to project, and between lenders. The larger the investment, typically the greater depth of diligence is performed. Normal sections of an IE report are:

- Project general details and overall description
- Technology review
- Key agreements review:
- Battery Energy Supply Agreement (BESA)
- Long-term service agreement (LTSA)
- O&M Agreement
- EPC Agreement
- Offtake agreement
- Technical review of interconnection studies
- Design review
- Construction schedule review
- Financial model review, looking at all key inputs used in the model and providing stress test cases
- Safety review
- Environmental compliance review

The major components of a battery project are some of those that are most scrutinized. To ease the process of financing a project using their technology, many battery integrators will hire an IE to conduct a technology review (also known as a bankability report), in which the IE summarizes all the key aspects of the technology. Typically, the IE has access to private information, and often conducts visits to manufacturing facilities, that allow them to evaluate the product relative to other standards in the industry. Since the OEM is funding these reports, they tend to include more positive than negative information about the product, but they give a more neutral view than the marketing materials that are provided by the OEM directly.

The IE's role is not to be a gatekeeper of the project, allowing some things and not allowing others. Instead, it serves to highlight the level of risk to lenders, so that they can make a collective decision about whether to live with the risk or require the project to mitigate it. For example, an IE report could identify that a transformer could have a 1year warranty but might note that a 3-year warranty is more typical of other large-scale projects.

Much of the IE's role is around making sure there is appropriate coordination between various agreements in terms of performance requirements/guarantees, schedules, and budgets on the project. The IE report also compares LD provisions. Although many battery projects have a guaranteed performance level, if they slip below that level, they may be allowed to pay LDs in lieu of repairing the system. Similarly, many offtake agreements have LD provisions around the key metrics of the project, such as energy capacity, RTE, or availability. If a project missed its capacity measure by 10 MWh and had a provision of $20 / kWh of LDs, it would owe $200,000 in penalty. If the offtaker's capacity was missed by 5 MWh, but had LDs of $30 / kWh, that would be acceptable, since the damages owed to the offtaker ($150,000) would be less than the damages owed by the battery integrator. This type of comparison is one of the key functions of the

IE report, helping the lenders evaluate various scenarios to see if the project is adequately protected.

Tax Equity

In the U.S. there have been extensive efforts from both federal and state governments to incentivize the deployment of renewable energy projects. Back in 1992, the U.S. congress enacted an incentive known as the Production Tax Credit, or PTC for wind power plants. This incentive provided a credit of 2.5 cents to be provided by the federal government for every kWh produced by a wind farm. This means that a 3 MW wind turbine, which might produce around 11.8M kWh (11,800 MWh, or 11.8 GWh) each year (depending on the location and type of turbine), could receive a tax credit of $295,000 each year. As a credit, this amount could be used to offset an existing tax liability but could not provide a direct payment to the project. The intention was to provide a financial boost to the economics of wind projects, with the intention of increasing investment, which was thought to reduce costs as the economy of wind power scales.

By most accounts, this approach was a success – the capital cost of wind turbines in the U.S. declined from approximately $2,750 per kW in 1990 to approximately $1,100 by 2005.

Total energy costs are often compared across one technology to another using the metric **Levelized Cost of Energy / Electricity (LCOE)**. This method considers all costs across a project's lifetime, including development, CapEx, OpEx, and eventually decommissioning. By dividing the net present value of each of these items by the total lifetime energy production, the resulting LCOE gives the all-in energy cost, typically expressed in US dollars per MWh. This allows for an even comparison of technologies with low CapEx but annual fuel costs (such as natural gas plants) to be compared alongside renewable projects with much higher initial CapEx, but low OpEx owing to not having to purchase fuel.

From 1990 to 2020, the LCOE of wind projects fell from approximately \$100/MWh (\$0.10/kWh) to below \$60/MWh (\$0.06/kWh), as shown in Figure 7-3 based on data from Berkeley Lab [49].

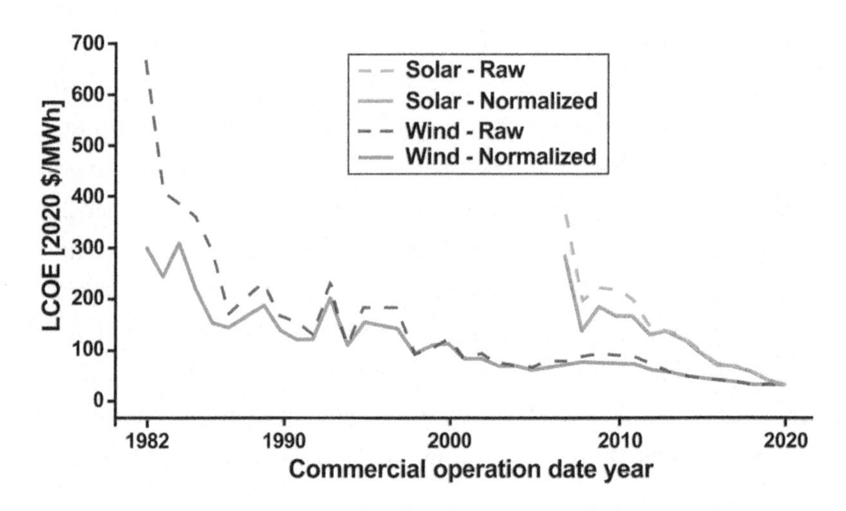

Figure 7-3
LCOE wind and solar values in the US [49]

While these cost declines were not exclusively driven by the incentives, they provided a strong boost to the wind industry in the US and accelerated the cost declines by improving processes and achieving economies of scale.

In 2011, the provisions of the PTC were modified to provide an alternative ITC, a credit based on the capital cost of the project at construction, rather than the energy production of the facility over its lifetime. The ITC was extended to solar PV plants, which similarly experienced a decline in the initial CapEx and the LCOE over their lifetime (also shown in Figure 7-3). The energy policy act of 2005 provided for an ITC of 30% by 2007, which provided a strong boost to the economics of PV projects.

When used to finance utility-scale PV projects, the ITC resulted in a large one-time tax credit, in some cases exceeding $100 million per project. The credit therefore created an incentive for the projects to involve investors who had a very large tax liability/appetite, as they are the only ones able to make use of the tax credits on large projects.

In general, the ITC tends to be more favorable for higher CapEx projects. This gave rise to the **tax equity (TE) investor**. Tax equity was needed to monetize the PTCs and was the common mechanism in the 2000s: a big tax liability was needed for ITC, but the PTC needed a very predictable, long-term liability (albeit lower for any given project in any given year).

Currently, TE investors are one of the main drivers of battery projects, contributing approximately 30-50% of the BESS CapEx. These investors fund a portion of the project's initial capital requirements but take advantage of all the Project's tax credits. There are several methods of doing this, with the most common being:

- **Sale-Leaseback**: In this structure, a developer sells the entire project to a TE investor. Then, the project is leased back to the developer over its lifetime. The Developer retains control of the project, but the tax equity investor receives the tax benefits as well as a share of the dividends produced by the project.
- **Partnership flip**: In this arrangement, the TE investor makes their investment up front, taking approximately 95% stake in the project. The contributed cash does not directly cover project expenses. Instead, the SPV allocates to the TE investor approximately 99% of the tax benefits (in the form of credits, losses, and depreciation), with the cash remaining with the SPV. Once the TE investor's IRR is met (typically 5 to 10 years), the

ownership stake "flips," at which point all of the tax benefits are fully monetized. If the project is successful, the sponsor will take most of the cash.

- **Inverted lease**: In this structure, the TE investor leases the project from the developer. The developer continues to own and operate the project. The TE investor receives operating cash flow from the project and the ITCs, but never owns the project assets.

Until 2022, the ITC was available exclusively to wind and solar projects. Even PV projects that had an energy storage component were only allowed to fund a portion of the energy storage investment via the ITC. However, with the passage of the 2022 IRA, the ITC was extended to be able to apply to energy storage projects [50].

For projects over 1 MWac, the tax credit is 6%, plus an additional 24% provided the project meets wage and apprenticeship requirements, under which a percentage of total labor hours are performed by qualified apprentices. Furthermore, there are bonus incentives available for projects located in specific areas:

- 10% for projects that meet domestic content minimums
- 10% for those in an "Energy Community," which are brownfield sites or those related to mining operations
- 10% for those in a Low-Income community or on indigenous land
- 20% for qualified low-income residential building project, or economic benefit project

While it is unlikely that any one project could qualify for all these incentives, many projects seek to have over 50% of their CapEx offset by the IRA incentives, providing a huge boon to investment in energy storage projects.

Another novel provision of the IRA is the allowance for transferability of the tax benefits from renewable energy projects. This allows for a secondary market for the tax credits, and as of late 2023 the first transfers of tax credits from BESS projects are starting to take place. Transferability is expected to drive even more investment in the sector. In 2022-23, the traditional TE market was approximately $20 billion / year in the US, whereas to realize the full potential of the IRA's provisions, this would have to grow to roughly $100 billion / year, representing a sizable portion of the US gross corporate tax receipts.

As of early 2024, there was concern about a global set of banking standards developed in response to the 2008 financial crisis. A provision of these standards, known as Basel III, would require TE investors to hold higher capital reserves against the cash invested in projects, which would significantly reduce their returns. Industry experts are optimistic that this will be addressed by US agencies, in order to avoid imposing this onerous requirement on investors.

In summary, tax equity has been one of the most significant drivers of investment in renewable energy projects. According to the American Council on Renewable Energy (ACORE) [51], over the past 20 years the availability of tax credits has facilitated over $485 billion of private investments into renewable energy projects. This investment is expected to exceed $50 billion annually by the mid-2020s.

7.5 BESS Asset Management

Once a BESS project has been constructed and started operations, it becomes a financial asset, just like an apartment building or a business. The asset produces revenue, in the form of payments from the various contracted and/or merchant value streams. The asset requires some ongoing operational costs such as O&M or service-agreement payments, insurance, debt service, and land lease payments. Occasionally there may be large one-time costs, such as battery augmentation, as elaborated in Section 3.9. The financial goal of the battery facility is to meet or exceed the returns expected in the financial model. This is done by keeping the operating costs to a minimum, while maximizing the returns expected of the project. The process of keeping any battery project operating profitably is known as asset management. This may be performed in-house by the owner of the project or may be through a third-party asset manager.

Revenue Optimization

Unlike other renewable technologies, BESS is completely controllable. Hence the revenue potential for BESS lies largely in the manner it is operated to tap on all possible revenue streams. If the BESS has a contract structure, like a tolling agreement, it will have a fixed monthly revenue stream that is pre-defined in the contract, so long as the availability of the system is kept at the guaranteed level (although there may be other requirements of performance).

However, the BESS owner can often reap wholesale market-based revenues by letting the BESS participate in market services such as frequency regulation, frequency response, and energy arbitrage (often in addition to the contracted revenues). These, as described in the Value Streams section of Chapter 7 in detail, are known as merchant revenue streams. The nature of pricing for these products on the wholesale market in most regions, especially on a real-time or short-term basis is often complex and highly volatile. For example, ERCOT, which monitors and operates the vast majority of the Texas grid, experiences massive unpredictable upward swings in real-time energy price on extremely hot days when exacerbated by units being offline unexpectedly – these can last for short time periods before being re-adjusted by the market. Figure 7-4 from Modo Energy shows an example of a real-time spike in ERCOT energy pricing, from June 20, 2023 [52].

Advanced software-based algorithmic capabilities, beyond the EMS and controls layers, can dynamically adapt the BESS's

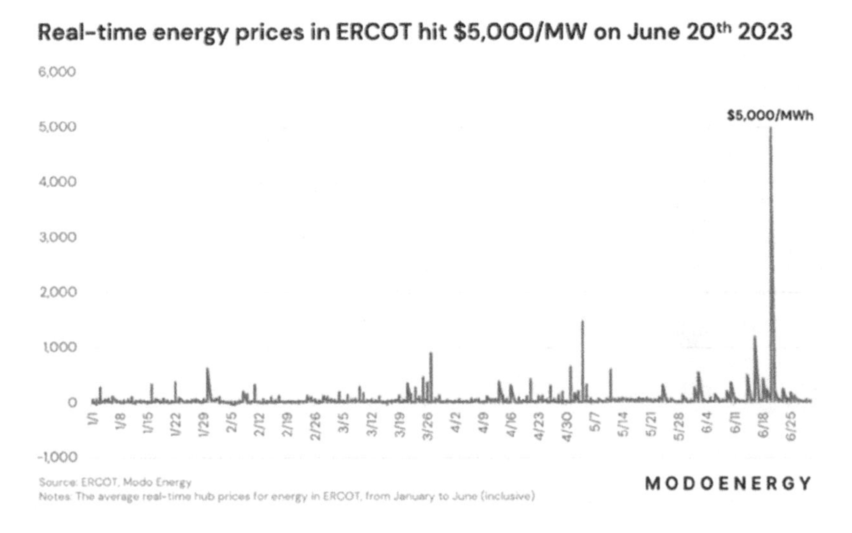

Figure 7-4
Real-time energy price spike in ERCOT in summer of 2023 [52]

response or dispatch to these market signals are critical to maximizing the BESS's revenues. These bidding engines are commonly built on data cloud systems and use sophisticated machine-learning and AI-based models. In turn, the models are built on large volumes of historical data and local real-time conditions like weather to forecast pricing behaviors that determine the BESS's dispatch. The models must also be fed data about the real-time battery conditions such as the current SoC, power limits and availability, as well as market participation rules to decide on the optimized outputs for the BESS. Without this capability, the BESS could operate on a pre-set pattern and make revenues from expected seasonal price behaviors but may lose out on dynamic pricing incidents – and the BESS is excellent at capitalizing on these events because of its fast-responding nature and potential for revenue-stacking. Many integrators wrap in this layer of intelligent bidding software as part of their offering. Some owners and operators opt to perform these types of simulations with in-house expertise.

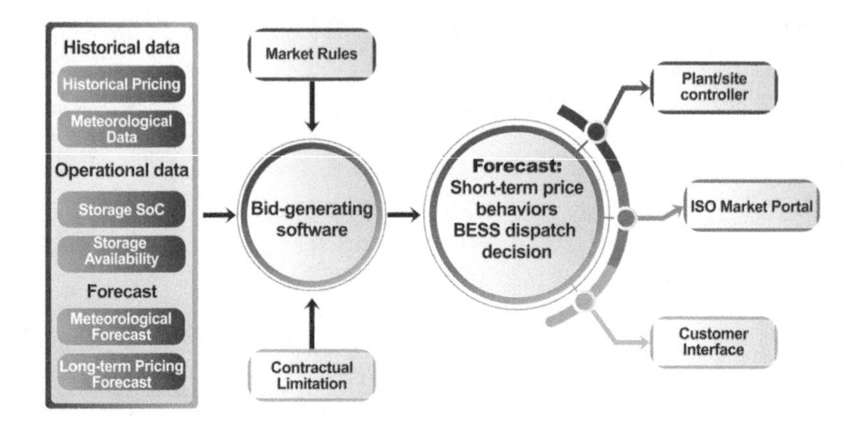

Figure 7-5
Overview of inputs and outputs of a BESS bidding software

The bid generated by this type of software needs to be communicated to the ISO market portal, if in an ISO-regime, according to the rules of the market service. For example, day-ahead energy participation bids, in the form of MWs of participation, are to be turned in or scheduled to the ISO typically by midday the day prior. Real-time energy participation bids, also in the form of MWs, may have to be scheduled at the top of the prior hour, or fifteen minutes prior depending on the ISO. This needs to be automated and can be accomplished via communication of the bid-generating software to other third-party software programs dedicated to ISO-compliant scheduling. Some large developers with extensive renewable fleets have in-house scheduling capabilities.

Cost Control

Battery systems' state of health (SoH) and useful life is heavily dependent on the nature of operation, as explained more fully in Chapter 9 on O&M. The SoH depends on the number of cycles per year, depth of discharge, average state of charge, as well as operating conditions at the site, such as cell temperature. BESS asset management is complex because the asset must be optimized to achieve the desired SoH over time as guaranteed by the supplier. This, in turn, involves carefully maintaining the operating conditions within constraints specified in the supply contract. Otherwise, the project could incur additional augmentation and/or replacement costs, or the useful life of the asset may be reduced. In some markets, it requires continuous evaluation of the current versus future value of the asset. For example, high frequency regulation prices may prompt aggressive participation and cycling of the BESS at the cost of reducing its overall lifetime, but the revenues may be realized earlier (sometimes even higher than estimated during planning).

In other cases, there may be higher energy-price volatility in later years for which the BESS state of health needs to be preserved. This is the challenge of operating and managing BESS in the most lucrative manner as market situations continually change. But high-level market trends can be created by monitoring the convergence of complex factors such as expected renewables deployment in the market and impact on afternoon and evening energy prices (duck curve), impact on regulation needs and electrical load growth (from EV charging, etc.).

7.6 Usual Suspects

This section lists some of the publicly available data of firms involved in utility-scale investments. Sponsors are not listed here, as the majority of sponsors of these projects are developers / IPPs and can be found in the Usual Suspects section of Chapter 6.

Most investors in this space are involved in the finance of many different renewable energy assets. Where information is available, their specific commitments to BESS-related finance have been highlighted.

Equity Investors

Blackstone Private Capital: Blackstone has made many renewable energy investments through a variety of funds. Recently, their $7 billion Green Private Credit Fund III (BGREEN III) set records as the largest private credit energy transition fund. They are focused on investments in renewable electricity generation and storage, as well as energy storage technologies.

Breakthrough Energy Ventures: Founded by Bill Gates in 2015, BEV has raised over $2 billion toward funds that are focused on addressing climate change. Although not typically an investor in storage projects directly, they have made storage-related investments, such as in iron-air storage firm Form Energy.

Energy Impact Partners: EIP manages over $2 billion in assets under management, including a diverse portfolio of investments across North America, Europe, and Asia. They have been leaders in deployment of capital toward renewable energy.

Macquarie Group: Macquarie is a global financial services and investment firm that has made significant renewable energy investments aimed at addressing climate change. In 2023, Macquarie launched Aula Energy, a renewable energy business aimed at developing utility-scale wind, solar, and integrated battery projects across Australia and New Zealand. Aula plans to develop over 4 GW of projects.

MN8 Energy: Formerly Goldman Sachs Renewable Power (GSRP), MN8 was spun off as an independent firm in 2022. MN8 now manages renewable power assets across 27 US states with a total capacity exceeding 2.3 GW.

Temasek: State-owned Singaporean conglomerate Temasek has a portfolio of over $280 billion March 2023. In April 2024 they announced a partnership with Blackrock to invest over $1.4 billion in clean energy investments.

Private Equity firms:

BlackRock: BlackRock, Inc. is a global investment management corporation based in New York City. The world's largest asset manager, Blackrock has made many climate and energy-focused investments across many of their funds. Blackrock is the owner of Australia-based global BESS developer Akaysha. Additionally, Blackrock recently committed $500 million to Recurrent Energy for development of projects including energy storage.

Foresight Group Holdings Limited: Foresight is a private equity investment manager based in London. They have invested heavily in renewable energy, including a stake in Foresight Energy Infrastructure Partners (FEIP) fund, which invests in European renewable energy projects. These projects will include wind, solar, storage, as well as grid infrastructure.

KKR & Co.: Founded in 1976, KKR has been a significant player in renewable energy investment. KKR's platform has seen investment of over $15 billion in wind, solar, and other renewable energy and climate-related projects.

Lenders

CoBank: CoBank is a key financier of renewable energy, having committed over $11 billion in loans to renewable projects. This includes both infrastructure project investments and support for corporate initiatives.

Deutsche Bank: Deutsche is a major investor in project finance of renewable energy projects in the US and other markets. Outside of this, they have committed to providing at least $3.2 billion in financing to developing economies and emerging markets by the end of 2025.

First Citizens Bank: First Citizens Bank, a US financial institution, has been a leading lender to renewable energy project finance.

Although the total investment is not publicly announced, they have provided senior debt financing for solar PV wind, and BESS projects.

Mizuho: Part of Mizuho Financial Group, Mizuho is a prominent Japanese bank that has made many investments in renewable energy projects, including energy storage.

Norddeustche Landesbank: NORD/LB is a prominent German bank that has taken a leading role in the finance of renewable energy projects. NORD/LB has been a lender to Plus Power's large-scale battery energy storage project portfolio, with over 2.76 GWh of BESS planned in Arizona and Texas.

Siemens Financial Services: SFS, a division of conglomerate Siemens AG, is a leading renewable energy equity investor. They have financed projects totaling over 27 GW of wind, 14 GW of solar PV, and 2.3 GW of other technologies including BESS projects.

Société Générale: SG is a major French multinational investment bank and financial services company. They have been involved in financing several large-scale BESS projects in Europe, the US and globally.

US Bancorp: US Bank has been a major player in renewable energy project finance, investing over $15 billion in renewable energy tax equity projects since 2008. Additionally they have done project finance transactions offering debt directly to projects. Through March 2024 US Bank has stated that they have facilitated the deployment of over 24 GW of renewable energy capacity.

Tax Equity Investors

Most lenders to energy storage projects tend to be large commercial banks, given the large ticket size of batteries. Here are some of the larger North American lenders that have been leading BESS financiers. Collectively Bank of America, JP Morgan, and Wells Fargo are often referred to as the "big three" of tax equity investors.

Bank of America (BofA Securities, Inc., previously Bank of America Merrill Lynch, or BAML): The investment bank is the second largest in the US Bank of America is one of the largest players in the tax equity market for renewable energy projects. They have financed approximately 41 GW of renewable energy project across wind and solar since 2015. They recently committed to mobilizing $1.5 trillion in sustainable finance capital by 2030 to support ESG-related investments.

JP Morgan (JP Morgan Chase & Co.): JP Morgan is another giant in tax equity finance (JP Morgan and BAML combine for over 50% of all tax equity investments made in the US). Similar to BoA, they have committed $2.5 trillion over 10 years to support the UN's Sustainable Development Goals (SDGs).

US Bank: A smaller player in the US Tax Equity market, US Bank closed 10 transactions in 2023 across solar, wind, and battery projects. Since syndicating deals in the renewable energy space in 2014, they have continued to grow, with a focus on transferability of BESS projects.

Wells Fargo: Through their Renewable Energy & Environmental Finance (REEF) division, Wells Fargo has provided tax-equity financing to hundreds of utility-scale wind and solar projects, as well as battery and fuel cell investments. Since their founding in 2006, Wells Fargo has financed projects across the US, with their first battery project in 2021.

Independent Engineers

DNV: Based in Norway, DNV is a leading global advisory firm providing assurance across several business sectors. Their Energy Systems business unit is one of the leading IEs in the renewables industry worldwide. They also serve as an owner's engineer and provide trainings and other services to developers, financiers and utilities. In 2023, DNV supported over 2 GW of energy storage project transactions in North America.

Fractal Advisors: Fractal Advisors offers strategic advisory services to clients in the energy sector, helping them navigate the complexities of the energy transition. Fractal's sister company, Fractal EMS, provides hardware and software control solutions.

Leidos: Leidos has provided IE services in over seventy countries and provides due diligence for over a hundred projects per year.

Luminate: a smaller IE firm based in Denver, Colorado, focused on technical due diligence of renewable energy projects.

Natural Power: an established independent consultancy in the renewable energy industry with over 30 years of experience.

PowerSwitch: A small energy storage advisory firm based in Portland, Oregon, PowerSwitch has experience as an OE, IE, and in providing training services to the energy storage industry. In disclosure, PowerSwitch is managed by Drew Lebowitz, one of the authors of this book.

UL Solutions: a global safety science and reliability engineering leader that offers a myriad of technical services including independent engineering.

8. ENGINEERING, PROCUREMENT, AND CONSTRUCTION

After months of land acquisition, financing negotiations, contract revisions, and due diligence, finally the day developers have long waited for: giving the green light to contractors and suppliers to move ahead with the construction of the project! Following such an arduous journey with minimal physical progress, starting the **engineering, procurement, and construction (EPC)** phase of the project can be a relief.

However, there are many challenges on the path to successful EPC of a utility-scale BESS project. The process involves dozens of stakeholders, and often tight budget and schedule constraints. Any delays, scope changes, or cost deviations can cost the project portions of its revenue, or even affect its financial viability. A smooth construction process often involves managing volatile factors such as commodity pricing, weather, and changing regulations.

Despite these concerns, a thoughtfully planned and executed EPC process can result in the successful completion of nearly any project. This chapter covers key considerations for the EPC of a utility-scale BESS project, from financial close (FC) through to the start of commercial operations (COD). While every project is different, we review the most common structures for EPC, typical issues, and the tools used to build the project on time and on budget.

8.1 Early-stage Engineering

At the time when a project reaches FC, it is common to have a **preliminary design package** of the plant, also known as a **conceptual design**. This outlines key details like quantities of components, interconnection point, and land boundaries, but does not contain many of the critical specifics of the project engineering. These details include elements such as conductor sizing, protection schemes, substation design, or communications diagrams. The process of moving from the conceptual design through to **Issued for Construction (IFC)** design package is an iterative process which involves several key stakeholders: the project owner, battery supplier, utility, and grid operator, among others.

Many projects perform their initial design work either in-house or with an **owner's engineer (OE)** at an early stage of the project. This level of design, often referred to as a conceptual, preliminary, or early-stage design, is a simplified version of how the system will be built and will operate. At this stage the final design of the project may not be known, but the key aspects of the system layout and design can be determined with preliminary versions. Some of this depends on the developer preference: some like to have their initial designs be more detailed, while others prefer to keep the early-stage planning to a minimum. At a minimum all conceptual designs have two critical components: a **site layout drawing**, which indicates major components and where they will be built on the site, and a **single-line-diagram (SLD)**, also known as a **one-line diagram (OLD)**, which shows how those components are electrically connected and how they feed into the grid.

Site Planning

The design starts with the land on which the project will be built. This includes a wide range of inputs, from visible aspects such as site boundaries, vegetation, and terrain, to studies and investigations such as geotechnical and hydrology engineering.

Prior to any preliminary design, most developers contract to have an initial survey performed, which often uses a scope described by the American Land Title Association, or ALTA. Often referred to as an **ALTA Survey**, this study assesses the high-level aspects of the parcel, its ownership, and its feasibility as a project site. ALTA surveys typically consider the following:

- Precise boundaries and lines of possession
- Any existing buildings or structures on the property
- Any easements, cemeteries, or water features
- All power, gas, sewer, and water lines or other utilities on the site
- Rights of way and access to public or private roads

Optional items which may be included in some ALTA surveys include wetland easements, topography, contour lines, off-site easements, parking, zoning and flood zone classification.

ALTA surveys help position the BESS correctly within the site by giving a clear picture of the features, constraints and risks that may come with a site. The survey may present red flags which make development impossible like flood zones and raise issues that present additional costs or burdens.

Geotechnical studies analyze the soil and geological conditions of a site to determine its suitability for construction. These involve evaluating the soil's bearing capacity and stability for supporting the foundations, identifying potential issues such as brownfields, and aid in developing effective civil engineering design for the equipment. **Hydrology studies** analyze the water flow patterns, and potential flood risks at a proposed site. Closely related are **hydrologic studies,** which assess the flow of water within physical systems such as trenches, culverts, and pipes. All three of these studies are crucial for determining the risk of flooding or water inundation, which could damage equipment, and designing appropriate drainage systems.

System Architecture

Following selection of a site, one of the earliest system design steps is design of the system architecture. This work is typically conducted in the conceptual design stage, as it will dictate some of the key aspects of the land use, costs, and revenues of the project. In planning the BESS architecture, the exact models of equipment, precise layout, or cable sizing are not required. However, it is important to have a broad idea of the type and quantity of battery units and other major components, and how the project will interconnect with the power grid. The way the system is designed and works together is known as the **system architecture**, which determines the basic layout and functionality of the project. For a BESS project this includes:

- Point of interconnection (POI)
- Ties to transmission lines (called "Gen-tie lines") if applicable
- Substation location
- Whether the BESS is being paired with new or existing PV/ wind or built as stand-alone
- Layout of the project
- Expected power and energy rating
- For PV-coupled projects, whether the system is AC- or DC-coupled, as elaborated below
- Whether the project will use an AC-block or DC-bock BESS product as explained in Section 4.2

Project use-case is also a critical factor in designing the system architecture. Use cases range from arbitrage, frequency response, peak shaving, or a mix of various use cases. These are covered in greater detail in Section 6.1.

The flexibility of battery systems and their ability to provide multiple services at once make them powerful tools on the power grid. Understanding the nuances of each use case and how those function in each market is critical to designing the system correctly. The developer may find in the design process that, for example, an AC-coupled PV+Storage plant system is more beneficial than a DC-coupled system. Other projects may find that the system is more profitable with a high-energy battery rather than a high-power battery.

Other determinants for project architecture include cost, location, regulations, available equipment, and constraints that may not be obvious until project location and application are known.

Solar Plus Storage (PV+Storage)

Building a BESS together with a solar PV plant has many technological and financial advantages. This type of installation was one of the first commercially viable use-cases of energy storage when first deployed from 2016-2020. There are a variety of use cases for PV+Storage systems, but the most common is to use the solar to charge the batteries during the daytime to capture higher energy pricing in the evening. This often provides the projects with increased revenue that justifies the cost of the construction of the BESS. Another common arrangement is to use the BESS to allow the system to capture the energy lost when the PV production exceeds the PV inverters' ability to convert the power to DC, known as **clipping losses**. These use cases are covered in greater depth in Section 6.2.

Besides the technical benefits, there are often significant tax incentives for pairing solar with storage. In the US, an **Investment Tax Credit (ITC)** gives a 30% tax incentive for building a PV-paired storage system, which serves as a large boost to the financial performance of the system by reducing the up-front capital investment. This credit was originally scheduled to wind down in the late 2020s, but the Inflation Reduction Act (IRA), passed in 2022, extended the credit and allowed it to apply to standalone BESS projects in addition to solar-coupled systems. Besides tax advantages, PV systems coupled with a BESS have several other advantages such as:

- Common land management
- Decreased project development costs, when compared with two separate projects
- Making the PV production more predictable and "firm," since the BESS can complement and balance inconsistencies in the PV production
- Reduced Balance of Plant (BoP) equipment in the case of DC-coupled systems

PV-coupled BESS are called hybrid when the PV and BESS share the POI and are controlled together as a unified system. Alternatively, some PV and BESS plants may be located adjacent to one another, and connected at the same substation, but are financed and built separately. The following section focuses on the two most popular PV+Storage architectures: AC-coupled or DC-coupled. Each has benefits and drawbacks that make them a good fit for certain types of projects.

AC-coupled PV+Storage

In an **AC-coupled PV+Storage system,** the batteries and the photovoltaic array each have their own inverter or set of inverters. The power from the batteries and the PV field are electrically connected on a bus located on the AC-side of the PV and battery inverters. The batteries can typically be charged from either the grid or PV, although some systems may limit the charging to PV only. This can be the case if the local utility requires that the batteries do not draw power from the grid since they can become large loads when the PV is not covering the battery charging needs, or in some cases where this is required by the tax incentives used by the project.

Essentially, an AC-coupled system is two separate adjacent systems: a solar PV plant and a BESS plant. However, since they are controlled by a single plant controller, and financed and executed together, they are able to serve certain revenue streams that each plant would not be able to achieve on their own. A detailed example of an AC-coupled PV+Storage project, the Fifth Standard project developed by RWE, is provided as a case study in Section 13.2.

DC-coupled PV+Storage

In contrast, **DC-coupled PV+Storage systems**, the PV and the battery plants are electrically connected on a DC bus behind a single set of PCSs. This architecture requires the storage to connect to that bus via a **DC-DC converter** to connect the batteries to the PV system, since the PV voltage is determined by irradiance (how much sunlight is received by the PV panels) and other nuances of how the PV panels are wired, and the BESS voltage is determined by state of charge (SOC) and string size. Since these voltages are independent of each other, for them to operate on a common DC bus, one of the voltages must be managed by an intermediate device to maintain a common DC voltage at the coupling bus. The converter monitors the voltages of the buses on either side of it, and if a charge or discharge command is sent, it will synchronize with the main inverter and make the power flow on the two buses equivalent. It does this by managing the current flow on either side in relation to the bus voltages. The DC converter can also provide galvanic isolation between the equipment on either side of it. This means that the component has a physical gap between the metallic conductors of the input and output of the device, which helps to mitigate short circuit issues that can occur when paralleling many batteries together.

Figure 8-1 shows a simplified schematic comparing the AC- and DC-coupled architectures.

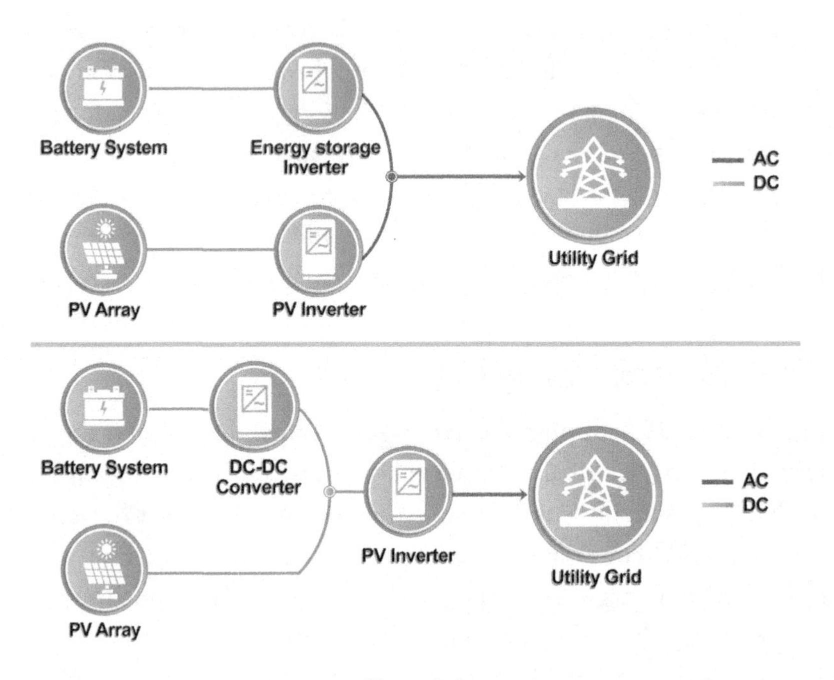

Figure 8-1
AC- and DC-coupled architectures

AC- vs. DC-coupling

AC-coupled systems are less electrically complex than DC-coupled systems since they are essentially two entirely separate systems (one PV and one battery) being built on the same AC bus. In contrast, DC-coupled systems must have specialized inverters and DC-DC converters to balance DC loads being fed onto the same bus at different voltages. However, DC-coupled systems are advantageous when the PV energy can be stored in the BESS before being discharged to the grid since the power passes through fewer components than in an AC-coupled system. In an AC-coupled system, the power first must exit the PV inverter onto the AC collection system, typically through a set of transformers, then pass through the BESS inverter to charge the battery, and ultimately pass through the BESS inverter once again at discharge. This decreases the net round trip efficiency of an AC-coupled PV+Storage project. One other drawback of AC-coupled batteries is that they are unable to absorb clipped energy that is lost at the PV inverter (unless the shared inverters are oversized compared to the POI power limit) This lost heat energy is common with projects that are built with a PV capacity larger than that of the inverters (referred to as having a high **DC-to-AC ratio**). DC-coupled projects can capture these losses, while AC-coupled projects cannot.

DC-coupled systems were viewed as the better way to couple BESS with PV especially because of rules placed by the ITC – for the PV+Storage project to jointly qualify for the ITC - the BESS had to charge from the PV for at least 75% of the time. However, since the passage of the IRA, which allows BESS to qualify for tax credits on their own, this constraint is no longer applicable in the US.

As of mid-2024, the majority of PV+Storage projects in the US are AC-coupled now owing to the simplicity of design and flexibility of this architecture.

Standalone BESS

Stand-alone BESS projects are installations that are not co-located with a renewable energy source like solar or wind. The key advantage with this approach is that these systems that need far less physical space than PV or wind plants, can be located closer to load centers in urban areas that tend to have high nodal congestion and price swings. Stand-alone BESS projects are typically used to provide grid stability services, balance intermittent renewable energy, defer costly grid infrastructure upgrades, and provide energy when solar resources cannot produce enough energy after the sun sets.

Utilities have begun to embrace stand-alone BESS as resources that provide power capacity akin to conventional generation resources via capacity or resource adequacy (RA) contracts. A few examples that are well known are the Moss Landing Energy Storage Facility in California (400 MW / 1,600 MWh) which is contracted under an RA agreement with Pacific Gas and Electric Company (PG&E). Goleta Energy Storage in Santa Barbara (60 MW / 160 MWh) is contracted to provide RA services to Southern California Edison (SCE).

There are numerous stand-alone BESS installations in Texas, or more specifically the region operated by the Electric Reliability Council of Texas (ERCOT). This is mainly enabled by the need for fast-responding ancillary services like frequency regulation that battery systems are very well suited to perform. The other reason can be attributed to often-witnessed dramatic swings in energy prices especially on extremely hot days with surges in demand for energy, as the network is limited in its ability to import energy from neighboring grids. These systems, at the time of writing, are usually of 2-hour duration and have been rapidly installed in the region since 2018. By the end of 2024, the installed capacity in ERCOT is expected to grow to 9.5 GW. The development of stand-alone large-scale BESS in the US has seen a significant increase in recent years. In 11 of the 30 large-scale BESS projects that came online in 2023 were stand-alone. By the end of 2024, the US is estimated to have around 18 GW of stand-alone BESS capacity online, according to the EIA [3].

Microgrid Configuration

Microgrids are electric installations that can operate independently (disconnected from the grid) or in parallel with the larger regional grid (connected and synchronized to it). Although the "micro" part of the name implies that they are small, microgrids may be large, as in the case of mining operations with generation that may exceed 100 MW.

Microgrids come in one of two configurations:

- **Grid-connected microgrids**, which are normally connected to the larger power grid, with the ability to disconnect or "island" themselves from the larger grid as needed
- **Off-grid microgrids**, which are not connected with a large power grid. These are typically in remote locations such as on military bases, islands, or mining facilities.

The ability for a BESS to form a microgrid requires a different hardware configuration, at the PCS control level, than one that operates in grid-connected mode. This is because most PCSs are designed to synchronize and follow the operational power grid, rather than to create their own self-supporting grid. This functionality is known as **grid-forming**, which contrasts with the typical configuration which is known as **grid-following**. This is covered in greater technical detail in Section 4.2.

In the case of grid-connected microgrids, the facility can serve as backup power, commonly referred to as providing **resiliency** services. Although this is what many people think of when they imagine batteries, given the very high cost of the batteries and the relatively high reliability of most grids, this is rarely the use case for large batteries. Instead, the systems are often used to redistribute the demand of the facility, or to absorb production from on-site PV generation.

Off-grid microgrids often use BESS to reduce dependence on fossil fuel generation such as diesel or natural gas, or to be able to shift the PV production for nighttime.

Most microgrid uses of a BESS are not considered utility-scale, since they are designed to cover a smaller area such as an industrial facility. However, utility-scale microgrid applications have been used at military bases, private developments, and in some cases by utility companies looking to provide backup power for certain critical sectors of the grid.

Interconnection

Interconnection is one of the key aspects of a BESS project, and a primary concern from the earliest stages of a prospective project. The **Point of Interconnection (POI)** is where the BESS ties into the grid. Applying for interconnection and receiving all approvals from the grid operator and utility takes a long time, is expensive, and is technically complex. Interconnection approval is currently the largest bottleneck to getting projects online in the US: ISOs and RTOs have long queues with interconnection requests that are greater than the capacity needs for their regions. There are two aspects to interconnection: the physical connection and the process for obtaining approval to inject power into the grid.

The battery and the interconnection work together to exchange power as needed by offtakers. The battery supplies energy at a given power level to the interconnection; the interconnection moves power across transmission lines of varying voltages. The physical interconnection itself is different for every project and made of several components:

- **BESS substation**: This is part of the project and steps up the AC voltage to the voltage at the utility substation, usually from medium (typically 34.5 kV) to high voltage (for example 115 kV). This substation is usually built with the project, although wise developers can sometimes reuse substations at decommissioned power plants.
- **Utility substation or switching station**: The larger of the two substations and owned by the local utility, the utility substation or switching station connects transmission lines and may have several lines leaving in every direction, from distribution level voltages up to 765 kV. Developers often compete for land near these existing substations or switching stations or pay large sums to build one for their project. More

common is that these substations require network upgrades so they can handle the additional load being pushed through to the grid.

- **Substation components:** The substation includes bus bar structures, breakers, disconnect switches, transformers, capacitors, reactors, power measurement devices, and a control room to house the protection relaying, communication, and SCADA equipment needed to monitor the substation and the BESS array.
- **Transmission or gen-ties:** When the BESS substation is not adjacent to the utility substation or switching station, additional transmission lines are needed to connect the two, often referred to as a "gen-tie." These can be on the order of 0.1-10 miles and usually run at a higher voltage than the project, for example 69 kV. They may be overhead lines, which need to navigate around, below, or above any existing infrastructure such as existing transmission lines, roads, and buildings. When overhead lines cannot navigate existing infrastructure, underground lines may be needed, which is more expensive, as construction costs are higher. The responsibility of the transmission buildout depends on the utility managing the region the project is in, though the scope usually sits with the project developer. Either way, the buildout must be coordinated with the utility and constructed by a specialized subcontractor.
- **POI:** This location varies from project to project but is typically located at the high-voltage utility substation or switching station. This is the point at which the system's power to flows onto, or off of, the grid. A revenue grade meter is installed at this location to measure the power exchange between the BESS and grid. The POI may also be referred to as the **Point of Connection (POC)** and may overlap with the **Point of Delivery (POD)**, which is the point where energy is being measured for the offtake agreement.

- **Metering:** Utility-grade bi-directional meters are used to measure the apparent, real, and reactive power, as well as the energy flowing into and out of the project. The primary meter is typically located at the POI, though other meters may be installed to measure the auxiliary load or contribution from each renewable source on the medium-voltage side.

The more complicated part of interconnection is the process through which developers are approved to connect to the grid. This may take from several months to several years to secure, depending on the jurisdiction. Developers must submit a detailed interconnection application through the appropriate ISO or RTO (and pay of course). Very specialized engineers at the ISOs/RTOs then study the impact of adding the power plant to the grid, which involves modeling real/reactive power injection, frequency, harmonics, congestion, and line capacities. These studies have typically been performed for single projects, which is inefficient and doesn't consider how multiple projects in one area may affect the grid. Recently some ISOs have taken a cluster approach to improve the process: instead of studying one project at a time, they study a cluster of several nearby projects and can greatly hasten the process. The results of these studies may show that network and/or facility upgrades, improvements in the transmission or substations, are needed to get the project online. These upgrades can be very expensive and have long lead times, and both price and timing are managed differently by different ISOs/RTOs. Understanding how the interconnection process is managed by the ISO or RTO is a critical part of BESS project development. The timeline and costs of interconnection often make or break a project's feasibility.

All interconnection applicants to an ISO or RTO in the US must wait in a virtual line known as the **interconnection queue**. This queue consists of all the developers planning to deploy a project online, though typically at different commitment levels. Some are ready to build and are waiting in line while others only have tentative plans to build but want to get in the queue to mitigate long delays if they do commit to building. The varying commitment level in the queue is a major problem for getting projects online, and ISOs/RTOs are trying to solve the problem by requiring a certain level of commitment or scoring projects based on how prepared they are. Higher scores get into the queue faster. The Federal Electric Reliability Council (FERC) released a set of significant reforms in 2023, called Order No. 2023, aimed to streamline interconnection processes and reduce the backlog.

It's important to note that interconnection is a collaborative process between the developer and the utility and demands a healthy professional relationship to succeed. It is critical to have technology, especially the steady state and dynamic PCS model with accompanying specifications, be fixed early enough to make the interconnection process smooth (in most cases, amendments to the interconnection procedure can be submitted to change the PCS technology if needed but this adds time to the process). Further, an interconnection reform may enable a faster energy transition, and parties at all levels, from federal to state down to developers, are working hard to reform the interconnection process.

Sizing the System

When sizing a system, it is critical to keep in mind the project and system goals. This is usually a requirement to provide power at a constant level for a certain duration. For example, many systems in California that are participating in the CAISO market and have power purchase agreements with one of the large investor-owned utilities are required to have a minimum duration of four hours. In this case a system should be able to deliver the required power (for example, 100 MW) at the point of interconnection for four hours. This means the design of the system needs to oversize the PCSs so that, after all losses between the output of the PCS and the point where the power is measured, the meter will read 100 MW, and the battery system energy capacity needs to also be oversized so that it can deliver 400 MWh at the POI. If the system is sized exactly to 100 MW / 400 MWh, then only 92.5 MW for 3.6 hours may be produced at the POI roughly accounting for losses, and not meet the requirements of the contract with the offtaker. Oversizing the batteries is referred to as overbuild and is covered in more detail in the next section along with augmentation as the two strategies for maintaining a contractual energy capacity.

System sizing is more like an art than a science, but companies can create tools to help keep track of the assumptions and requirements when sizing a system. There are many commercially available tools for sizing battery systems, but we have found that developing proprietary tools, or working with an integrator that has a lot of experience in sizing can be valuable.

Elements to keep in mind when sizing a battery system:

- Power required at the POI
- Duration required at full power
- Real estate footprint – this can force interesting layouts to fit high energy densities in small parcels
- Local fire-compliance rules can also impose additional spacing and orientation requirements
- Limits on the number of battery containers that can connect to the power conversion system – depending on the products being used and their current ratings, the designer needs to be mindful of current limits, symmetry across PCSs, etc.
- Lifespan and augmentation plan – the degradation curve for batteries can have an impact on initial oversizing and augmentation cadence. Also be aware of space required for augmenting and how this will connect to the existing system
- Maximum and minimum operating temperature (major equipment such as the batteries and power conversion will be forced

to derate at higher than optimal temperatures and altitude, depending on their specifications

- Auxiliary power source – if self-supplied the system will need to be oversized to account for this
- Losses of each component
- Required power factor at POI
- How DC converters will interact with upstream PCSs
- Control and metering scheme
- Networking connections
- Use cases, how they stack, and how they will impact system health
- How the system will be installed and maintained

The above is not an exhaustive list – there may be requirements specific to the site, permitting authorities, and offtake agreement that need to be considered. A good battery system designer will keep their own growing list and take notes on what was missed in previous projects and add them in. Each project will also have its own main focus where some items are more critical than others. The earlier in the process the requirements and nuances of a project are understood, the design can focus on what is critical for that project and drive to success. Typically, the system design would need to account for about 7-10% efficiency losses in one way of power flow.

Augmentation and Overbuild

As discussed in Section 3.9, degradation is the loss of energy capacity over time. For projects in certain markets, especially in CAISO, the energy capacity needs to be maintained at a contractual level for a prolonged number of years. There are two ways to accomplish this: overbuild and augmentation.

Overbuild is a strategy of deliberately building a project with an energy capacity much greater than the contractual capacity. This allows the project to meet the contractual capacity for many years, even when accounting for the degradation of the battery over the life of the project. This usually means overbuilding the energy capacity by around 30 to 40% for a 20-year project. By the time the project reaches the end of its useful life, the energy capacity will be approximately 65% of the capacity at beginning-of-life (BOL). Although the batteries may still be able to cycle at this point, in their degraded state they create significantly more heat, meaning that they may start to become a safety risk. Given that battery costs are expected to continue declining, paying an additional 40% up front is typically uneconomical. Because of this, overbuild is typically only used where returning to a site may be difficult or costly, such as a BESS project on a remote island location.

Instead of overbuilding, projects typically choose to deliberately build a system that will, after a few years, require fresh batteries to be added to meet its capacity requirement. This strategy is known as **augmentation**. This scheme anticipates that new battery enclosures be added to the project at certain intervals throughout the project lifetime – typically from 2 to 4 occasions throughout the project's life. Augmentation helps reduce upfront costs relative to overbuild, allowing the project to capture savings from declining battery prices. Augmentation must be planned during project design, so that when the contractor is mobilized to add new batteries, they can be added with minimal additional work. This includes things such as leaving extra concrete slabs for the new enclosures, or spare circuits on the collection system to accommodate the new containers. The costs of adding the new battery units, such as electrical integration and BOP, must also be factored into the project's financial model.

Figure 8-2 shows an example of a system that is designed to have a minimum of 20 MWh DC available energy. At construction (year 0) the system was overbuilt to have 24 MWh of capacity. As the system degrades, it anticipates performing an augmentation in year 7, to avoid going under the 20 MWh DC minimum requirement. Another augmentation is planned for year 14, which will maintain the project's capacity through to the end of the useful life, in year 20.

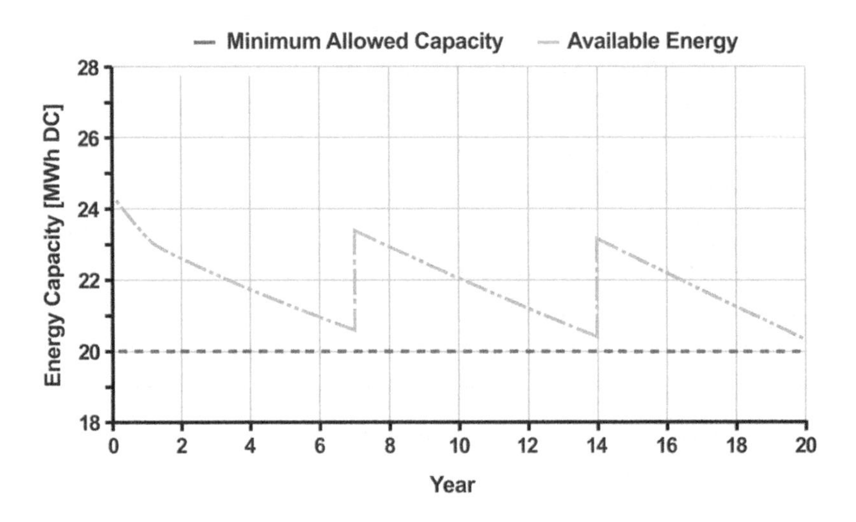

Figure 8-2
Capacity augmentation over the project's life

There are two ways to add additional energy capacity:

1. AC augmentation is connecting additional batteries and PCSs at the low or medium AC voltage side of the project, usually between the medium voltage transformer and PCS. Advantages with this approach are that different battery products can be added with ease. This is very relevant now with the high degree of evolution of batteries – it is highly likely that the original battery may no longer be available or supported at the time of augmentation, so a different battery or supplier is used. Moreover,

battery cells are easier to operate when they are at the same state of health (SOH). By connecting on the AC side, cells can be at completely different SOHs and not need complicated controls to work effectively. Finally, connecting on the AC side gives more flexibility to layout as DC connections require additions to be adjacent.

2. DC augmentation is connecting additional batteries on the DC side of the existing PCS. Note that in the case of most of the products as they stand now, this involves adding new battery containers (or DC-blocks), without making changes to the modules in the containers installed at the beginning of the project life. However, it does require recalibration of modules such that the old and new units are "normalized" and calibrated with the controls system. This method has the advantage of lower cost as it involves fewer PCS units, as well as avoiding changes to the interconnection agreement (which is typically warranted when more PCS need to be added). At the time of writing this book, DC augmentation is more prevalent because of the lower cost and easier installation, but this could change as the market evolves.

The decision between needing augmentation or overbuild is a decision that needs to be taken with careful consideration of the project's use-case and contractual requirements – in most cases, merchant projects will not have augmentation planned, nor overbuild, as the owner has flexibility on how they're using the system. But augmentation or overbuild will be needed when the project has contractual obligations to deliver a certain amount of MWh. Additionally, if augmentation is planned, space needs to be reserved and the overall design of cabling and BoP elements needs to accommodate the incremental units. Owners tend to avoid excessively frequent augmentation, since the process requires site mobilization, permits, additional labor, and the commissioning of new hardware.

Augmentation is most often planned for periods of low market revenue – for example, projects in ERCOT would be wise to avoid planning their augmentation in the summer, when BESS capacity is most critical. It's also wise to perform regular maintenance in parallel with augmentation while parts of the project are offline.

Typical SLD / Site Layout

Communicating the electrical architecture is critical for the design and construction of a BESS. This is done by means of a Single Line Diagram (SLD), which shows the electrical components and electrical connection or "lines" that connect them. The SLD helps develop contract documents by identifying stakeholders, and documenting and defining scopes. Figure 8-3 shows the SLD for an AC-block, which includes the BESS, PCS, and MV transformer for a 5 MW / 20 MWh block, which may be repeated throughout the project.

A site layout is a diagrammatic representation of the project with all major equipment depicted. It is usually generated using AutoCAD software that allows for exact dimensions of the site boundaries, equipment, and clearances to be laid out. For a conceptual layout, the CAD depiction is overlaid on an aerial imagery of the site so that the exact dimensions and constraints of the site can be visualized and captured in the design. This is a crucial aspect of project design as the first step is to verify if the site, with all its constraints considered, can physically hold the required number of battery containers, PCS units, and substation equipment to achieve the desired nameplate MW/MWh rating. This is often not as straightforward as it may seem initially – the layout needs to consider buildable area constraints such as wetlands, ecological features, zoning/permitting setbacks, clearances required from roadways, clearances required for each of the equipment, laydown areas, stormwater allocated spaces, grading restrictions, O&M dedicated space, practical constructability, space for augmentation units, and safety-related measures that may need to be incorporated. Figure 8-4 shows a sample of a typical project layout.

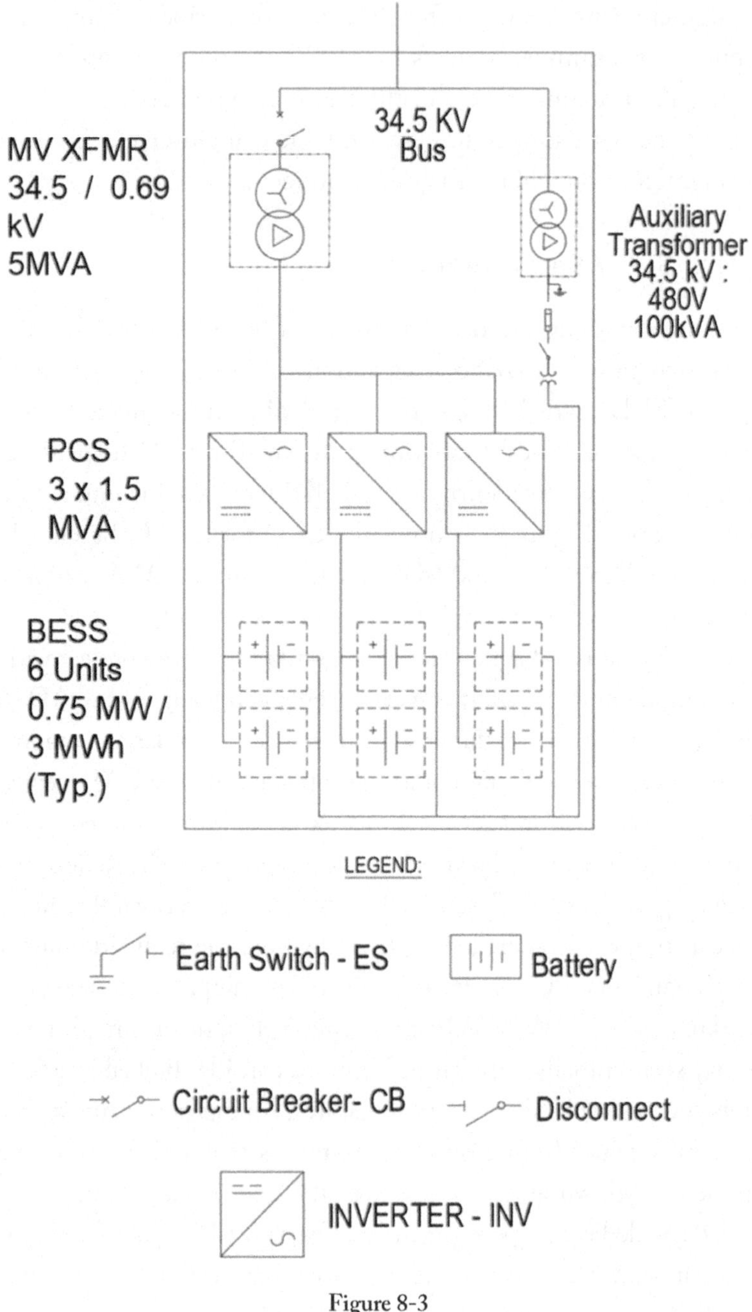

Figure 8-3
30% Single Line Diagram (SLD)

As with the broader BESS plant design, SLDs and layouts are typically refined and detailed through various stages of the project – the following are typical SLD and layout design phases.

The pre-design phase develops project goals, design criteria, and concepts, and this phase identifies stakeholders, potential risks, and regulatory requirements.

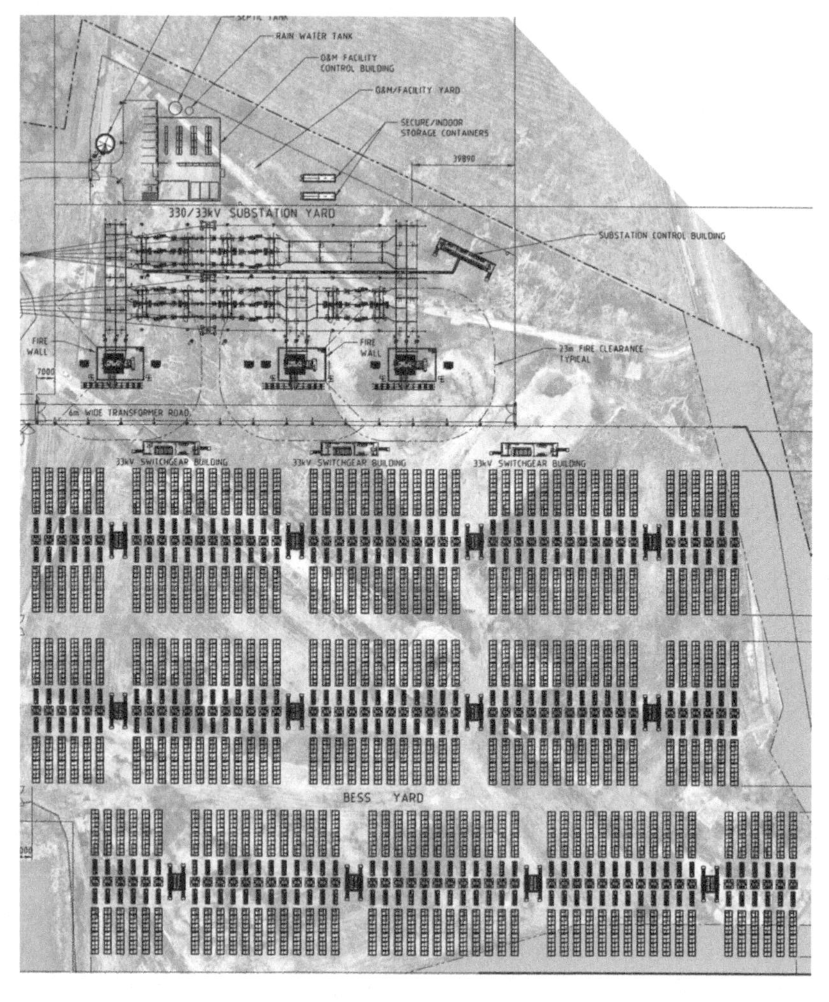

Figure 8-4
Typical BESS layout [53]

System Integration

BESS are composed of multiple building blocks – the batteries themselves as DC units, PCSs, transformers, and controls systems. One entity needs to take the responsibility for making sure that all these components are connected correctly and function synchronously. This is usually in the scope of the integrator if one is involved. Otherwise, it can be split between the project owner and the EPC companies. It is imperative to ensure that system design, right from determining quantities of BESS and PCS units, ensures the physical and electrical compatibility between equipment. Making all these pieces work together, achieved in the commissioning phase post construction is a complicated task and needs clear stakeholder management. Service technicians and engineers from the BESS and PCS suppliers need to be present during commissioning for testing and quick troubleshooting of their respective equipment.

8.2 Complete Design Package

Once the system has reached financial close (FC), the EPC contract is signed and an **Engineer of Record (EOR)** firm proceeds to work on designing the final system. Typically, up until this point the design will have been completed to approximately 30%, meaning that it has the overall structure, layout, and components described in the previous section, but specifics such as conductor sizing, foundation design, or interconnection details have not yet been defined.

The process of moving from the 30% design to the final set, known as **Issued for Construction (IFC)** set, is iterative, meaning that several sets of design packages will be issued over the weeks or months that the design team requires to build the system. This approach allows the design to proceed concurrently with early stages of construction, vastly compressing the overall timeline of the EPC process. There are several different methods of contracting structures used on BESS projects – the major ones are covered in Section 8.5.

Some EPC firms perform design and engineering of the system using their own in-house engineering teams, while others use a subcontracted design firm to perform the design work. In either case, the design is performed by a team of engineers, with each discipline (electrical, civil, and structural) having a licensed Professional Engineer (P.E.) who will stamp the individual drawing package and studies. This engineer must be licensed in the jurisdiction where the project will be built. In the US this is typically a P.E. or the EOR. Since the EOR is putting their professional license and judgement behind the project, they and the EPC sub-contracting them are ultimately the authority behind the design decisions and the decider if there is ever any question about the appropriate means or methods to build the project.

Components of the Design Package

The BESS drawing package, as well as the technical specifications, are divided by subsection, or discipline, to cover all aspects of the facility. The typical design package includes the following categories:

- **Civil engineering**: This broad set of designs includes all aspects of the surface earthwork, subgrade work, land development and built environment of the project. Sub-disciplines include:
    - Geotechnical engineering: Studies the subgrade soil that support the system. This is one of the earliest steps in design as it will determine the foundation system required and inform the grounding and ampacity studies for the BESS and substation.
    - Land development: Includes surveying, clearing and grubbing, grading, cut-and-fill, site fencing, and security. Roadway design and planning is included under this section.
    - Structural engineering: Includes engineering of the foundations and structures used to support buildings and equipment

on the project. It considers all expected dead load, live load, seismic loading as well as potential stressors such as hurricanes, snow or frost heave events, or earthquakes.

  - Hydrology: Studies the storm water, drainage, and infiltration resources on the site, including potential flooding risk, water quality issues, flow, and distribution of water on the site.

- **Mechanical, electrical, and plumbing:**
  - Electrical engineering: This is the heart of the battery design, including how the components are connected to one another, how the system will feed power to the grid under normal conditions, and ensuring that the system is designed to protect itself in the case of abnormal conditions. Electrical design includes subsections such as:
    - **AC and DC collection systems**: these are the design of how all DC components combine to connect from the battery product to the PCSs. On the AC side the collection system includes all design from the PCSs up to the high-voltage transformer.
    - **Communication and controls**: A subset of the electrical design, communications plan includes all low-voltage wiring such as ethernet, fiber optic, and radio communications connections. It also includes how these systems interact with external sources such as first responders, the grid operator, and the monitoring system of the project.
    - **Substation / interconnection**: This covers both the new HV substation on the BESS project as well as any transmission lines, and integration with, in most cases, an existing utility-owned HV substation.
    - Protection: This critical section includes all fuses, switches, disconnects, breakers, circuit switchers, and relays used to

protect the components of the system on both the DC and AC sides.

- **Mechanical:** This considers any systems such as compressors, pumps, fans, heating, ventilation, and air-conditioning (HVAC) equipment. Although most battery products typically have these systems self-contained within their enclosure, there are always some additional systems required, especially for larger sites.
- **Plumbing:** Although most BESS sites do not have extensive plumbing systems, most have some small water and wastewater systems, either tied into the existing water and wastewater service lines, or independent systems. Any water tanks and pipes for fire suppression are designed within this section.

- **Fire Protection Engineering (FPE):** The design of all components and systems related to fire protection (detection and suppression) such as fire alarms, smoke detection, gas monitoring, and first integration with first responder systems. This section is closely aligned with communication / controls and ensures that the fire protection systems are robust and compatible with one another.
- **Architectural plans:** Most battery systems have few architectural concerns since they are industrial facilities, but those located in urban or suburban locations may have architectural concerns such as landscape, walls, and appearance from the street. If there are any permanent structures built as part of the facility, their architecture is addressed in this section.

The full design of each of these sections includes a plan set showing the construction in visual form, including relevant tables and schedules indicating technical information needed to build the project. Additionally, the design includes a technical specifications package which outlines the performance requirements of each component in the project.

While some components may be explicitly named in the plans, such as the battery enclosures, others may be designed to meet certain specifications, and it is the contractor's responsibility to select a brand and model of the equipment. Typically, there is a **submittal** process whereby the proposed equipment must be approved by the engineer stamping the drawings or by the owner of the project.

Iterative Design Stages

Note that developers use different approaches, definitions, inputs, and outputs for their design stages. The following discussion is for typical design stages and not an industry agreed to standard.

The conceptual design is typically considered a 30% design. This level describes the quantity and size of major components – the BESS containers, PCS, transformers, and substation – and lines to indicate the general architecture of the BESS. The intention is that all parties review and agree on the architecture before moving into a more detailed design. This will include determining any fatal flaws and finalizing the design criteria.

The 60% phase finalizes the expectations and objectives of the project, confirms constructability, determines permitting, implements acceptable engineering requirements, and identifies preferred equipment and materials. The SLD and layouts have more details on the equipment such as junction boxes, breakers, and switchgear for a viable plant.

The 90% phase is near complete and shows the design as intended to be constructed on site, including detailed plans and specifications. Following this package, the **Issued For Construction (IFC)** drawing package is generally the definitive set that will be used for the project. For some projects, an intermediary set is issued for permitting, known as an **Issued For Permitting (IFP)** set.

After construction, an As-Built drawing package will be issued that includes any changes from the IFC drawings that were implemented on site, either due to problems encountered during construction or errors that were missed in the reviews through IFC.

8.3 Codes and Standards

Every market has specific codes and standards that define the legal standards that a project must comply with. These requirements will vary greatly based on the location and the county's AHJ where it is being built, so they must be studied by the designer to ensure that there are no issues with compliance. The US market, being as large as it is, has defined many of the codes and standards which are adopted by or adapted in other jurisdictions.

Note that a standard defines specifications, while a code is a standard that has been adopted by a governing authority and becomes requirements. This section deals primarily with codes and standards related to construction; the relevant safety standards for each product are covered in greater depth in Chapter 10.

Codes

The construction code is the most important standard used in each project. It is typically defined by the local jurisdiction in which the project is being built, often written in the laws of that jurisdiction. Some nations have a single code that applies across the country, while other markets, such as the US, have regional (state- or province-level) codes that govern; those in turn often reference a national code with some specific modifications.

The **International Fire Code (IFC), International Building Code (IBC)**, and **International Residential Code (IRC)** are some common international standards the govern construction across several countries – some national and state codes simply reference these codes. IFC is issued every 3 years – at the time of this writing, the applicable version is 2021, and the 2024 version is to be released soon.

For energy storage, the most applicable portion is Section 1207 of the IFC, covering Electrical Energy Storage Systems (ESS).

The **National Fire Protection Agency (NFPA)** in the US is one of the most referenced codes. NFPA 70, the **National Electric Code (NEC)**, has extensive requirements for the installation of indoor and outdoor battery systems, as well as for the supporting collection systems and communication networks. The **National Electric Safety Code (NESC)** is used to govern electric utilities and may also be adopted by state regulatory commissions for **investor-owned utilities (IOUs)**.

Other prominent NFPA codes are:

- **NFPA 68**: Explosion protection by deflagration venting
- **NFPA 69**: Explosion prevention systems
- **NFPA 70E**: Electrical safety

The most prominent standard for battery construction is NFPA 855 – Standard for the Installation of Stationary Energy Storage Systems. This standard, and the others above, are covered in depth in Chapter 10 on Safety and Environmental Considerations. Many guidelines in that code are related to the layout and engineering of the system, and so they should be considered by the design engineer for the project.

Within the US, New York City and California have been market leaders. The notable aspects of these codes are listed below.

- **New York City Fire Department (NYCFD)** As of October 2019, the New York Fire Code was updated to include Rule 3, which governs Outdoor Stationary Battery Systems. This code references Underwriters Laboratories (UL) 9540A, NFPA 855, UL 9540, UL 1741, UL 1973, and defines permits and supervision requirements, installation approval roles, remote monitoring guidelines, emergency planning, and several other requirements.

- **The 2022 California State Fire Code** (which is based on the 2021 IFC) includes updates to its section on Li-ion batteries (Section 1207). Additionally, the California Public Utilities Commission (CPUC) has Rule 21 which applies to grid-interactive requirements of PCSs, referencing IEEE 1547 (Distributed Energy Resources) and UL 1941 SA, which imposes special requirements on PCSs to comply with this standard. Many PCS manufacturers have complied with this across their entire product lines, while others offer SA compliance as an add-on feature.
- Other states may maintain their own fire code and construction standards for BESS projects.

Standards

Product standards in the battery industry may govern individual components, integrated battery systems, or the installation as a whole.

UL is one of the foremost testing labs in the US, and many codes have adopted UL standards. While some competing laboratories argue that UL standards and testing make for unfair competition, their codes have been adopted so widely that they have become the main benchmarks driving performance and safety in the industry. The specific UL standards pertaining to battery systems are also covered in Chapter 10 on Safety and Environmental Considerations. In brief, the main standards are:

UL 1741: Power Conversion Systems (Inverters)

UL 1973: Li-ion battery products

UL 9540: Integrated lithium-ion battery systems

UL 9540A: Methods for Burn testing of Li-ion battery cells, modules, racks, and systems

Note that the transport of Li-ion batteries is described in the United Nations standard **UN 38.3**, and this compliance is listed on nearly all battery storage products.

It is notable that while UL has defined the standards with which battery testing must comply, it is not a requirement that the UL organization test them directly. Other organizations may test to UL's standards, provided they are an acceptable facility – one relevant accreditation is the designation of a **Nationally Recognized Testing Laboratory (NRTL)**, which is provided by the US **Occupational Safety and Health Administration (OSHA)**. Although this is a US-focused standard, many international labs have elected to be certified to this standard. Besides UL, some common testing agencies are TUV Rhineland and Canada Standards Association (CSA) Group.

In some cases, a battery manufacturer may conduct in-house tests which are witnessed by an NRTL.

8.4 Procurement

Procurement of equipment for a BESS project is one of the most critical stages in a project. A project must ensure both the best price as well as the level of quality and performance required. Also critical are lead times, potential international shipping logistics, tariffs, and many other potential roadblocks in getting the needed components to site safely and timely.

Multiple parties are involved in building a battery system and it is challenging figuring out who is responsible for what and when. Many project owners like to have a single point of accountability, which makes warranty and O&M claims a lot simpler, albeit at a generally higher cost. Divisions of responsibility, scopes of work, and other tools are helpful for procurement. The major entities involved in procurement for a BESS project are:

- Battery OEM or Integrator
- PCS suppliers
- EMS suppliers
- Transformer suppliers
- Engineering, procurement, and construction (EPC) companies
- Engineer of Record (EoR) companies

There can be several combinations of scopes that each entity covers, depending on the supplier or company in question and their offerings.

Battery systems themselves tend to be the most dominant capital expense for utility-scale BESS. It is interesting to note that the scope of supply/services of battery providers can vary based on the type of entity the developer works with, broadly categorized as:

- **Battery Original Equipment Manufacturers (OEMs)** - Who typically manufacture cells, modules, and racks. Based on current market trends, it is more common that OEMs provide a containerized DC-block that is designed for plug-and-play. The PCS has to be procured separately in this case (or as an extension of scope for the battery OEM). The developer takes responsibility to ensure design compatibility between the battery blocks, the PCS, and other components, addressing voltage balancing, number of battery containers per bus bar and fuse protection aspects.
- **Integrators** – Who provide a more holistic solution, typically battery containers coupled with PCS and other elements like

EMS and transformers (depending on the scope). Traditionally, integrators have purchased individual components from OEMs, with the value-addition to the end-customer being the guaranteed responsibility and warranty of an integrated solution at low-voltage or medium-voltage level.

These classifications are blending, and the value chain is shrinking across the industry as the market evolves and everyone wants a larger share of the pie. Products are more standardized, and some large integrators are even venturing into battery module production to further streamline their product. OEMs that were focused on the battery portion (DC-blocks) are now increasingly offering an integrated solution with PCS and EMS. This is done as a scope expansion, meaning they purchase the subcomponents from other suppliers and are responsible for the supply and performance of the integrated product. Similarly, some of the large PCS suppliers are expanding their scope to include battery systems by either purchasing from traditional battery OEMs or venturing into battery module production themselves. While integration of the major components – batteries, PCS, and EMS – is a critical aspect for the asset owner to manage, lately the OEMS are increasing their BESS scope and diminishing the cost of integration. The downside of this evolution is that the project owners and operators may have less direct relationships with the specific OEMs, making it harder to troubleshoot issues fast and track product evolution. There will be many hiccups and lessons learned as this approach becomes more common, but if done correctly it should enable more efficient project execution.

Regardless of the type of battery solution provider, typical equipment scope from the battery supplier includes:

- Battery cell, module, racks, and all subcomponents
- Network and control hardware (module, rack, and bank-level BMS), up to a specific line of demarcation

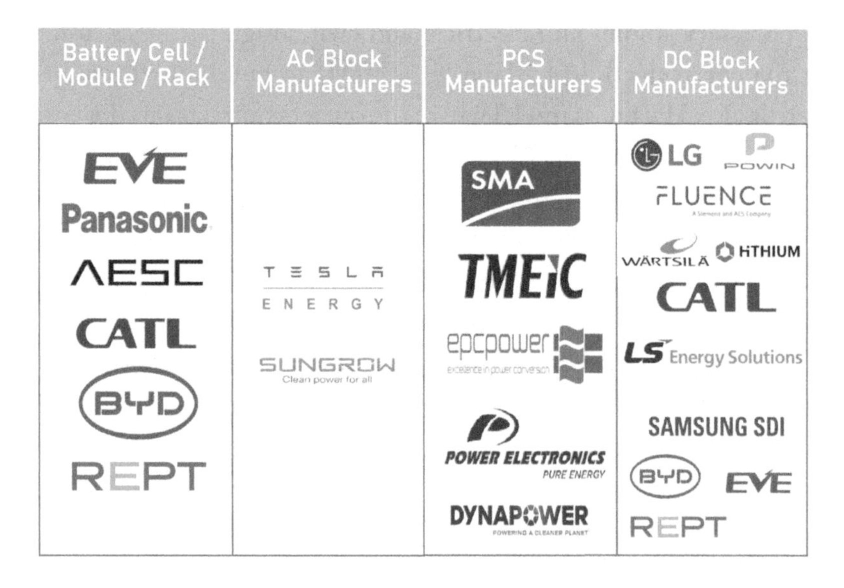

Figure 8-5
Major suppliers and market segments they cover

- All ancillary equipment required for safe and optimal operation of the batteries (thermal management system, fire detection, fire suppression, etc.)
- DC combiner bus and disconnect switches
- Enclosure or container

In the case of integrated systems, the scope of supply typically includes the following, on top of the DC battery units: the PCS and EMS for the BESS.

Currently in the case of most utility-scale BESS projects, the developer usually procures the major equipment: batteries and PCS, major HV equipment, and sometimes the transformers as well. An EPC firm is brought on board to help design the entire project to be construction-ready, procure all necessary equipment to build out the project and construct it. Many EPCs work with an EOR to complete the necessary studies and electrical and civil design packages to flesh out the complete Bill of Materials (BOM) to build out the project. There is also a business model where the owner contracts out the entire scope – that is including selection and purchase of the battery system and PCS – to an EPC. However, this is not economical because of the layers of margin involved, except for special circumstances, such as a lack of in-house expertise to manage all aspects of the project, as in the case of very small developers, utilities, or pilot projects.

Equipment typically provided by the EPC (or in combination with the developer/customer) includes:

- DC cabling
- AC cabling between the PCS and transformer
- AC cabling from transformer to substation
- Low-voltage to medium-voltage transformers
- Medium-voltage to high-voltage transformers, more commonly known as Main Power Transformers (MPT)
- Required substation equipment (except for major equipment)
- Auxiliary transformers
- Auxiliary panel board
- Communication/networking hardware and fiber cable

The scope of work can vary depending on the type of project, developer's resources, and the capabilities of the contractor. An accurate scope helps to ensure that everyone is aware of what is and isn't within their scope. This can be managed with a Division of Responsibilities matrix, which is covered in Section 8.5, and by checking the compatibility of different equipment and delivery timelines. All these items will help to simplify a complex project, minimizing errors and finger pointing. Concise, clear, and comprehensive documentation is critical for a successful procurement process.

Request for Proposal

Most procurement activities conducted by Developers and EPCs follow a set process with procurement guidelines, typically beginning with a **Request for Proposal, or RFP**.

An RFP allows the developer to get down on paper the particulars of the system components they want to procure/have built, and for multiple counterparties to provide their offering in terms of products and pricing. The RFPs can range from one-page overviews to hundreds of pages, detailing and specifying every component of the system. Multiple companies will typically receive or have access to the RFP and will be allowed to bid for the aspects of the project depending on the type of RFP. In this way a project developer can go to the market for project partners and get competitive pricing.

RFPs are usually issued for major equipment like batteries and PCS, and EPC contracts. Battery and PCS RFPs typically include the size and location of the system, specific design parameters like expected cycling/use-cases, system losses, temperature ranges expected, required power factors, etc. EPC RFPs should include details on the required scope of services and a preliminary layout other than the basics of the project. In general, the more open-ended the scope of work definition is, the more high-level the quoted price for services is. It is beneficial to have the battery solution decided and at least 30% civil design package ready in time for the EPC RFP to get certain price estimates.

RFPs allow the project owner to set the rules of the game up front. They are also often the best tool for enabling a level playing field amongst all the respondents and are even often required to satisfy procurement guidelines that companies have. They provide an open and transparent way for the auditors of the RFP to ensure that all responses can be quantified in the same manner, typically with the goal of the least cost, best solution response being chosen.

Battery Energy Supply Agreements

For a utility-scale BESS project, the battery supply agreement typically represents the highest cost, as the battery system is one of the key pieces of major equipment procured, usually by the project owner/developer. The battery supply agreement is known by different terms by different suppliers, such as Energy (or Equipment) Supply Agreement (ESA), Battery Energy Supply Agreement (BESA), or Sale & Purchase Agreement (SPA). This contract governs all aspects of the battery supply, from pricing, to schedule, payment terms, testing, commissioning, and performance guarantees. The contract typically is for the battery system – racks or containerized systems – and associated performance metrics. It can also cover the PCS and other ancillary equipment, depending on the scope or supplier.

Commercial Terms

Critical commercial aspects to be reviewed and agreed on carefully by both the buying and selling parties are pricing, payment milestones, delivery milestones and liquidated damages (LD) for any delays in delivery or performance, and insurance. Pricing for the product can be for delivery to site (called "DDP" which stands for Delivered Duty Paid terms), where the supplier assumes responsibility, costs and risks for shipping of the goods and delivery at the project site, or on Free Carrier terms (FCA) where the hand-off of the goods and responsibility for it transitions from supplier to owner at a mutually agreed upon point such as the port of entry to the destination country. This decision on DDP vs. FCA needs to be made balancing risks and costs in mind. In the PV world, we often hear of situations of PV modules being detained by US Customs authorities at the port of entry for significant periods of time causing uncertainty for project schedule and financials. Suppliers from outside the US are generally better equipped and experienced to manage this type of risk (which would otherwise fall on the owner in the case of FCA terms) and cover insurance for the product until delivery. Conditions for storage of equipment, if applicable, and delivery of the product without any damage during shipment need to be carefully discussed and laid out in the supply agreement as battery storage units are particularly sensitive to temperature and humidity. At the time of writing, containerized BESS are touted to be as high as 5-6 MWh per container implying a significant increase in weight (typically 45-55 tons). This may warrant diligence on road transportability of the equipment to the site, especially if it is remote, and specific permits to be sought from the Department of Transportation in the US.

A key commercial risk to carefully manage is potential volatile import tariffs especially for products manufactured in China, which is the current global leader in battery cell supply by a large margin. For instance, the Biden administration imposed 25% tariffs on battery cells manufactured in China in mid-2024 to take effect starting in 2026. This is done to prevent artificial lowering of prices, via substantial subsidies by the government, and the inundation of supply in international markets, more commonly called "anti-dumping" measures as seen with PV modules. One of the strategies to hedge against these kinds of risks, beyond diversifying supplier options, is to "safe harbor" products by purchasing and sometimes even importing them well ahead of project schedule. This way, the products will be subject to known rules at the time of purchase (and/or import) of equipment, thus circumventing volatile rules at the risk of storing batteries for a prolonged period in warehouses where they can degrade. While such tariffs boost supplier diversity beyond China and hence make global supply market more robust, they invariably result in the increase of overall costs of building BESS projects.

At the time of writing this book, the total cost of a four-hour duration utility-scale BESS is roughly between $250-300/kWh. The bulk of that cost is from the cost of battery procurement – battery system costs are anywhere between $150-215/kWh depending on the supplier and battery chemistry. This price varies widely based on the market share of the supplier, as emerging suppliers may offer bullish price to sell more products. The next biggest shares of the project cost are the PCS and EPC cost. Beyond the cost of production itself and supply-demand dynamics, pricing quoted by suppliers can embed strategies based on several factors such as volume of sale, strategic partnership, market entry, production demand, etc.

Liquidated damages associated with delays in delivery are particularly important when the project has an offtake agreement and associated milestone obligations which may carry penalties if not met. This way, the project can offset some or all the penalties that may be incurred to the offtaker in the case of delays in product delivery from the supplier. There are usually caveats for uncontrollable circumstances, such as natural weather events and pandemic circumstances – called **Force Majeure** events in legal terms – that the supplier can use to be exonerated from LDs.

Suppliers generally offer the price of battery systems on a fixed $/kWh basis. They may include an index component to this price to hedge against mineral price fluctuations, as was seen in 2021-2022. Payments are paid based on a mutually agreed upon milestone schedule with a sizeable downpayment, much like a house purchase (typically 10% of the overall contract price). The decision around the downpayment itself, which is easily several million dollars, is a major financial commitment for the project. The total contract price is paid out when delivery and site acceptance are completed satisfactorily.

Technical Terms

The supply agreement lays out important technical details and performance guarantees from the product supplier. It will list the binding quantities and specifications of the exact product being purchased, delivery schedule, spare parts arrangement for commissioning and maintenance, and the scope of support for the supplier during commissioning and testing, which is further elaborated in a later part of this chapter. The quantities of units are decided by sizing analysis – usually collaborated on by both the supplier and owner – for the desired application of the BESS. Spare parts for most of the key components may be at an extra cost but are critical to have to mitigate in a timely manner and maintain smooth operations. The recommended quantities and storage conditions for them are specified by the supplier.

When the battery systems arrive on site of the project, a visual inspection for damage and defects is performed by the owner. After installation and commissioning of the units (partially or wholly), a capacity test is performed to ensure that the entire MWh purchased by the owner and guaranteed by the supplier has been received. Since the batteries need to be charged and discharged for this exercise, the overall plant and substation need to be energized to perform this test. Depending on the scope of supply, the capacity is verified at either the DC or AC levels.

Some of the most important technical aspects needing diligence and negotiation are product warranty, performance guarantees beyond the delivery window, and conditions for operation needed for warranties and performance guarantees to apply. The product warranty is similar to the warranty that comes with most major appliances – it is a protection against product or workmanship defects that cause the product to function at less than their promised capacity/performance. These can be caused by production quality issues or damage during shipping that the supplier agrees to rectify. Typically, product warranties are offered as part of the supply agreement for three to five years, with an option to extend them for longer times at incremental costs. Whether the extended warranty is procured or not depends on the risk appetite and resource availability of the operator of the project.

Classic performance guaranties for battery systems include system availability, degradation or SoH guaranty which is tied to specific operating constraints, and round-trip efficiency (these can be on DC or AC basis, depending on the scope covered). Usually, availability guaranties range at 94-98%, which provides contractual assurance that the system will be online and available at full capacity for at least 94% of the 8760 hours in one year, apart from Force Majeure events. Round-trip efficiency at the beginning-of-life on the DC-level is typically around 94%, and about 87% for AC-scope, considering other losses like PCS, transformers, line losses and auxiliary loads.

The degradation or SoH is a key metric for Li-ion batteries as it significantly impacts the economics of the project. The performance guarantee is usually laid out as the usable percentage DC or AC MWh energy (depending on the scope) for each year. These are commonly provided for ten years as default as part of the supply agreement with an option to extend at incremental costs, but this can vary between suppliers and is a point of negotiation. If there is a need to maintain the MWh of the project through a period, typically the case if there is an offtake agreement, the degradation guarantee forms the basis for calculating how much the system needs to be augmented in future years, and accordingly how much of an overbuild is required in the first year, which is typically the most expensive. Hence it is critical to have a firm guarantee on the expected SoH of the system. Even if there is no need to maintain the MWh capacity of the system, as in the case of merchant projects that participate in wholesale markets without offtake obligations, the SoH curve determines the usable energy in the system and hence the revenue potential for the project. There needs to be clear understanding, and ideally a guarantee of rated power output through the operating SoC range of the system. It is common to have decreases of power output at the tail ends of the SoC range during both charging (on the high end of SoC) and discharging (on the low end of SoC) – the system needs to be sized adequately if there is a requirement to discharge at constant power for a set duration of time.

The degradation guarantee is generally contingent on a number of complex intertwined factors: the C-rate, depth of discharge, operating temperature, number of cycles per day or throughput per year of the system (total energy discharged), resting and average SoC that the battery is expected to be maintained at in the planned use-case, as further elaborated in Chapter 3.9. So, the suppliers set boundary limits on these metrics, meaning the operation of the batteries must ensure that limits set by the supplier are always adhered to. These conditions are typically on the operating temperature, total cycling or throughput and maximum power of charge or discharge.

As explained in Section 4.3, containerized BESS are generally designed and equipped with thermal management systems such as HVAC and chillers that need to be continually powered to maintain desired temperature and humidity levels (which is the responsibility of the operator of the BESS) - these power requirements are collectively called auxiliary power for the battery systems. It is paramount that a strategy for providing uninterrupted auxiliary power be worked out by the owner/operator of the BESS as it is critical to retaining the performance guarantee. Otherwise, the performance guarantee may be stepped down using pre-set formulas, or in worst-case scenario, entirely voided. Similarly, the constraints around cycling/throughput and average SoC (if applicable) should be maintained as boundary conditions while operating the BESS so that the performance guarantee is retained.

The degradation curve is usually set at the time of contract execution with whatever expectation of the application of the battery is at that time. In more recent times, as the understanding around battery degradation in utility-scale ESS systems grows, suppliers are offering more flexibility around constraints for SoH guarantees, and evolving SoH guarantees instead of static ones over the term. This is especially valuable if the use-case of the BESS is not fixed or may vary in future years. Flexibility is provided in the form of a variable SoH guaranty depending on the way the BESS is cycled – varying most commonly as a factor of equivalent cycles/total discharged energy. This is very helpful as BESS operations need to be dynamic and adaptable to evolving market conditions and rules.

An annual capacity test, either as a physical test or by calculation via operational data, is typically performed by the operator of the plant to check if performance guarantees around rated power/energy capacity and RTE are upheld. As mentioned in the case of delivery delay LDs, failure to meet performance guarantees should have penalties or LDs associated. It is favorable to the project to lay out clear terms of non-performance, and the method of remedy or value of compensation in case a performance guarantee is not met. Usually, for availability non-conformance, a cost penalty is stated in the agreement. For not meeting SoH guarantees, suppliers may have the option to repair, replace, or compensate with agreed charges at the time of non-conformance – these will be agreed upon mutually with the asset owner. The ideal practice is to have penalties for such performance metrics with the supplier equivalent to those the project may incur to an offtaker as part of contractual performance requirements and associated LDs. Even if not for a one-to-one backup of such penalties, having clear and measurable metrics and reimbursement methods will save time and money spent in dispute resolution when an issue occurs. If there is a need to raise warranty or performance guarantee claims, varying levels of data to support the claim are required – ranging from hourly EMS data to second-level data on temperature and voltage from the BMS. Hence it is important to understand and enable data-logging needs to uphold performance guaranties, as well as raise the claim as soon as issues are observed.

Battery Supply Chain

Most of the Li-ion batteries in the global market are currently manufactured in Asia-Pacific – by the end of 2021, 90% of the manufacturing was done in China [54]. There is a global surge in announcements of battery gigafactories around the world because of the anticipated surge in demand for Li-ion batteries – primarily driven by the electric vehicle market accounting for about 80% of the demand now.

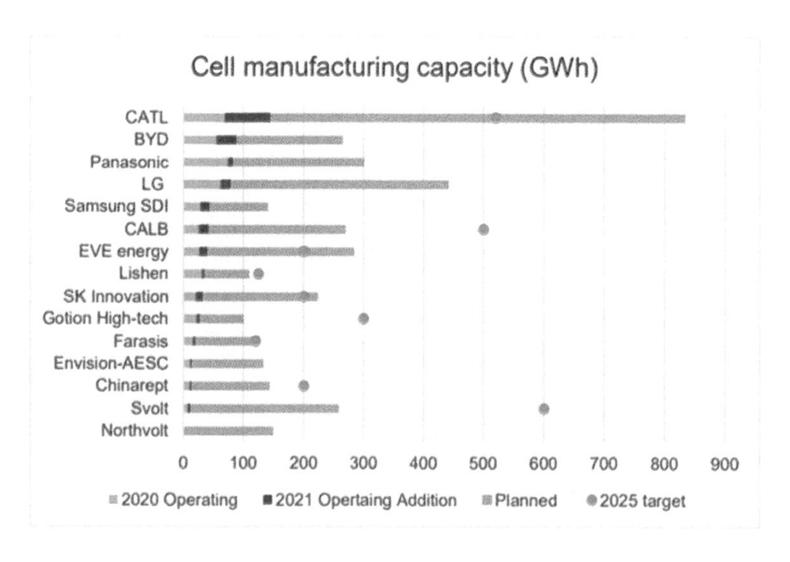

Figure 8-6
Global lithium-ion cell manufacturing leaders [54]

By the end of this decade, however, the Asia-Pacific region's share of manufacturing is expected to reduce to 69%. There are intensive plans to expand manufacturing in Europe by some of the leading battery cell manufacturers. This in-part is owing to the sensitivity of goods imported from China and the potential for tariffs to be imposed on them, similar to what had been observed in the PV module industry as touched on earlier in the chapter.

There is a noteworthy surge in battery manufacturing plans in the US: North America's share could expand 10-fold by 2030. This is largely driven by the Inflation Reduction Act (IRA) of 2022 which provides incentives in the form of subsidies and tax credits for domestic production of all components of BESS, as well as additional tax credits for developers to deploy home-made solutions (the IRA provided investment tax credits for stand-alone BESS for the first time). With this, US manufacturing is set to outpace Europe by the end of this decade, though Asian-Pacific countries will still lead the race.

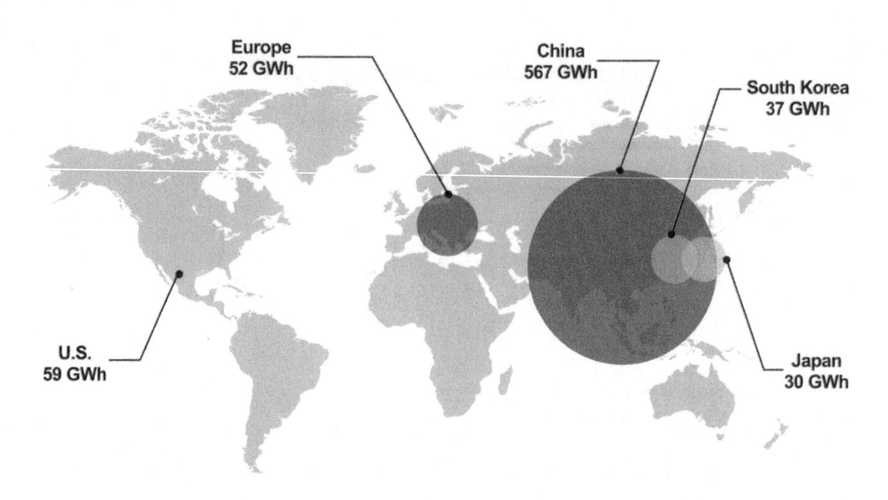

Figure 8-7
Lithium-ion battery cell manufacturing by country, 2021 [55]

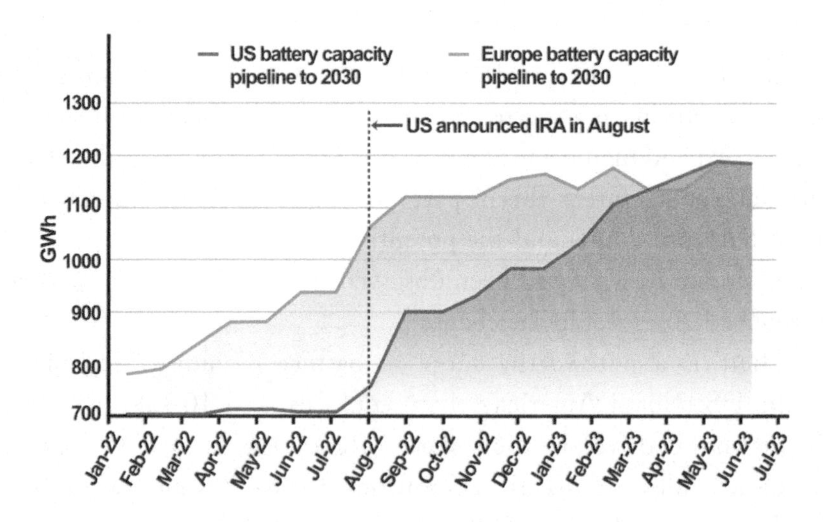

Figure 8-8
US lithium-ion battery manufacturing capacity increases post-IRA [56]

Like any industry, there is a dynamic market-driven balancing of supply and demand. The price for Li-ion batteries (and the cost of BESS) has steadily declined since 2010 as the industry matured, very similar to what was observed in the early days of solar modules. For the last few years, because of the successful surge in electric vehicle adoption (the same suppliers and factories cater to the EV battery market) as well as growth in the stationary storage industry, there was concern that battery supply may not be able to keep up with the anticipated demand. COVID-19 threw these dynamics into unprecedented spins with a sudden drop in buying of consumer vehicles and difficulty in availability of raw materials at expected prices. Over 2021-2022, lithium-carbonate, the primary mineral from which lithium is derived, saw an extremely sharp increase in cost over a short period of time. This led to a surge in Li-ion battery price – an increase year-over-year was unheard of up until that point. Suppliers indexed the price of batteries as a variable on raw material prices, which made BESS project economics unexpectedly complicated.

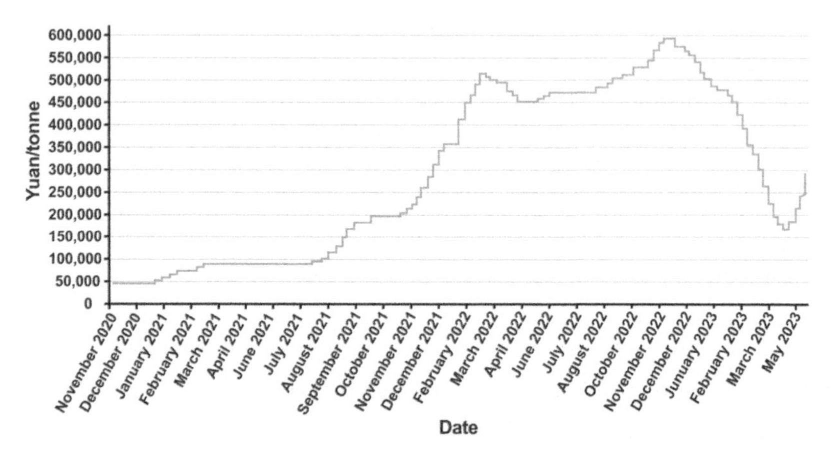

Figure 8-9
Lithium-carbonate spot price 2021-2022 [57]

This situation has since stabilized and lithium mineral prices have come down just as dramatically. Broadly, battery prices are expected to continue to decline because of swelling of planned supply – the market is expected to be flooded with batteries with competitive pricing given the production volumes being planned, and an extremely optimistic demand from electric vehicles.

Planning is the key to any project's success, and building BESS with myriad components and parties involved is no exception. This has come even more to the forefront in recent times as the renewables industry overall grapples with dynamic equipment lead times – something that was not quite the case before the COVID-19 pandemic. Battery systems – the containerized systems that most of the suppliers offer – require placement of orders between twelve and fifteen months prior to when the equipment needs to be delivered on site. As more BESS projects are being announced and developers are building out ambitious pipelines, this could grow quickly, so securing production slots with desired product providers will be crucial. The story is worse for BoP equipment such as transformers, medium-voltage switchgear, and control enclosures; this equipment does not get as much attention because they are fairly standardized but are indispensable to every BESS project. They have evolved quickly from having traditionally predictable lead times in the range of multiple weeks into multiple months and even years, throwing project schedules into disarray. At the time of writing this book, some equipment like high-voltage breakers is being quoted with lead times as high as five years. Supply constraints induced by the pandemic are a part of the puzzle that led to this situation. In the wake of the IRA, along with developers rushing to procure equipment for their project pipelines, utilities maintaining the transmission and distribution networks are also said to be bulk-buying equipment to prepare to integrate these projects onto the grid they manage. There isn't yet a clear path to alleviate this crunch for the industry as manufacturers of such equipment do not have enough incentive to scale up manufacturing to meet such a demand swiftly. To manage this situation, developers and EPCs with deep pockets are procuring these pieces of equipment in bulk to meet the needs of some of their projects in development. The challenge with this is that projects do not become firmed up until late stages of development as all the pieces fall together, and in some cases need customized equipment. Hence the inventory of secured/stocked equipment needs to be continually balanced with the dynamic project pipeline. Developers must start planning out construction schedules and procurement of equipment and contractors as early as three years before planned CODs and get creative with sourcing some of this equipment.

8.5 EPC Contract Structures

One of the earliest decisions to be made by the developer is which firms will be used to perform the EPC work on the project, and what scope each will perform to bring the project from financial close through to commercial operations. The options range from a **turnkey contract**, in which a single contractor performs 100% of this scope to a more distributed approach with five or more individual contracts for different portions of the EPC scope. There are a variety of different structures in between these two approaches. This section presents some of the more common contract types and the pros and cons of each.

A note on terminology: Whether turnkey or split contract, some in the BESS industry refer to the firm handling this scope as the **EPC Contractor**. This can be somewhat confusing, since in a turnkey contract, EPC includes all aspects of the project, while in a split-contract, the EPC contractor does not perform some critical aspects of the procurement, such as of the batteries themselves. For clarity, this book uses the alternate term **Balance of Plant (BoP) Contractor**, which refers to an EPC contractor that is performing a portion of the scope outside of provision of the major equipment such as the BESS, PCS, or transformers. It is common for BoP contracts to include provisions for certain large equipment to be procured directly by the project owner, known as **owner-supplied equipment**. This allows the contractor to negotiate directly for these components, avoiding paying a markup to the BoP contractor for this scope. However, the downside of this approach is that it requires the owner to take added responsibility for ordering this equipment correctly and ensuring that it arrives on time.

Part of the contracting structure decision has to do with economies of scale. Large developers who manage portfolios of larger projects tend to pursue multi-contract approaches, driven by their larger purchasing power and staffing. Smaller projects, or those built by firms with smaller teams, tend to use a turnkey approach.

Given that they often are contributing large portions of the funds to the project, large investors and lenders may have influence on the contract structure. Typically, their bias is toward using a large contractor, as it reduces risk on the project but may account for a larger cost. Certain lenders, such as the IFC exclusively finance projects that use a single EPC contract. Most financial institutions, especially those engaging in tax-equity transactions, accept the multi-contract approach so long as reputable contractors and suppliers are used.

Common Contracting Types

This section presents the three most common contracting methods: turnkey, EPC+BESS Supply (dual-contract), and multi-contract.

Turnkey

A turnkey contract is all-encompassing: The contractor takes responsibility for all aspects of the design, purchasing, delivery, installation, and commissioning of the project. This contractor may perform some of the work with their own staff (known as "**self-performance**"), but typically they will engage subcontractors for most of the work. This type of contract is relatively rare for utility-scale battery projects, since it results in the highest cost project, but has the advantage of having a single point-of-contact for the owner to hold the contractor to account. Although the contractor must balance their many suppliers, subcontractors, and other conditions to ensure an on-time delivery of the project, from the owner's point of view there is a single point-of-contact when things go wrong, or delays happen on the project – it is always on the contractor's shoulders to work this out.

In the early days of the BESS industry (roughly 2015-2020) it was common for battery integrators to offer turnkey EPC contracts. In practice this meant that the battery integrators supplied the batteries (the single biggest expense on the project) as well as hired a subcontractor to do the BoP work on the project. As the industry has expanded, and more contractors have gained experience in BESS installation, it has become rare for any battery supplier to take on the risks associated with turnkey projects. As of this writing, where turnkey arrangements are currently used, it is typically a large-scale construction or engineering firm that acts as the prime general contractor, with the battery integrator serving as their subcontractor.

The downside to the turnkey approach is that it nearly always results in a higher cost, with lower risk to the owner, as compared with a dual-contract or multi-contract approach. This is because the contractor always marks up the services provided by their subcontractors, to account for the added risk of performing the scope. For many projects, the increased cost is not worth the benefit.

In a turnkey contract, the owner may not have full visibility into all the contracts of the project since they are not a direct party. However, the prime contractor may voluntarily share the terms of their subcontracts, often in redacted form, to provide assurance to the owner of the quality and technical merit of the product.

BoP+BESS Supply Agreement

As of this writing, the most common contracting approach for BESS projects is to have two separate contracts covering the EPC of the project: a BESA for the purchase, delivery, installation, and commissioning of the batteries, and another contract for all other aspects of construction of the BESS. This approach has the advantage of simplicity – a project can manage two major contractors, limiting the points of failure, while also allowing the project to negotiate directly for the purchase of the batteries, which comprise the largest single line-item in the overall cost of any Project.

For projects using a DC-block BESS, it is common for the battery supply agreement to also include the PCS, MV transformer, and EMS, since many battery suppliers have standing supply agreements with transformer and PCS manufacturers that they have successfully integrated with in the past. If using an AC-block product, such as the Tesla Megapack or Sungrow PowerTitan 2.0, the PCS supply is already included as part of battery supply, since it is integral to the battery container.

The BoP contract in this structure includes an extensive scope, such as:

- All design work outside of the BESS components themselves and other equipment furnished by the BESS supplier
- All permitting and compliance work that has not been performed by the owner
- Civil, electrical, structural, and mechanical BoP work
- Communications systems to interconnect the BESS, PCS, controls system, and power plant controller (if applicable); also interconnecting these systems with other SCADA systems, if applicable
- Continuous coordination with the BESS supplier to ensure aligned schedules and scope alignment

The design and construction of the new battery substation, aerial or underground transmission lines, and any upgrades to the interconnection substation is specialized work requiring HV-specific equipment, gear, and expertise. This scope may be included under the BoP contract for smaller scale projects (50 MW and less), while for larger projects there is often a separate agreement. Even if the scope is included within the BoP contract, it is typically performed by a specialized HV subcontractor. The HV contract also encompasses the critical scope of coordinating the substation work with the utility and/or ISO/RTO.

The biggest risk of a project with multiple large contractors is that a delay or issue with one contract can have large effects on the other. The nature of contracting is that if there is ever a chance that a contractor can blame someone else for their cost overruns or delays, they will usually try to do that. Because of this, it is important for the owner to take a proactive approach to ensure that there is proper alignment between the contractors. This is accomplished by:

- Open and continuous communication between all parties from the contracting phase through to substantial completion
- Prior to contracting, alignment of milestone and overall project schedule so that the expected date of delivery of the BESS units and other key components is aligned with the BoP schedule, including some buffer to allow for minor delays or acceleration of the project schedule
- Delay damages for the most critical dates to provide a strong disincentive for the project to go over schedule; in some cases these can be coupled with incentive bonus payments for early delivery
- Clear definition of responsibilities, including for items which need to be conducted jointly (covered in depth in Section 8.5)
- If and when delays, problems, or incidents occur, addressing them early and in a forthright manner to maintain trust and get the project back on track
- Prioritizing a safety-first culture to avoid near-misses, incidents, lost-time injuries, and fatalities on the project; in addition to the trauma of these incidents, they also may have grave effects on the project budget and schedule
- A third-party construction management firm overseeing the EPC by the Owner or for the Owner

The most common schedule for the dual-contract approach is for the BoP contractor to complete works through to the civil phase, including the foundations for the BESS, PCS, and MV transformers. At this point, provided the schedules are properly aligned, the BESS units and other equipment supplied by the BESS contractor are delivered. On smaller projects this typically happens all in one phase, while on larger BESS installations the deliveries may be tiered to arrive in multiple groups as the foundations are completed.

When the major equipment has been delivered, it passes into the hands of the BoP contractor for them to connect the units and ready them for commissioning. Since this milestone involves a change in responsibility, it is critical that there be a full inspection of the units after delivery by all parties – this is because if there is any damage caused to the units in shipping, this should be identified prior to handing them off, since this could be blamed on the contractor rather than on the shipper.

Following the connection of all of the major equipment, the major suppliers return to the site to begin the process of commissioning, which is covered in detail in Section 8.7.

Multi-contract

While the dual-contract approach has become popular, some developers find that a large EPC contract can be broken up still further to result in additional cost savings for the project. Some of the common divisions with multi-contract approaches are splitting out the civil BoP, electrical BoP, and substation / transmission work. Some developers go so far as to contract for the enclosure in which they will locate the battery. This approach may result in the lowest cost project, since it allows the developer to negotiate directly with each supplier to get the best pricing. In practice it is more commonly done by larger developers, since they have the purchasing power with each supplier to sign supply agreements that cover multiple projects. The downside of having several contracts is that there are more parties involved, internal and external, and an increased number of risk points for the project, and more interfacing between contracts which could result in disputes or scope gaps. From a finance point of view, some lenders balk at having so many parties involved, since it increases the overall failure points to the project.

As the industry matures it is likely that more large developers may go the route of using the multi-contract approach, as projects get larger and lenders become more comfortable with the associated risks. At the same time, however, BoP contractors are becoming more competent at building energy storage projects and have more experience in how to successfully execute them.

Division of Responsibilities

The engineering, procurement, and construction process on a utility-scale BESS project involves hundreds of individual tasks and may have dozens of contractors working on site at a time. Even if the turnkey EPC contract approach is used, there are many other stakeholders involved such as the Owner, Owner's Engineer, Independent Engineer, AHJ, offtaker, and landowner.

To avoid any confusion around who is responsible for what tasks, it is common to have a matrix listing all major tasks, which is known as a Division of Responsibilities matrix. This is important both when a project is being bid, as well as throughout the execution of the project. Ideally an identical version of the DoR is included as a formal attachment to all the contracts used on a project, so that no one can claim that they were unaware of what was expected of them.

In some cases, the DoR is known as a **RACI matrix**, which stands for Responsible, Accountable, Consulted, or Informed. The simple DoR has only an X indicating who is responsible (and in some cases it may be multiple parties, for tasks done collaboratively), while the RACI matrix has different letters marking the various levels of involvement on each task. Where possible, it is preferable to have a single "A" per line-item, so there is no doubt about who owns each task. It may be common to have multiple parties who are consulted or informed at each step.

Many DoR matrices involve multiple parties and must be performed collaboratively between various entities. For example, the performance testing of a BESS is usually coordinated by the BoP contractor, but must be closely coordinated with the battery supplier, the offtaker, the IE, the Owner, and possibly the Owner's Engineer. In those cases, it is best to mark the primary party as responsible, and the others as Accountable, Consulted, or Informed (if using the RACI structure). Most DoR matrices have a Remarks or Notes section, in which further breakdown can be given for shared tasks.

Most battery manufacturers have a DoR matrix for their projects, while other Owners and BoP contractors may have their own standard version. One of the first steps in the scoping phase of a project is to select the DoR that will be used, and to walk through that document with all parties to ensure that everyone is aligned about who is doing what.

Appendix A shows a typical DoR that outlines high-level tasks for a battery project using a dual-contract approach. These vary widely based on the project, jurisdiction, and developer.

8.6 BESS Construction Process

The construction stage is the last in a long process of development work that culminates with a project entering commercial operations. It involves many parties working together to perform a successful execution – contractors, subcontractors, OEMs/suppliers, financiers, and others. As with many other parts of development, proper preparation is key to successful construction of a project. The more time in the development process spent assessing the site, negotiating comprehensive contracts, and thoughtfully engineering the project, the more smoothly the construction will proceed.

That said, the construction process will inevitably run into snags. Unexpected subsoil conditions, adverse weather, and delivery delays are just a few of the many issues that can arise on a project. The management of these conditions requires an experienced construction firm with a proven track record and on-site team.

A typical battery project's construction may last over 6-12 months, sometimes longer depending on MW/MWh size, involving a few prime contractors, and dozens of subcontractors, and may require extensive planning to stay on budget and schedule. While every construction project is unique, this chapter lays out the typical steps for contracting, mobilizing, building, testing, and commissioning an energy storage facility.

For simplicity, this section assumes that the project is contracted with a dual-contract approach, as described in the prior section.

BoP Contracting Process

The BoP scope may vary widely from project-to-project, as discussed in Section 8.5. The following description assumes that the BESS OEM provides the BESS enclosures and PCSs, while the BoP (EPC) contractor performs all other work on the project.

The first step of any successful BoP contract is the negotiation of the contract documents. Most large construction firms have draft construction contracts that they use, some of which are hundreds of pages long, covering all the administrative, legal, and commercial terms of the contract. Some developers may have their own versions of these contracts. These contracts include provisions that apply to all construction projects, such as mobilization, scheduling, weather delays, payment, acceptance, and contract modifications. In many countries outside the US these provisions are known as **Employer's Requirements (ER)**. Most contracts are structured so that these provisions are in the body of the contract, while the project-specific aspects are contained within the appendices to the contract.

Many volumes have been written on construction contracting, which is a specialization within the construction, legal, and project management fields. Here are some of the most important aspects of a successful construction project:

- **Scoping**: The most critical aspect of any contract is the scope. This describes precisely what is required of the contractor, and what will be handled by other contractors, or by the owner themselves. Scopes may be as simple as a bullet point list or may consist of a spreadsheet which details all of the steps and sub-steps required for the contractor to perform. Since many scopes are performed collaboratively between other parties, such as the battery integrator, the utility, and the Owner, it is common to have a DoR matrix to determine how these scopes interact. The DoR is covered in depth in Section 8.5. A clearly written scope will help to avoid many issues down the road, from pricing to contractor relationships to costly procurement mistakes.
- **Owner-supplied equipment or owner-furnished equipment**: As mentioned in Section 8.5, it is common for the contract to define any equipment that will be furnished by the owner, meaning that it is to be furnished and delivered to site by another party. In most battery projects, the battery itself falls in this category. Other common owner-supplied equipment includes the PCSs, high-voltage transformers (HVT, MPT, or GSU), MV transformers (MVTs), the SCADA/controls system, and the substation major equipment.
- **Payment terms**: The time and amount of payment for the contractor is a critical element of any contract. Most contracts include some down payment for the work, payments throughout the construction process, and a retainer payment to be paid some months after final acceptance of the work. Some projects

may have the construction payments invoiced on a monthly basis, while others may be invoiced only upon the completion of certain milestones in the construction process. For larger projects, there may be a requirement from the lenders that an IE perform site inspections.

- **Scheduling provisions:** All utility-scale projects should be managed using commercially available scheduling software to track the hundreds or thousands of individual steps in a battery project. Two of the more common scheduling platforms are Microsoft Project and Primavera. It is common for large contractors to employ a full-time scheduler to track the progress of the contractor. The schedule helps to determine the **critical path**: the series of steps on the project that, if delayed, will result in delay of the overall delivery of the project. This allows the contractor to appropriately prioritize the design, procurement, and construction phases of the work to minimize delays.
- **Force Majeure (FM)**: This provision describes extraordinary events that may occur during the project which are outside of the contractor's control. Typical FM provisions include acts of god, war, natural disasters, pandemics, or government intervention. If the contractor invokes Force Majeure, they request that the project not be held liable for the delay, so the acceptance of this requires formal acceptance by the owner and is often the subject of extensive discussion.
- **Excused delays**: Most contracts contain provisions that allow the contractor to push back the schedule if certain events occur, outside of those listed in force majeure. This could include excessive adverse weather, delays in delivery of certain components, or delays in the Owner providing a critical decision or piece of information. Excused delays typically require written notice from the Contractor and must be accepted by the Owner.

- **Delay LDs** : In the case of late delivery of the project, in the absence of FM or an excused delay, the contract may have terms to penalize the contractor for the delay. These are typically assessed in dollars (or equivalent) per day. While the contract gives the Owner the legal right to withhold these payments, in practice it is common for the Owner to forego delay LDs, or to reach a deal in which they receive some other benefit from the contractor in lieu of LDs.

- **Performance LDs**: Similar to delay LDs, performance LDs are a penalty assessed to the contractor to account for reduced performance of the system. In energy storage, it is common to see performance LDs that are pegged to shortfalls in energy capacity, power or round-trip efficiency (RTE).

- **Testing and commissioning**: Since the testing and commissioning of the system is one of the last elements of a system to be conducted, it's critical that the testing process be well-defined. In energy storage, this typically means the provisions for the energy capacity testing, RTE testing, or network performance tests. This section requires close collaboration between the provisions of the battery supply agreement (BESA) and long-term supply agreement (LTSA)

- **Substantial completion (SC)**: This section describes what the final requirements are for the contractor to receive payment. Beyond testing and commissioning, there is often a long list of requirements to reach SC, such as the SC for certain subcontractors, acceptance by the utility/ISO/RTO, IE certificate or site inspection, and site cleanup and mitigation. It is common that even after SC, the contractor may have a punch list of items remaining to finalize construction, but these are typically minor tasks not central to the operation of the project.

Construction Works

The construction process is typically considered to begin when the project issues **Limited Notice to Proceed (LNTP)**. Not all projects use this intermediate step, but it is often granted to allow the contractor to proceed on critical steps prior to mobilization to the site. It may include provisions for the contractor to begin placing purchase orders for key equipment, moving the design process forward. The design work is usually covered under the BoP contract, which is discussed in detail under section 8.1. LNTP is often granted immediately after the project reaches financial close (FC).

Following LNTP, the project typically receives Notice to Proceed (NTP). This is the green light that allows the contractor to mobilize and begin site work. The first steps are typically securing the site with fencing, clearing and grubbing the site, and performing cut and fill work if required. A field office is commonly erected on the site for the prime contractor and main subcontractors to work from. Once these measures are put in place, the battery plant is ready to be built.

Civil

Civil construction for a battery plant is focused around preparing the base that will be used for the energy storage containers, PCSs, transformers, and other equipment. Foundations will vary based on soil type, but the most common structures used to support the BESS units are concrete slabs and driven piles. For concrete slabs, the area is first excavated to remove the subsoils, and then fill material is brought in and compacted to serve as a solid base for the concrete slabs. Proper compaction of this subbase is important to ensure that there is no settling of the battery units, which are quite heavy, over time. Where driven piles are used, there is typically no requirement for sub-base preparation since the civil design will consider the driving requirements for the piles in the existing soil.

After the sub-base is prepared, conduits are run in their required locations, and trenches are cut where cable will be buried. This work is critical since any inaccuracies in the placement of conduit or trenching can result in large delays later in the project. This is because most conduit is encased in concrete, making it very difficult to modify once built. Wherever possible, it is recommended to check the conduit and trenching plans multiple times with all stakeholders to ensure all parties approve.

Following the conduit work, the concrete slabs can be poured. Some battery integrators prefer individual slabs for each battery storage container, while others may call for several large slabs which hold many containers. Where possible, large slabs are more efficient since they allow for larger concrete pours.

Some projects use a single civil contractor, while others may use separate subcontractors for earthworks, concrete, and trenching. For the substation, there are many concrete slabs required, as well as some galvanized steel structures used to support the aerial conductors, rigid buses, and other equipment. All of this is civil work, although some contractors prefer a single subcontractor to perform all work within the BESS substation, including the civil scope.

Although they are becoming rare, some BESS projects are built in purpose-built buildings in which battery racks are installed. Although this approach was more common prior to 2020, safety concerns have tended to make outdoor-mounted enclosures the preferred method of construction. For projects that use a purpose-built building, there are several added complexities, such as building HVAC and fire control strategies.

As of mid-2024, some BESS projects have begun to be built on compacted backfill, rather than on piles or slabs. This method is feasible only where the geotechnical, hydrology, and seismic design conditions allow, and where it is permitted by local code. Where it can be used, it reduces construction costs by eliminating concrete, pilings and even the underground conduit, which can be replaced with UV-rated surface conduit or raceways to connect the various components. As of this writing, this approach is still rare among utility-scale systems.

Delivery of Major Equipment

When the foundations have been completed, the next major phase is the delivery of the BESS equipment. Most dual-contract structure projects require the battery integrator to deliver the storage enclosures to the site and unload them onto the foundations prepared by the BoP contractor. This interface of scope is critical, as it represents a passing of custody from the battery integrator to the BoP contractor. Typically, the BESS units will have some method of tracking their acceleration during shipping – this may take the form of a data logger, or a sensor which can change color if a significant shock is incurred. It is also common for temperature tracking to be attached to some of the battery units, to record the state in which they were shipped. Once the battery manufacturer and BoP contractor have both agreed that the enclosures have been delivered in good order, the BoP contractor must secure them in place. This may be done by bolting the enclosures in place, or in some cases welding them to the foundation.

At this stage of construction, the BESS units have not yet been back fed with power, nor has their auxiliary power or communications structure been connected. The acceptable conditions vary between manufacturers, but typically the containers can withstand high ambient temperatures (commonly up to 50°C) since they are neither charging nor discharging. Most manufacturers have a certain period in which they can stay in this state without energization. In some cases that time can vary if the units are back fed with auxiliary power or communications.

At this stage the battery integrator is not involved in the construction, as it is the responsibility of the BoP contractor to connect the units, at which point the integrator can return to begin their commissioning.

It is important to note that while BESS units are frequently shipped lower than full SOC, even a BESS at 5% SoC often has voltages between rack terminals of over 1000 VDC. These products are designed to make the terminals far apart to minimize safety risk, but within the battery modules and racks, there is always some element which cannot be fully disconnected. For safety purposes, battery units should always be treated as if they are fully charged, with proper personal protection equipment (PPE) and procedures to minimize the hazards to workers.

In addition to the batteries, all other major equipment is also delivered on site at this stage of construction: PCSs and MV transformers (which may be delivered together on a single skid), control enclosures, auxiliary transformers, and GSU transformers are the large equipment typically included, as well as several other smaller components. Each of these pieces of equipment has its own acceptance process which must be followed to ensure that when the BoP contractor takes control of the equipment, all parties acknowledge the state it was in when delivered.

Electrical

Following equipment delivery, the next phase is to connect all the various components to work as a fully functional BESS. The most important of these is pulling conductors through conduit and terminating those conductors at the appropriate locations.

While this may seem a simple task of plug and play, it must be performed in alignment with codes and standards, the EOR design and specification, manufacturer's recommendations, and in a way that will be safe for everyone involved. The primary electrical work is the connection and termination of the various cables which are run through the conduits into the sections of the various pieces of equipment.

Most systems include both DC and AC conductors, although with systems that include an integral PCS, no DC conductors may be required.

Although it is common to think of a battery as being able to power itself, most DC battery enclosures are not designed in a way for them to be energized if they are not back fed with power. This is because the electronics which monitor and control the system (the BMS and EMS) are run on low voltage (typically 120-900 VAC) power which must be interconnected before any of the systems come to life.

All electrical connections must be made without power present – to be safe, some contractors require the conductors be grounded to avoid any inadvertent connections.

Fiber communications wiring is typically performed by a subcontractor specialized in this field. Most battery manufacturers have detailed requirements for the procedures of landing the communications wiring, and at this stage they will have signed off on the communications wiring plan, including interconnection with any SCADA terminals or communications with outside entities, such as the ISO.

The AC conductors between the output of the PCS and the MV transformers are typically designed to be as short as possible. This is because they carry high power at low voltage (typically 600-800 volts), and which can result in losses of power. These are typically less than 1 m, and sometimes are done through solid or braided busbars rather than through traditional conductors.

Typically, an underground MV collection circuit will feed many MV transformers. This circuit emerges within the BESS substation to feed a circuit that connects to the HV transformer, where the voltage is stepped up (or down, depending on charge or discharge) to high voltage. The MV lines which connect the MV transformers are often referred to as **home run** circuits, or simply **home runs**.

The terminations of cables are all common failure points in the connection of a system. All terminations require meticulous care to ensure they are done in accordance with manufacturer's instructions. This includes things like using calibrated torque wrenches to reach the required torque, application of dielectric grease between fittings, and in some cases working in tents to ensure that dust from the construction site does not contaminate the connections. Most subcontractors require systems to document the connections, through pictures or in some cases automated reporting from their tools. Additionally, the connections are tested, and the results recorded and verified. All these procedures require diligent quality control (QC) and quality assurance (QA) to ensure that they are performed correctly, even when repeated hundreds or even thousands of times throughout a project.

Mechanical

Batteries have few moving parts, but there are a few critical mechanical assemblies that form part of the battery system. These are primarily associated with the battery enclosures, PCSs, and transformers.

Except for the rare indoor BESS projects, the HVAC components are typically installed as part of the integral system. Some products have them more visible, such as air-cooled units which are mounted on the outside of battery enclosures, while liquid-cooled units have their compressors and pumps concealed within the enclosure. Any mechanical systems within a piece of equipment are typically the responsibility of the manufacturer, but it is important that the project be aware of these systems, given their importance to correct operation of the plant.

For battery products that utilize liquid cooling, some manufacturers ship their units full of coolant, while others may ship them without coolant, or at some partial fill level. If they require filling, the systems may need to be fed with auxiliary power before this can happen.

PCSs are typically cooled with forced air provided from fans, while some models incorporate hybrid air and liquid cooling. Transformers may have combined air and oil cooling depending on their size and function. These methods are covered in depth in Section 4.3.

Battery systems are most concerned with the cooling part of HVAC, since components produce large amounts of heat when operating near their working limits. However, in areas with cold winters, heating systems may be required to ensure that no components fall below their allowable temperatures. Battery cells can have drastically reduced performance at low temperatures, and permanent damage can be inflicted at extremely low temperatures, which may risk voiding the warranty (typically at -30^{0}C or below). In these locations batteries and other products will be provided with integral heating systems to ensure they are kept within their operating temperature range. In some cases, these may be augmented by heaters within buildings or built into concrete equipment pads.

Besides HVAC systems, other mechanical components of a BESS plant include the following:

- Water pumps, water towers, and water pipes which may be required as part of the fire suppression system on site
- HVAC units for any occupiable enclosures, such as control rooms or spare parts storage areas
- Mechanically operated switchgear (typically only in large substations)
- Mechanically operated gates

8.7 Testing, Commissioning, and Start of Commercial Operations

The final step in the construction of a battery system is achievement of COD (Commercial Operations Date). A wide variety of tests and checks must be performed to ensure that the system can reach COD. These include tests of individual components, tests of the system to check overall capacity and integrity, and tests required by the offtaker, ISO, and/or interconnecting utility. Testing is often a critical portion of the financing agreement for a project, as it is often a milestone which will allow for payments to flow or for the utility to accept a project. Conversely, should a project fail a test, this can result in LDs, delays, or costly remedies.

Testing is typically a collaborative process between the suppliers, owner, EPC, offtaker, grid operator, and other stakeholders. Testing procedures vary widely from project to project, but the following six categories represent the typical steps used to commission a utility-scale BESS project.

Mechanical Completion

Mechanical completion (MC) is a milestone which indicates that all components have been physically delivered in place, connected with their respective conductors and communications wiring. Any mechanical systems such as water pumps, HVAC units, or gates will have been installed and tested at this point. For conductors between pieces of equipment: ensuring that all point-to-point testing has been performed on AC and DC circuits, all labeling is done appropriately, and all documentation is submitted and approved by the BoP contractor.

MC often has a specific legal definition that is referenced in the financing agreement and BoP contractual documentation. This often requires that the Contractor and/or IE submit a certificate attesting that the system, or portion of the system, has all equipment installed and all required documentation has been provided. Although MC is common on wind energy projects, which include more mechanical components, many BESS projects do not include an MC milestone.

Cold Commissioning

Once MC is reached, the controls systems for each component can be brought to life to check that the unit is in good working order and ready to connect to the power grid. The process of performing these checks is known as **cold commissioning**. This name refers to the fact that this testing is done prior to energizing, or **backfeeding**, the system from the main grid power. Despite the name, cold commissioning of BESS units involves many components functioning at dangerous voltages that could easily harm workers on the site if proper precautions are not taken. For the BoP equipment, cold commissioning is conducted by the BoP contractor, or often by their subcontractor. For the individual components such as batteries, PCSs, or controls system, this is typically the stage where the suppliers return to the site to begin commissioning their individual components. For cold commissioning of batteries, the enclosures typically require auxiliary power. This can be fed one of two ways: via the grid, through auxiliary transformers or via temporary power provided by on-site generators. Grid power is always preferable, since it is more stable and requires less temporary work; however, in some cases the construction schedule can be compressed by using the temporary power option. In all cases, the battery integrator should be consulted to ensure that the process complies with their requirements and recommendations.

Nearly every step of the cold commissioning process requires a large amount of documentation which shows the certified results of the tests, signed off on by a representative of the manufacturer or contractor. Despite the name, this step requires some components to be energized and still must be conducted in a safe and careful manner.

The specific steps taken during cold commissioning varies from project to project and among different manufacturers, but typically includes the following steps:

- Verify that the unit is in good physical condition
- Verify that the unit does not show any faults or defects as a result of shipping or installation
- Confirm that the units have internet connectivity
- Confirm that the unit's cooling system is in good working order; for liquid-cooled systems, this includes checking the level and circulation of coolant in the pipes and chillers

- Conduct basic health checks such as ability to circulate current, verification of SOC, and SoH of the system
- Check that the system has sufficient cells that are appropriately balanced across the modules, racks, and enclosures
- Verify that the PCSs are communicating correctly with the EMS and BMS, and that all portals are up and running
- For the communications system: Confirm that all individual components are connected correctly, that the system is connected to the internet portal, and ensuring that any outside connections (such as to the ISO or fire department) are working properly
- For the substation: Confirm that all components are commissioned, safety checks are complete, and communications and metering are completed
- Perform electrical test procedures such as grounding, MV cable checks, MPT transformer checks, CT/VT checks, and relay checks

Hot Commissioning

Following the cold commissioning of all components, the system may be energized by the grid. This normally happens first on the AC side, where the MPT is energized, energizing the MV bus. If it hasn't yet been provided, auxiliary power is fed to the auxiliary transformers. Then, one by one, each MV feeder circuit is energized, feeding power to the AC side of the PCSs. On the BESS side, individual racks are closed into the DC bus, which typically operates at 1500 VDC nominal for utility-scale systems. Each DC bus is then closed into the DC side of the PCSs, at which point the PCSs come to life and begin the process of synchronizing with the grid.

Throughout the energization process, the BMS and EMS are all checking to ensure that there are no fault codes triggered when the connection has taken place. Given the complexity and size of these systems, there will inevitably be some issues that arise. Common issues are quality issues in parts, missing components, and damage during installation. These may often require an interdisciplinary approach between all parties to resolve them, so solid communication between all suppliers, contractors, and subcontractors during this phase is critical to avoiding delays. It is common that it may take several weeks until all faults are cleared, and the system can be energized entirely.

Once all components are safely connected, the system can begin to execute some of its commands. The EMS typically is instructed to begin sending small amounts of power charging and discharging the grid. This process is conducted jointly with inputs from the BESS OEM, PCS OEM, BoP contractor, ISO, utility, and owner. Eventually the system will be able to execute commands at full power, demonstrating that the nameplate power rating of each battery circuit can be held in the charging and discharging direction. This is typically first done in stages – first with individual PCSs and all the batteries behind them, then together with an entire feeder, and lastly by the complete system.

Commissioning is a complex step of project execution which requires clear coordination between the battery technicians, PCS technicians and BoP contractors to ensure all the equipment is connected and synchronized correctly. The time it takes for commissioning and issues that can arise are often underestimated – it is important to allocate at least three months to complete this step rigorously.

Final Performance Testing

Once all individual components have been tested on the system, the final performance testing can be conducted to demonstrate that the system is able to do what it was designed to do – to fully charge from the grid until the system reaches 100% SOC, and to discharge it at nameplate power to the power grid. Testing is critical for many reasons

– first, it will indicate the system's ability to perform in the markets it is operating in, which is the key driver of project revenue. Second, every battery supply agreement has clear requirement for the system to demonstrate its performance metrics; without reaching these requirements, the supplier must either repair or replace components of the system, or in some cases pay LDs, as prescribed by the contract. Last, most offtake agreements require the project to demonstrate certain performance metrics – most commonly these include discharge of the maximum power level for the contractual duration of the battery, and in some cases includes demonstration of the RTE.

Even if test procedures are described in the contracts, it is common for the BoP contractor to submit a test plan to both the Owner and offtaker 2 to 3 months prior to conducting the test. This allows for a mutual agreement on how the test will be conducted, discuss any side effects of the test, and for the grid to prepare to receive the energy from the test. For larger grids this is typically not an issue, but smaller grids may need to take special precautions to ensure that the large power and energy of the test does not disrupt grid service. In some cases, systems are tested with one portion of the BESS discharging and another simultaneously charging, to minimize the effects on the grid; however, this approach should only be used if necessary, as it does not allow the system to demonstrate its ability to simultaneously discharge all battery units.

Test procedures vary widely from project to project, but tests are a variation of the following procedure, which is used to test the energy capacity, the power capacity, and the RTE of a project:

1. System is discharged down to 0% SOC, with proper balancing to ensure that all racks and enclosures are at similar voltages
2. The system charges at its rated power from the grid, typically measured at POI, until the SoC reaches 100%. Typically, this charging duration is longer than the contracted duration of the battery, since BESS losses require the system charge up with

additional energy to ensure that it can discharge the full power for the contractual duration.

3. The system rests for one hour.
4. The system discharges at its rated power, as measured at the POI.
5. The system is run until it reaches 0% SoC (which is defined by when the voltages of all battery enclosures drop to their lowest limit for discharging)
6. The system then records the discharged energy at the required metering point. For the offtaker, this measurement is adjusted based on the agreed-upon losses between that meter and the HV point of interconnection.
7. RTE is calculated by dividing the discharged energy by the charged energy measured at the same point. Depending on the system architecture, this may need to be adjusted to compensate for auxiliary loads both charging and discharging.

Using the formulas and equations defined in each contract, the test results in an energy capacity and RTE for the system. Sometimes multiple contracts reference the energy and RTE percentage at a certain point in the system (most commonly the POI), while in other cases different contracts may reference different points. Regardless, the contracts define a minimum passing criteria for the system to be determined as functional. If the system has met this requirement, then it can proceed to finish the other remaining requirements and move toward declaring COD.

In the case that the system has failed a capacity test, it is normally given a chance to retest. If the system missed its performance target by a small amount, sometimes the test can be re-run without updates to the hardware, minor changes to the balancing or controls to optimize performance. If the system missed the target by a larger amount, the contractors must consider the root cause of the miss and develop a plan to re-test the system.

In some cases, the project is allowed to pay LDs for a missed target, normally defined in the contract in $ / kWh or $ / percentage-point of RTE below the required values. In these cases, there are typically back-to-back requirements for LDs, in which the BESS supplier and/or BoP contractor will owe damages equal to or exceeding the values assessed by the offtaker.

Substantial Completion

The final step in the EPC process is the project's achievement of **Substantial Completion (SC).** This is a term typically defined in the financial agreement governing the project, as well as the EPC agreement and possibly some other contracts governing the project. When substantial completion is released, the project has typically reached COD, all required approvals from offtakers, AHJs, and the IE have been received, and the path is clear for the project to transition into full-scale operations. There may be some final minor steps, such as site clean up, or perhaps some required documentation that needs to be received. These final items are gathered in a punch list which is agreed to by all parties as the final list of things that are required for all parties to agree that work has been completed. Typically, the terms of SC will stipulate a period (such as 30 or 60 days) in which all punch-list items must be completed.

There are often cash flows associated with the achievement of SC, such as the project's being able to draw on its last tranche of debt, or a large payment flowing to the contractors on the project. Some provisions of contracts include a "pay-when-paid" clause, in which the final payment will only be received by a given contractor or subcontractor when the payment is received by the entity higher up in the hierarchy of the project.

SC as defined in the financing agreement is often a difficult milestone to achieve, as it may reference many other documents – grid requirements, the EPC agreement definitions and their subsequent subcontracts, and close-out documentation such as as-built drawings or volumes of test reports. It is common to create a spreadsheet to track all the requirements of SC, which can periodically be checked on as they are completed.

Once SC is reached, most of the on-site crews will have demobilized from the site, although there may still be a couple individuals on site to complete the punch list and assist in the transition to operations.

Transition to Operations

After reaching SC, the project transitions to managing an operating storage asset. This may vary widely based on the given project, but typically involves some amount of training – from the battery supplier's team to the service provider and operator, and the operator to the fire department. There is often a transition period where there is some close coordination between all parties.

Batteries, as with many complex systems, often have issues described as a bathtub curve, which implies that most issues happen within the beginning of operation, or the end. The initial portion of this period is known as a **breaking-in, or teething period**, and often plays out with nuisance trips or minor issues that require both hardware and software fixes. In a project that goes well this period can last as short as 30-60 days, while in other projects it may go on as long as the first year of operation. Some issues are always to be expected, and the project will benefit from an amicable relationship between the key parties to resolve these things as they arise.

In most EPC contracts there is typically a provision for some form of retainer to be paid, which happens from 1 to 3 months after SC. This retainer is to give the owner some leverage for any final construction-related issues that may arise that were not identified when the punch list was composed. Retainage is normally around 10 to 15% of the contract payment for the work. Once retainage is paid, any claims need to be made under the workmanship or equipment warranty provisions of the contracts, or under the long-term service agreement (LTSA) if applicable.

Chapter 9 reviews in-depth the requirements for the successful operations and maintenance of a utility-scale BESS project.

8.8 Usual Suspects

This section provides a list of some of the more well known EPC firms operating in the US and Canada. They are divided between turnkey and BoP contractors, although some may provide services across both of these sectors.

Turnkey EPC firms

Major companies that undertake turnkey projects in the US and Canada are listed below. The statistics on projects completed are as reported by the companies on their websites.

AECon: One of the leading Canadian electrical contractors, Aecon provides end-to-end services for commercial and residential BESS and growing experience with utility-scale systems.

Burns & McDonnell: An EPC firm that has been in operation since 1898 serving multiple industries, Burns & McDonnell has over 3 GWh of energy storage project experience as of 2024. They are able to provide end-to-end services for energy storage projects.

Bechtel: A multi-national contractor that has also been in operation since 1898. Bechtel is able to provide complete permitting, engineering, procurement, and construction services, and has a growing arm that can execute turnkey renewable energy projects.

Kiewit: A leading global construction firm, Kiewit has led some of the world's largest infrastructure projects. Their energy practice provides engineering and construction services for various storage technologies including wind, solar, battery energy storage, pumped storage, and hydrogen.

Mortenson is on of the top 25 builders in the world. They have divisions that build projects such as sports facilities, data centers and healthcare, in addition to renewable energy. Specific to energy storage, they have completed over 13 GWh worth of BESS projects.

RES America is a company focused on clean energy and transmission and distribution infrastructure construction. Beyond construction, RES also offer asset maintenance and performance management services post-COD. By the start of 2023, RES had built over 23 GW of new renewable energy capacity globally.

Balance of Plant Contractors

The following contractors are also large firms but are more often used as a BoP contractor, installing all components outside of the BESS, and possibly other major equipment.

Blattner: A diversified energy storage contractor, Blattner has capabilities on project feasibility, engineering, pre-construction, and construction of renewable energy projects. They have business units that can cover wind, solar, hybrid PV+Storage, and standalone storage BoP. Blattner is owned by multi-national construction engineering firm Quanta Services.

Black & Veach: B&V is a global EPC firm with an extensive renewable energy practice. They have experience in wind, solar, and BESS projects, and are a global company with operations around the world.

Blue Ridge Power: Blue Ridge Power has completed over 7 GW of solar projects, making them a significant BoP contractor player in the US solar market. They are primarily operational in the Southern US.

Borea: A leader in the Canadian renewable energy industry, Borea is responsible for the realization of nearly one-third of all renewable energy development opportunities in Canada. They are actively growing their US practice.

Dashiell: is a leading national provider of high and medium voltage electrical services to the electric utility, power generation, industrial, and renewable; the company has recently started taking on the full construction scope of BESS projects.

Jingoli Power is an EPC company offering a full spectrum of services with experience building solar and energy storage projects in the US.

McCarthy is a full-scope contractor that has completed over 4 GWh of BESS projects. The company also offers operations and maintenance support. They are based in St. Louis, Missouri and active across the US.

Mill Creek Renewables: MCR is an engineer and EPC firm focused on utility-scale solar PV and BESS projects across the United States. MCR's team has worked on over 4.0 GW of photovoltaic (PV) assets.

Moss Construction: Moss is an experienced BoP across solar and storage utility-scale projects. They are active across the US but primarily in the South.

Rosendin: Rosendin is largest electrical contractors as well as a full-service contractor with over 2 GWh of BESS under construction. They are one of the leading US electrical contractors in the California market.

SOLV Energy: SOLV is an EPC contractor specializing in large-scale solar and energy storage projects. SOLV has built 293 projects and operates across 31 states in the US.

9. OPERATIONS AND MAINTENANCE

Completing the development, finance, and construction of a BESS plant is difficult work that culminates with the plant's reaching Commercial Operation Date (COD). From this point onward, the plant's role is to perform the services for its offtakers that it was designed to provide – and in so doing, to generate revenue that services the project's debt and provides a return to equity investors. For the plant to do this, it must be in good working order. Collectively, this work is known as Operation and Maintenance (O&M).

A mutually agreed upon long-term maintenance plan can help keep all stakeholders on the same page with the required system upkeep. This often includes both scheduled maintenance activities, as well as contingency plans for what to do when the system experiences unexpected issues and requires partial or total shutdown to fix the problem. Contracts will often reference "system availability" and will have liquidated damages (LD) tied to downtimes that exceed this guaranteed uptime, which is calculated annually. Intelligently scheduled maintenance periods, smart system design with engineered redundancy and the ability to isolate issues, and a well-kept spare parts list will all help minimize downtime and associated LDs.

O&M is a combination of preventative and corrective maintenance (sometime called Scheduled and Unscheduled maintenance). Preventative maintenance (PM) is the regular care for each component of the system, performed at regular intervals. Corrective maintenance is the work undertaken when unexpected problems are encountered. Even in ideal conditions, all plants will have some corrective maintenance items. The art of O&M is not to avoid these expenses, but rather to be prepared for these events, and to have good systems in place to manage them efficiently and cost-effectively, minimizing system downtime.

When done well, O&M is invisible, allowing for the plant to identify problems early, address them, and continue operating to provide grid services. However, when O&M work is neglected, the results can be costly downtime that may result in unsafe situations for the operational team. Although the O&M of BESS plants lack the mechanical complexity of wind farms or the large land areas of PV, it is still a critical task that will be a key determinant of a project's profitability.

The most visible O&M work at a battery plant is the maintenance of the physical infrastructure of the plant, which is a critical task. However, the O&M of a battery plant also includes the controls, HVAC, communications infrastructure, PCS (inverters), fire detection, fire suppression, warnings, electrical contractors, alarms, and data analysis tasks, which is equally as important as the physical battery. These unseen aspects result in a significant fraction of downtime and lost revenue. Figure 9-1 shows a typical BESS maintenance visit.

Figure 9-1
Maintenance personnel inspecting a BESS [39]

This chapter is dedicated to explaining the basics of the O&M process for a utility-scale battery plant. Although every plant's O&M process will differ slightly, the following six elements are hallmarks of any well-tuned O&M process:

- Ensuring safety: Safety comes first, as accidents can be both tragic and costly. Regular maintenance, including inspections, helps ensure early detection of safety issues. This helps reduce lost time injuries and maintain the safety reputation of the project and owner.
- Managing costs: Proper O&M practices are the appropriate balance of preventative work and corrective work – helping avoid unexpected repairs and replacements, minimizing the total operational cost of the plant.
- Maximizing performance: O&M activities, such as monitoring and parameter optimization, keep the project operating at maximum performance and efficiency, to meet or exceed the operational revenues anticipated in the plant's financial model.
- Maintaining high availability: Staying on top of repairs and replacements keeps availability high by avoiding unexpected downtime of failed equipment, especially when replacements come with a long lead time.
- Extending equipment lifetime: The lifetime of every component in a BESS can be improved by regular maintenance, thereby reducing how often replacements are necessary. This improves availability, replacement costs, labor expenses, and administrative costs for the project.
- Fostering a culture of continual maintenance: An effective O&M procedure starts by assigning maintenance tasks an important role. Where a culture of continual maintenance permeates an organization, it tends to generate both positive results and a positive reputation with stakeholders, investors, customers, and regulatory bodies.

9.1 Common O&M Arrangements

The contracts governing the O&M of a BESS plant can be structured in many ways. Some firms prefer to perform O&M in-house, while others prefer to hire third parties to do this work. On certain projects the battery integrator provides support throughout the entire life of the system, while on others that support is only provided for the first several years and is later taken over by others.

This section reviews some of the more common arrangements used by utility-scale BESS plants to operate and maintain their systems.

O&M work is commonly divided among tasks that are performed on the battery itself, and the balance of the plant (BoP). The following sections detail the more common arrangements used for these elements.

Long Term Service Agreement

The most common arrangement for the maintenance of the BESS itself is for the owner to sign an agreement with the battery integrator to perform the preventative and corrective maintenance of the BESS units. Battery and PCS OEMs provide a "product warranty" that covers defects of the product caused during manufacturing or shipping. This typically lasts 1-3 years during which claims can be made if issues are observed with the components covered by the warranty. Even during this warranty period, the OEMs expect appropriate maintenance to be carried out for the warranty claim to hold good. Maintenance procedures are usually lined out by the OEMs for their products.

Plant owners and operators can opt to manage the maintenance of the units in-house if they have the resources or hire a qualified firm to do that. Battery integrator terms vary, but most manufacturers will allow external firms to maintain their products only if they are certified as installers or maintainers. These installers are typically existing electrical firms who have elected to take the courses and go through the accreditation process with the OEM.

As an alternative to contracting with these accredited firms, most OEMs offer their customers a separate, all-inclusive agreement for the maintenance of the BESS unit or the whole plant. This agreement typically covers the project from delivery of the products to the end of the project's life (15 to 20 years). The most common name is Long Term Service Agreement, or LTSA, although some OEMs prefer other names such as Battery Service Agreement. Care must be taken to understand what is covered in this type of agreement.

The overarching purpose of an LTSA is to ensure that the battery supplier is responsible for their own product – since they designed and installed their product, it makes the most sense for them to maintain it. In exchange for an annual payment, the supplier is responsible for performing all routine maintenance of the BESS, as well as any corrective maintenance. For a DC-block BESS, this would include the battery itself, native controls platform, and any associated software, but typically excludes the PCS and any other equipment upstream. For AC-block systems, the LTSA typically covers the PCS since it is an integral part of the enclosure.

Beyond maintenance responsibilities, LTSAs may have provisions for a Performance Guarantee of the BESS, under which the OEM commits to achieve certain metrics for each year that the LTSA is active. The most common Performance Guarantee metrics are the available energy capacity – often referred to as State of Health (SoH), round-trip efficiency (RTE), and availability. Some values have a minimum threshold, like availability, while others often decline year-over-year, such as the capacity and RTE. Notably, many manufacturers now include a capacity or RTE guarantee as part of the base BESA, or at an incremental cost but tied to the BESA. But preventative maintenance is critical to uphold the guarantee.

The LTSA remains valid so long as the Project Owner continues paying for it, usually on an annual basis (this could also be done upfront as a lump sum) and abides by the technical conditions. This may include things like having to de-rate the system in certain heat conditions, or not overriding any of the safety protections built into the battery controls. In general, most integrated battery products, and their controls, are designed in a way to protect against any conditions that would negatively affect the warranty.

The terms of LTSAs vary significantly between manufacturers, but one important term is that the annual fee normally covers routine maintenance, but not corrective maintenance. This means that the standard inspections, cleanings, and consumables replacements are performed at no added fee, but incidents like PCS replacement, swapping BESS cells, or a broken controls system may incur a fee at the hourly rate charged by the OEM.

This may seem unfair – if a Project is paying the annual fee, why should they be responsible if the project breaks? The principle is that as long as the system is in good working order, the OEM is responsible for all work to keep up the performance metrics – for example, if the system discharges without any error codes, but cannot meet its capacity goal under the performance guarantee, the OEM is responsible for all costs required to pass the test.

However, if any systems are broken and require repair, the OEM will fix them (often within a maximum response time), but the cost of these repairs is borne by the end user paying for the LTSA. Especially if the OEM is responsible for an availability guarantee, they are incentivized to take corrective actions to fix issues, albeit at an additional cost. During negotiation of the LTSA, it is best to ask specific questions about what is covered and what is excluded to ensure the project does not end up incurring unexpected costs. Larger capital costs, such as augmentations or replacement of PCSs are typically not covered by an LTSA, unless there is a Capacity Maintenance agreement, covered in depth below.

Many LTSAs have terms around LDs for each of the key performance metrics. Ideally, a project will negotiate to have whatever LDs are incurred by the project be passed on to the OEM, minimizing the project's risk exposure. However, LTSAs often have a damages cap, often equal to the annual LTSA fee.

Although they may be expensive, the LTSA is a key feature for de-risking a project for performance and availability over its lifetime. If the developer/project owner decides to take on operation and maintenance in-house instead of an LTSA by the OEM, which is elaborated further below, the project bears the risk of the system under-performing, which is especially important to consider for projects with contractual obligations from an offtaker.

Some OEMs offer different levels of service, such as a low-, medium-, or high-level option, each with a differing price. At each level, there is a slightly different balance of involvement between the OEM and the Owner, so that a balance can be struck between minimizing costs and ensuring the project will be in good working order.

For Projects that have a Performance Guarantee, the OEM is incentivized to provide technical support and troubleshoot issues, since missing the performance metrics often results in costly LDs.

BoP O&M

Beyond the LTSA, the BoP also requires O&M services. This includes all components besides the battery, such as the AC and DC collection systems, substation, controls system, MV transformers, and PCS (if it is not integral to the battery enclosure). Some PCS manufacturers are starting to offer LTSAs on their products, but since the usable life of these products is typically less than the project life, many PCS makers are unwilling to extend a warranty beyond the initial workmanship coverage (typically between 3 and 5 years).

The controls system is one that often falls outside of the LTSA and is a critical component that requires upkeep. Although the controls infrastructure is built to last throughout the system's life, the soft aspects of the plant require continuous attention: software updates, security audits, and data maintenance are all examples of tasks that are critical to the operations of a project. These are covered in greater depth in Software and Network Components under Chapter 5.

Many project operators contract with third-party firms to perform the BoP maintenance – these firms are often the contractor that did the installation in the first place. As with most maintenance, it is preferable to keep a close eye on all components to identify and address problems early.

Owner-provided O&M

In the case of many of the large independent power producers (IPPs), O&M activities may be done in-house, especially if the company already has experience doing this for solar and wind projects. BESS projects are considered to involve lower levels of maintenance relative to solar and wind, since they have fewer moving parts and less land area to maintain. While BESS projects are less labor intensive in their day-to-day operations, it is important that companies devote sufficient personnel and resources to the ongoing BESS maintenance tasks. For owner-operated systems, the OEMs often provide documentation (product and operational manuals) as well as conducting training as part of the commissioning process. The manuals often detail the maintenance requirements specific to the project, which vary from project to project. Some OEMs and integrators, such as Tesla and Fluence, have intensive training courses that can qualify the operators as certified OEM contractors; however, in this scenario the contracts must be carefully crafted to avoid creating a conflict of interest where the owner could be both the holder of the warranty and the firm tasked with maintaining the system.

Augmentation

As noted throughout this book, batteries degrade over time, most notably decreasing both their energy capacity and their RTE. Some projects choose to accept this declining performance, but many prefer to or have contractual obligation to boost the system capacity throughout its lifetime which is currently most achieved by augmenting the system at different points. The alternative is to overbuild the system to compensate for degradation upfront. While this is invariably a costlier strategy, it avoids having to disrupt the project operations and the mobilizing, electrical installation, and labor costs.

Augmentation is an important part of managing operations of a BESS plant. It typically happens only after several years of the battery being operational; however, it should be planned for from the start of the project by designing the plant to accommodate the addition of more batteries and/or PCSs – in terms of space, civil and electrical design. Some developers choose to leave concrete slabs and conduit for future installations, while others leave only bare land, preferring to complete the augmentation design when the time comes.

Electrically speaking, augmentation can be done by two different methods: AC- or DC-augmentation. This is described in more detail in Section 8.1. In a nutshell: In an AC option, the project adds additional AC-blocks to the system, meaning that new battery enclosures and PCSs are added. In some cases, they can be added to existing MV feeder circuits, while in others, new MV feeders and/or transformers must be installed to accommodate the augmentation. AC-augmentation has the advantage of not mixing new and old batteries behind a given PCS. Additionally, AC augmentations can be installed with relatively small disruption to operational equipment, resulting in less system downtime. However, AC augmentations are more expensive because they require new PCSs, which are expensive components. This is a significant expense considering that augmentations normally do not need to change the power capacity of the plant, only the energy. Further, the addition of PCSs may trigger interconnection updates that can be costly and time-consuming.

In contrast, DC augmentation involves addition of DC battery blocks behind existing PCSs. However, this mixes new and old battery units, which can be problematic since these products have different states of health. This means that electrons have easier pathways through the newer cells than the older ones, which can result in unequal wear on the batteries. To remedy this, DC-augmented systems often use a DC-DC converter between the new and the old batteries. In this approach, the number of PCS units stays constant throughout the life of the system, while the number of batteries behind each PCS increases. Additional slabs or pilings must be left for the additional DC-blocks while planning the site layout. Figure 9-2 shows a typical arrangement of a new system in which space is left for future augmentation.

Figure 9-2
BESS blocks on foundation pads [40]

Planning augmentation during the project design phase is critical for engineering and finances. Knowing where the additions will be and how they connect into the existing system will save massive headaches when it's time to roll the new equipment in for the operations team that will typically have ownership of the project during this phase. In most cases, equipment called a DC-DC combiner would be required to connect new containers to existing PCS units. Some OEMs allow for this by leaving space within the container for additional modules. Augmentation design and strategy needs to be laid out at the time of initial design by consulting with the OEM in question, as it is likely that the augmentation units will be from the same OEM. In today's rapidly evolving market, products available for augmentation may differ from those installed at the beginning-of-life, as battery integrators unveil new products over the years. It is critical to maintain an ongoing relationship with the supplier to ensure smooth augmentation and continuous operations, especially in the case of owner-provided O&M.

It is important to figure out the timing for augmentation during the year as it involves the plant being offline, at least in phases. So, the times when the plant needs to be kept fully available such as summer months, which are usually most valuable monetarily, need to be avoided.

The division of responsibility for this work must also be clearly squared out between the entity managing operations of the BESS and the supplier of the equipment (typically same supplier as the initial capacity). When projects sign up for service agreements with integrators, as described below, the responsibility for capacity maintenance and augmentation is typically passed to the integrator. In the absence of that, the entity that owns and operates needs to arrange for this in a timely manner. Financially, augmentation years will be costly compared to non-augmentation years and getting accurate estimates of these impacts are necessary for understanding the project's costs.

Capacity Maintenance

For project owners that want to outsource as much as possible to the BESS supplier, some OEMs offer a Capacity Maintenance Agreement (CMA). Under this arrangement the OEM covers the maintenance of the plant, but also ensures that the plant will have a minimum available energy capacity. This means that the supplier is responsible for determining and carrying out the augmentation strategy. CMAs are typically quite expensive, and as developers have become more sophisticated, many do not want to pay an additional amount to cover something they can do in-house, reducing costs. As of this writing, the majority of BESS projects do not opt for CMAs on their projects.

If opted for the project, as of this writing, rapidly falling battery costs mean that most OEMs offering CMAs will plan on two or more augmentations, rather than overbuilding the system. The distinctive aspect of a CMA is that the OEM takes responsibility for the capacity, whether that is through augmentation, overbuild, or some combination. Although more expensive than traditional owner-managed augmentation, the CMA reduces project risk since the OEM is required to maintain the capacity throughout the project's life.

9.2 Maintenance

This section discusses the different aspects of BESS maintenance that need to be arranged for by the project owner/operator for smooth and efficient functioning.

Preventative Maintenance

Preventative maintenance functions are most often performed either by the battery integrator themselves or by one of their authorized subcontractors. After an initial period following COD (often referred to as the 'teething' phase), systems most often are visited every six months for routine maintenance, and annually for capacity testing and maintenance of the more complex systems. Other maintenance functions happen every several years, such as the replacement of PCSs or the swapping of battery modules.

Preventative maintenance is often performed when expected revenues are lower, to avoid missing high-profit periods. These may be during certain seasons of the year or times of the day.

Although the service regime will vary from product to product Table 9-1 shows the more typical categories of work performed as part of preventative maintenance.

Table 9-1
Typical preventative maintenance work

Task	Description
Regular cleaning and visual inspection	Visually check for damage, corrosion, or leaks
Harness inspection	Verifying that power and cable harnesses are still intact
Rodent infiltration	Checking for chewed cables, nests, droppings
Cooling system	Filter cleaning, testing the HVAC units, checking cooling fluid pressure, level, ventilation cleaning and condition of cooling tubes.
Battery connection verification	Verify that all terminals are well-greased, connections properly torqued, and no connections are loose
Spot checking bolt torques	For any mechanical connections, verifying that bolts have not loosened by checking the torque on some sample of them
Temperature logs	Keep the system operating within stipulated temperature limits, verify if there have been any localized high temperature events over time
Ancillary equipment	Verify functionality of smoke sensors, gas sensors, vents and alarms

Software updates	Implement bug fixes, ensure adequate tracking of the system's performance, and ensure compatibility with the supplier's service protocol. This is typically done in real time as a part of the cybersecurity program.
Network equipment maintenance	Verify connections and functionality of all communications gear
Fire protection testing	Often conducted in concert with local first responders, verify that the protection systems are alert and ready to respond to any incidents
Environmental compliance	Check if there have been any spills, noise events, wildlife encounters, or other environmental protocols.
Inventory check	Wherever parts have been used for preventative or corrective maintenance, verify that all stocks have been replenished
Capacity test	Plan for capacity test as per conditions stipulated in the BESA and/or the offtake agreement, if applicable
Market dispatch strategy	Verify that the system is responding as designed to all events and is complying with the intended market optimization strategy. This is often conducted by the asset manager.
Data historian	Checking to ensure that proper backups are being performed, that any data anomalies are investigated, and that no unexpected data losses have occurred.

Corrective Maintenance

Corrective maintenance happens when any unexpected event occurs, such as a fault or external damage to the system. Key causes for BESS outages include failures of components such as PCSs or compressors, overheating events, overcharging or mechanical abuse of modules. These events can seriously hamper a project's ability to earn revenue, since they are unannounced and may coincide with high market price windows. It's important for the service provider to have experienced, capable, and fast-acting service providers in the area who can rectify issues quickly. Most O&M contracts include provisions for required response time for corrective maintenance events, some of which may be under 4 hours for critical events. Downtime here correlates directly to decreased system availability and can lead to LDs. Figure 9-3 [58] shows an example of a corrective maintenance replacement of a battery module. A **Root Cause Analysis (RCA)** is then performed to ensure the source of the issue is found and reoccurrence can be avoided.

Figure 9-3
Replacement of a battery module [58]

Major tasks under corrective maintenance vary widely based on the events experienced, but some of the more common categories of events, with examples, are shown in Table 9-2.

Table 9-2
Typical corrective maintenance events

Event	Example events
BESS unit corrective maintenance	Cell / module overheating, voltage imbalances, sensors not functioning or capacity degradation
PCS unit corrective maintenance	Ventilation system issues, low availability, or slow response times
MVT unit corrective maintenance	Cooling system failure, de-rates, or catastrophic transformer failure
BoP corrective maintenance	Tripped breakers or fuses, conductor insulation issues
Communications/network corrective maintenance	Inability to read sensors, slow reaction times, cyberattacks
Substation corrective maintenance	BoP issues or failures, replacement of key components,
Tripped fire alarms	Nuisance trips being sent to first responders
Warranty claim processing	Verification of any potential Defect or Serial Defect provisions within a battery warranty
Root cause analysis	Verification of the cause of some critical failure event

Spare Parts

Many corrective and preventative tasks require parts, so many projects elect to keep some of the more common consumable parts on site. This reduces down-time, avoiding lead times and shipping by having parts ready for replacement. Which parts are kept on site, and who pays for them, are negotiated as part of the BESA or LTSA, typically between the developer/owner and BESS Integrator. Critical spare parts for a BESS may include battery modules, PCS skids, and components of HVAC systems. While helpful in maintenance activities, storage of spare parts increases upfront costs and site management duties since spare parts must be stored and maintained per their storage specifications. For example, battery modules often have a temperature at which they must be stored, which requires HVAC for the storage area. Due diligence informed by experience will help negotiate what and how many spare parts are maintained, balancing cost, management, and having spare parts readily available.

Recommended spares and quantities can vary greatly depending on the product in use and OEM recommendation. Some of the more common spare parts stored on site are:

- Battery modules
- Battery Control Platforms (Battery Control Unit or Battery Management System unit)
- HVAC filters and belts
- Networking equipment Repeater
- Internal and external Cabling
- Fuses
- Switches
- Module busbars
- Smoke and gas detectors
- Dampers
- Horn strobe
- Coolant fluid
- HVAC or chiller
- UPS batteries

Figure 9-4
Typical spare parts container [59]

Product Documentation

It is the responsibility of whoever provides the O&M of a project to maintain all available documentation on a given product. That includes the product manuals, warranty documentation, as well as the data storage of the operations of the plant. These documents and data are all typically kept in digital format with secure cloud storage to ensure they are safe and cannot be destroyed.

On the project site, the most important documentation is around safety. For each component on the site, a materials safety datasheet (MSDS) document will list all components as well as their percentage amounts present in the cells. It is the responsibility of the O&M provider to keep and maintain all the MSDS available and accessible.

Any other equipment in the system that contains potentially toxic chemicals (such as certain types of breakers, oil-filled transformers, liquid cooled systems) should also have associated MSDSs, which should be filed with system documents. This helps AHJs and first responders have a better idea of the potential environmental and health impacts of the system if it experiences certain types of failures.

Beyond the standard safety documentation, many AHJs will require on-site remediation equipment in case of equipment failures, such as oil containment and liquid absorption mats.

Software and Network Components

Since battery systems are totally dependent on their controls systems, maintenance of controls software and hardware is as important as the maintenance of physical infrastructure.

Most of the communications components for each portion of the system are maintained internally – for example the PCS has its own proprietary components, as do the battery EMS, fire alarms, and meters. These may require periodic software updates. The BESS's control system may be subject to frequent updates by the OEM for both performance and safety reasons. This can be done virtually via the internet-connected EMS by remotely patching updates to the software code. But this action often faces opposition especially for BESS contracted by utilities, as it requires the BESS supplier to have direct online access to the system which can introduce cybersecurity-related concerns. More commonly, these types of software patches are deployed manually in the field by staff sent out by the supplier. Changes may be made to the operational code, BMS, or logic used in the systems. As the OEM learns of security flaws in the battery system, these updates also ensure that the system controls are protected from cyberattacks. The overarching hardware and software that connects the individual elements, such as power plant controllers, SCADA systems, on-site network switches and routers, are the responsibility of the O&M provider to ensure they are operating correctly.

Neglecting BESS software updates can expose the system to cyberattacks, which have become increasingly common. These attacks are varied but may be ransomware attacks focused on holding the data or operational capacity hostage or provide methods for bad actors to affect the battery system or the power grid as a whole. BESS cybersecurity is covered in depth in Section 5.5. Software updates are also critical for maintaining system performance and reliability. Software updates often include improvements in the control algorithms, which result in the battery being able to perform better, degrade less, and ultimately make better returns for the project.

9.3 Operations

The operation of a battery comprises the ongoing processes and functions that keep the project running as efficiently and profitably as possible. Given that a functional battery has few moving parts, the focus of operations is on controls and strategic dispatch – this section presents three of the critical software functions of a BESS: system monitoring, data analysis, and optimization.

System Monitoring

A robust BMS and EMS form the portal through which the battery is operated. These systems provide real-time data about the system, and allow the operator to make operational decisions, address any potential risks, and provide all the necessary operational information about the system. This is typically performed through a graphical interface provided by the operator of the EMS, whether that is the battery integrator or a third-party controls platform. This is most often accessed via a secure web portal which allows the remote operator to see all the relevant conditions of the operational system. Figure 9-5 shows a screenshot of a typical operational BESS monitoring platform.

Critical data that is monitored through the platform includes:

- The current flowing through each cell, module, enclosure, and circuit. This provides the basis for the system to calculate the power flows that charge or discharge the system at any time.

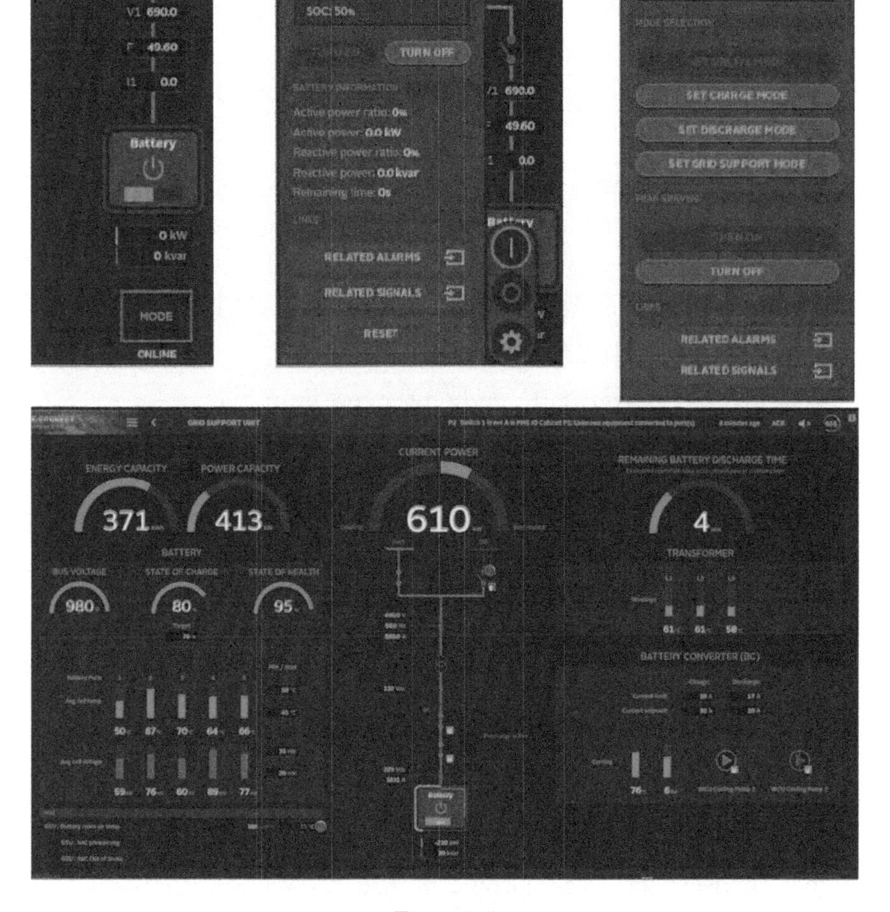

Figure 9-5
Screenshot of typical BESS controls platform [142]

- The voltages of each individual component of the system, along with comparisons to ensure that similar voltages are maintained throughout the system.

- The SoC of the batteries: although this is typically analogous to voltages, many systems also use other methods to count the amount of energy which enters or leaves a cell. SoC monitoring helps prevent discharge beyond permissible or recommended limits to retain the warranty terms, and damage from over- or under-charging
- Operating temperature conditions: most systems monitor temperature at thousands of locations throughout the system, which is critical information to determine if the cooling systems are functioning properly and to provide early identification of potential safety hazards
- Auxiliary power status and power consumption

The monitoring system commonly shows these values in real time, as well as maintaining their historical values in a large database which is usually uploaded immediately to a cloud platform. Given the vast amounts of data, the monitoring portal often highlights the minimum, maximum, or average values for the relevant portion of the BESS. For example, the temperature reading might show the highest cell temperature reading for each enclosure, or the minimum state-of-charge of a rack in an enclosure.

If any values that are shown fall outside of the bounds of normal operation, this can trigger a warning. Examples of such triggers are a higher-than-expected cell current, or a rack-level SoC that is out of step with the other racks in the enclosure. The warnings often have various levels of seriousness depending on the value:

- The lowest level conditions are considered nuisances, and may be flagged to be reviewed or dismissed the next time an operator reviews warnings
- Mid-level conditions may trigger a notification to an operator who is on call to receive these and investigate the causes and potential effects.

- Urgent conditions will cause immediate notifications to the system operator and responsible party for the site, to form a rapid response to mitigate the issue. If sufficiently severe, these conditions may also trigger communications with the local first responders.

Larger and more mature battery project owners may have in-house capabilities to provide their own on-site and remote monitoring. For large sites this may include on-site staffing, while for smaller projects it is often done remotely via a regional control center for all the assets. Smaller companies may elect to hire a third party to manage the monitoring of the facility, and many asset operators and monitors are available to provide this service.

Enabling comparisons between past and present states can often highlight a new problem or latent issue that may not have been obvious from a snapshot view of the system. Most performance guarantees that come with the supply and/or service agreement with an OEM or a provider require continuous data monitoring and recording at the minute or even second level data; these requirements need to be well understood. While monitoring data from the BMS and EMS is accomplished on site, most projects these days also have remote monitoring, so that personnel can access and analyze data from anywhere in the world using cloud-based servers.

Data Analysis

Large battery systems often contain over one million cells, and the monitoring of their conditions, which often happens every few seconds, means that batteries are producing many gigabytes of data every hour, resulting in massive amounts of data. This topic is covered in depth in Section 5.2.

This data can be difficult to manipulate given its volume, but it is a wealth of information about how the system is performing, and how it can perform better. For example, certain signatures of voltage flicker in the data can indicate where there may be a loose cable connection on a battery rack. Alternatively, if a rack is consistently having a temperature imbalance in the same location, this may be indicative of a cooling system malfunction that the HVAC controls system may not have identified.

The process of drawing conclusions about battery systems from operational data via data analytics is a burgeoning field. This process often involves the use of machine learning (ML) and AI tools that attempt to draw conclusions about the system in terms of expected SoH degradation and possibility of safety-related events in a predictive manner. Most battery integrators now advertise this as part of their scope offering, but many Owners opt to hire a third party specializing in data analytics to perform this review as an ongoing service. The third party may be the same firm that provides the system controls, if one is used, or may be another firm that specializes exclusively in such data analytics. There are many startups in this area – the more common firms working in this space are listed in the Usual Suspects section at the end of the chapter.

Performance Tracking

Battery systems can run unattended for extended periods of time, but industry standard operating procedures require that the data from the battery be periodically reviewed, for both safety and performance reasons. This is a separate process, outside of the standard tracking of warnings discussed in the prior section. The review is normally conducted monthly and involves working through the available data sets with two primary goals: safety and performance.

For safety, all warnings are reviewed to determine if there is anything more serious that should be verified. Through data analytics, as described in the prior section, any potential issues, even those that may not have caused a warning, are investigated. If warranted, the analysis may result in corrective action, such as dispatching a field team to investigate an issue. For less serious issues, an action item may be added to a list of tasks to be performed at the next upcoming field visit.

For performance, the earnings of the system are reviewed to ensure that the system is performing up to its maximum technical capabilities. This includes things that pose no safety risk but may result in accelerated degradation. For example, if a battery routinely has incorrect SoC measurements, this may mean that there is still remaining energy that the system can't access at 0% SOC, or still remaining room to charge the battery when the SoC is reporting 100%. Correcting these issues is an ongoing task which involves coordination with the battery integrator, review of market conditions, and learnings from operational projects. Individual modifications based on performance reviews may result in small gains from month-to-month, but the overall effect can have a large impact on the project's performance, health, and ultimately its ability to earn revenues in the markets in which it is operating. The financial performance of the battery is a separate task known as Asset Management – this is covered in depth in Section 7.5.

9.4 Energy Capacity Testing

Nearly all utility-scale battery projects require energy capacity testing throughout the project lifetime. These tests are first conducted when the project is built and are a critical element of the performance testing required for a system to come online (see Section 8.6 for a complete discussion of this process). As the battery ages, the system often reports an SoH value, which indicates the performance ability relative to the system's initial characteristics. This value ranges from 100%, which indicates the system meets or exceeds its nameplate capacity, to 60%, which is widely considered the lowest possibly value to which lithium-ion BESS should be allowed to degrade. Along with capacity, the system's RTE also typically declines. Values vary widely, but the high-voltage level of RTE often declines steadily from an initial value of 80-90% at COD to a value that may be 75 to 85% at end-of-life.

The specific minimum values of energy capacity and RTE are normally detailed in the BESA and/or the LTSA. There may also be requirements for these in the offtake agreements. Other values such as auxiliary power, response time, or thermal performance may also have guarantees, but these are less common.

To demonstrate the project is meeting its required performance values over time, the system undergoes a performance test, most often conducted on every anniversary of the project's COD. The performance test procedures are normally spelled out in detail in the BESS contractual agreement and are most often identical to the tests that were conducted at COD. The specific procedure varies between projects and from one battery integrator to another, but the basic steps of any capacity test is to start with the project at 0% SOC, fully charge the battery at full power to 100% SOC, wait some time, and fully discharge the battery at full power back to 0% SoC (the full, step-by-step process is detailed in Section 8.6). Constraints around temperature and time of the year are often added so that the test conditions are controlled. The resulting energy discharged is the total energy capacity of the system, while the amount of energy lost in the process allows the RTE to be calculated. These values must meet the values stipulated in the applicable contract.

The capacity test represents the hardest the battery will ever have to work, and some systems never experience the stresses of a capacity test during normal operations. Because of this, beyond energy capacity and RTE, other critical systems such as battery cooling and PCSs are also put to the test. It is good practice for the operator to review the full operational data set during the test to identify any potential issues in the system's operation.

Although the offtake agreement may require the system to meet some fixed capacity or RTE value, in Li-ion batteries, it is preferable for the system to be discharged fully to 0% SOC, even if that involves discharging more energy than is required by the contract. For example, if a 40 MW / 160 MWh project is built with 190 MWh at COD, the year-2 capacity test might demonstrate that the system can discharge 167.2 MWh, indicating it is at an 88% SOH. If the test is stopped at the minimum offtake capacity (160 MWh), the precise value of the system's energy capacity would remain unknown.

Some battery integrators require the system to come offline for some time prior to the test, so that the system can undergo module, rack, or enclosure-level balancing, or for some maintenance to be performed before conducting the test. The BESA or LTSA often stipulates the maximum time allowed for this process, but it typically does not exceed one day. Tests are usually performed during off-peak times to minimize revenue loss or availability deficiency in the case of contracted projects.

If the system meets its performance obligations, the system can continue operating in the marketplace, whether it is for a utility (contract) or an ISO. However, if the system fails the test, the contract dictates the required course of action. Some contracts provide an opportunity for the project to remedy the situation and perform a re-test once repairs have been made. Other contracts allow for a deficiency in energy capacity or RTE but require the project owner to pay LDs if the values are below the contractual minimums. If it is a merchant project participating in the wholesale market, the capacity bid needs to be adjusted according to the available capacity to avoid penalties of non-performance.

Energy capacity can be maintained by augmenting the system capacity, as described in Section 9.1. Most systems have a planned augmentation schedule at COD, which indicates in which operational years they intend to add battery racks, enclosures, and/or PCSs. Based on the capacity testing results, the augmentation may have to be performed earlier if the system degrades more quickly than anticipated. Alternatively, if capacity testing indicates less-than-expected degradation, augmentation may be able to be delayed.

RTE deficiencies are harder to remedy since these are typically the result of the internal resistance of the battery cells. Augmenting the system may cause a marginal improvement in RTE, but most systems experiencing failing RTE tests elect to pay LDs, if allowed by the contract. Because of the difficulty in improving RTE after COD, it is best practice to provide a reasonable buffer between the expected and warrantied RTE values wherever possible.

9.5 Battery Warranties and Guarantees

Given the large capital costs of batteries, the owners of these assets want to be assured that they will perform to a certain level throughout their life. To provide this assurance, nearly all major battery integrators offer warranties and guarantees, some of which extend throughout the life of the battery, which is currently 20 or 25 years for most utility-scale systems. This section outlines some of the more common terms and provisions of these agreements, and how they can be enforced.

Warranties

Warranties cover the product if it experiences defects and workmanship for a fixed period after it has been constructed. Like many technical products, battery plants tend to follow a bathtub curve, which is illustrated in Figure 9-6. This implies that there is an initial period of early failures (sometimes called the teething period) where faults become prevalent following commissioning, a period of normal operation, and eventually a wearing out of the system toward the end of its life.

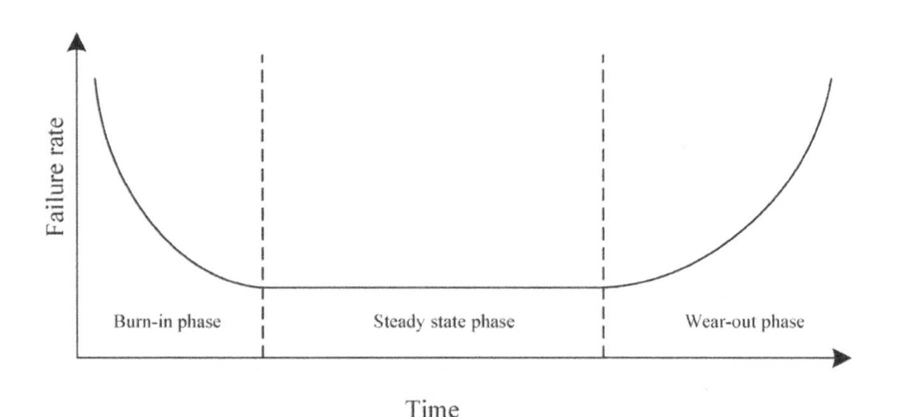

Figure 9-6
Typical product bathtub curve [60]

A BESS workmanship warranty is intended to cover this early failure period, and usually lasts between one to three years from COD. In some cases, OEMs may offer options for an extension at additional cost. These warranties cover any physical defects, such as loose connections or water ingress, but also cover defects in battery controls, which are a frequent source of issues early in a battery project's life. The warranty commonly includes a definition of how a defect is defined. In the case that many parts of the system experience the same defect, there may be a special clause to designate a Serial Defect.

When a defect is identified, the warranty often defines a minimum time for the supplier or their contractor to address. This may be through remote means, such as clearing system faults via web portal, or through on-site interventions, such as resetting tripped circuit breakers. The LTSA often has clear definitions of these response times, ranging from days or weeks in the case of minor issues, to response within minutes for critical safety events.

If an issue is identified with the system that is still within the workmanship warranty, the owner can file a claim with the battery supplier to return and address the issue. This may be as simple as putting a ticket into the supplier's service portal to report the issue. However, if these issues persist and remain unaddressed, legal notices or eventual litigation may be required to enforce the workmanship warranty. Most project owners seek to maintain an amicable relationship with the battery integrator to facilitate quick and easy fixes during the warranty period.

Performance Guarantees

While the warranty is focused on addressing problems in the functionality of the battery, performance guarantees (PGs) are intended to address a minimum level of performance for a functional system. The most common PG metrics are degradation, availability, and round-trip efficiency. The levels of performance at COD are usually defined in the BESA, while the guaranteed values throughout the years of the project's useful life are most often contained in the LTSA.

The most common term for the PGs is the full useful life of the BESS (20-25 years), although some OEMs offer a limited PG period (10-15 years) which can then be extended by mutual agreement. Table 9-3 shows some typical values of performance guarantees for a utility-scale BESS. Note that capacity declines significantly, while RTE declines slowly, and availability often stays constant throughout the life of the system.

Table 9-3
Typical performance guarantee table

Year	Capacity	Round-trip Efficiency (RTE)	Availability
0 (COD)	100.0%	88.0%	98.0%
1	95.1%	87.7%	98.0%
2	92.6%	87.5%	98.0%
3	90.8%	87.2%	98.0%
4	89.1%	86.9%	98.0%
5	87.6%	86.7%	98.0%
6	86.8%	86.4%	98.0%
7	86.0%	86.2%	98.0%
8	85.1%	85.9%	98.0%
9	84.3%	85.7%	98.0%
10	83.1%	85.4%	98.0%
11	81.9%	85.1%	98.0%
12	80.5%	84.9%	98.0%
13	79.3%	84.6%	98.0%
14	78.1%	84.4%	98.0%
15	76.8%	84.1%	98.0%
16	75.3%	83.9%	98.0%
17	73.5%	83.6%	98.0%
18	71.8%	83.4%	98.0%
19	70.1%	83.1%	98.0%
20	68.4%	82.9%	98.0%

The PGs are vital to help the project hedge against the risk of underperformance, especially if the project's offtake agreement includes its own performance obligations and associated LDs.

The LTSA has extensive conditions that must be met for the battery integrator to enforce the warranty. These vary across suppliers, but some of the more common terms are:

- The system cannot be serviced by anyone except the battery integrator or their authorized agent
- The system must never be operated outside of the allowable temperature range
- If there are any interruptions to the connectivity of the system, they must not exceed certain durations or frequency
- The deficiency must be reported to the battery integrator within a certain period following the operator learning of the issue
- The defect must not be caused by a force majeure event, as defined in the contract

Enforcement of Warranties and Guarantees

Although warranties and PGs for battery systems are important protections that enable the system to limit its exposure risk, their enforcement is an ongoing process between various stakeholders. In the early days of a plant, there are bound to be issues as the plant starts its operations. The plant operator and BESS integrator are typically in close communication to address these issues. In the first months, it may be common for availability numbers to not hit the required values. In some cases, LDs may be waived by both the offtaker and by the owner of the plant.

In the case of most deficiencies, the process can be worked out amicably through communication. For example, if an operator notices that a given battery container has gone offline without reason, they can contact the integrator and begin a collaborative process of investigating the cause and solution. Building strong relationships through training, open lines of communication, and site visits is critical to maintaining a strong O&M network for the system. Only when normal lines of communication have broken down must the system resort to legal enforcement of the terms of the warranty or performance guarantee. Although many systems reach this point, the goal of a successful O&M program is to foster the communication that provides many offramps for amicable resolution prior to the costly effort of legal enforcement or LDs.

While the BESS is the key component of any BESS, there may also be workmanship warranties or performance guarantees on other components, such as the PCS, controls, system, or the BoP components. In many cases, a response to an O&M issue encountered on site may involve coordination between several different stakeholders to ensure the issue is dealt with accurately.

In practice some of the more common O&M issues encountered are:

- Annual capacity testing issues such as excessive balancing time, non-compliance with established test procedures, or failure to meet minimum testing levels
- Controls problems, such as PCSs failing to communicate with the EMS, BMS, or plant controller.
- Nuisance trips, in which alarms or faults are activated without any actual incident.
- External communications issues, such as delayed response time to an ISO signal, or inability to communicate with the local fire department
- Individual battery cells or modules which experience temperatures that approach or meet the maximum allowable values

Each of these events requires a unique intervention, both to address the issue at hand and to avoid it occurring in the future. Managing this process is one of the key responsibilities of the O&M manager for the site.

9.6 Market Participation

At the time of writing, most large-scale BESS plants materialize with an offtake agreement with a utility where the utility decides the manner in which the BESS is operated. However, there are many projects, especially now in Texas (ERCOT region) where BESS projects are merchant and free to participate in the wholesale market as the operator deems strategic. This is also the case with a Resource Adequacy contract where the BESS gets paid to be online and available as capacity and must provide services in the wholesale market as seen in California. ISOs or RTOs have stringent rules on how to interconnect into their systems and how to bid into their market structures. The controller used on the project is often integrated with the ISO's secure communications portal, allowing it to receive commands, report on its operational status, and provide safety information to the grid operator.

To bid into a market, there are many complex requirements to meet to become certified as a bidder and to place bids. Some project owners become certified by themselves, particularly if they are large developers with large portfolios of projects. Instead, some elect to hire a scheduling coordinator to act on their behalf. Some of the more active firms in this space are listed in Section 9.8. Experienced schedulers have market experience and can help provide the required hardware and software for interfacing with the market control and settlement structures.

Whether done directly or through a scheduling coordinator, the system must determine its dispatch strategy and market-bid decisions that are fed to the ISO. These bids are relayed electronically to the

ISO through scheduling coordinators that interface with the ISO. The scheduling coordinator serves as the gateway for the project system with the ISO, submitting bid information including the power setting and hours of dispatch.

The bidder could also receive signals from the ISO, in the form of an automatic generator current (AGC), for the system to follow in the case of services like frequency regulation and response. The scheduling coordinator typically monitors the asset remotely around the clock and is responsible for reporting system anomalies or outages.

While each market is different, in general the BESS will be bidding into the market and will get dispatched by the system operator based on bid acceptance. For example, a BESS may bid into the day-ahead market at various price points for each hour for an ancillary service, such as frequency response. If the price bid for a certain hour is below the clearing price the system will win the bid for that hour and will be expected to respond to frequency dispatch commands for that hour.

These commands may be updated every 4 to 15 seconds and the system must be available and responsive to get paid for that hour. There may be times when the system wins the bid but isn't called on, allowing it to collect without charging or discharging.

Chapter 6 includes greater detail on the merchant markets available to battery systems.

9.7 End of Life / Decommissioning

A battery system typically reaches its end of life by one of two methods: either the system has degraded to a point where it is no longer financially viable to operate in the market, or its internal chemistry has reached a point where it is unsafe to continue operating. At this point, the system owner must consider several questions: Will the system continue under a new contract? Will it be decommissioned and thrown away, recycled, or reused? Can the system be deconstructed and transported safely? Can portions of the system be reused in some secondary application?

As of this writing, the useful life of batteries ranges from 15 to 25 years, or approximately 7,200 cycles, whichever is reached first. When a BESS reaches these conditions it may still be operated, but the warranties and performance guarantees which covered the system may no longer be available. At this point, the battery cells' internal resistance will be higher than it was when originally commissioning, resulting in excess heat generation and lower round-trip efficiency (RTE). The higher heat may harm the battery, and in some cases may increase its risk of catastrophic failures such as fire or explosion. Safety considerations are always the highest priority when evaluating an end-of-life system.

The owner or operator is responsible for scheduling and executing decommissioning activities, which typically begin within 12 months of the project ceasing operation. The activities involve the removal of above and below-ground components of the project and restoration of the site as per local environmental requirements. Components of the BESS facility with resale value may be extracted and sold, while components with no wholesale value will be recycled or disposed. Potentially toxic components, such as the battery modules themselves, must be dealt with in accordance with both the manufacturer's recommendations and all applicable codes and standards. Battery recycling is discussed in more detail in Chapter 10.

The specific practices for BESS decommissioning may vary based on local regulations, project size, technology type and other factors. It's important for BESS owners to develop a comprehensive decommissioning plan that addresses all relevant considerations to ensure a safe and environmentally responsible process. Although this plan may not be fully formed when the project comes online, the project's financial model should plan to set aside a reserve amount to be used for the eventual decommissioning of the plant.

9.8 Usual Suspects

The largest battery O&M providers are the integrators themselves, as listed in Section 3.8. Additionally, many large developers perform their own O&M work in-house – some of the larger developers are listed in Section 6.7.

For projects that use a third-party O&M firm, they are often firms that have a strong presence in the region where the BESS is located. The following are some of the more common larger US battery O&M providers.

Third Party O&M Providers

Consolidated Asset Management Services (CAMS): manages approximately 4 GW of wind, solar and BESS projects in North America for leading private equity firms and financial institutions. CAMS provides O&M, financial services, compliance services, and consulting.

Fortum: an O&M provider specializing in operations of renewable energy plants, with a growing practice in energy storage. They are based in Sweden and service projects across Europe and India.

IHI Power Services Corp.: a subsidiary of IHI Corporation, who offers full-service O&M to both fossil and renewable power plants. They are based in Tokyo with a presence in over 20 countries.

Rosendin: one of the largest US electrical contractors, is an O&M provider in addition to being an EPC for battery plants. They are a full-service contractor that can be used to service BESS project sites, with over 2 GWh of BESS under construction.

Scheduling/Market Operators

The firms that provide these services are listed under Section 5.7, the Usual Suspects section for Controls. They are listed here for reference:

- PCI Energy Solutions
- OATI WebTrader
- Customized Energy Solutions (CES)
- Boston Energy Trading & Marketing (BETM)

Performance analytics

Accure Battery Intelligence: An energy storage analytics company whose software and AI-driven solution assists manufacturers and operators in monitoring the overall health and safety, currently serving over 2 GWh of projects world-wide.

Twaice: Provides predictive analytics software that offers a battery analytics platform, simulation models, development solutions, energy solutions, and fleet solutions. It serves automotive, energy storage, battery manufacturing, and commercial vehicle industries.

Zitara: Zitara is a provider of AI-enabled battery management software to verify the accuracy of native BMS products, monitor battery performance, and make operational decisions to enhance project profitability. Zitara offers both live BESS data monitoring through its Zitara Live product, and one-off advisory work through its Zitara Studio offering.

10. SAFETY AND ENVIRONMENTAL CONSIDERATIONS

Outside of Phoenix, Arizona in April 2019, the McMicken battery facility quietly hummed along, a single container with 2 MWh of energy storage, with a maximum power of 2 MW. The battery was working to smooth out fluctuations in the electricity grid from solar PV and shift excess energy from daytime peaks to evening hours, after the sun had set. Without anyone realizing, deep inside rack 15 of a containerized battery enclosure, small needle-like dendrites had been growing inside of one of the cells. At around 4:56 pm, the dendrites, growing on the anode, made their way to the cathode of the cell. This caused a short-circuit in the cell, resulting in a large amount of current crossing from anode to cathode. The temperature of the cell skyrocketed, going from its normal operational temperature of 30°C to over 700°C in under 10 minutes. The electrolyte chemicals and lithium within the cell had begun to burn, even after an aerosol system within the container had discharged. The fire quickly spread to all the cells in the rack, releasing clouds of toxic and flammable gases. Luckily, the fire did not spread into other racks.

Approximately an hour after the initial short circuit, firefighters arrived on site. After waiting for an additional two hours, the smoke that had been coming from the container had subsided. The firefighters opened the side container door of the container, thinking that the fire had cooled. Within three minutes of opening this door, the gas within the container spontaneously ignited, scattering debris, and blowing the doors open violently. Four firefighters were injured and immediately brought to the hospital. Fortunately, all survived their injuries, although two were seriously wounded. Following the incident, the battery system was disconnected from the power grid and inoperable, and an investigation conducted by DNV jointly with APS commenced that would take more than 18 months to conclude. Figure 10-1 shows some pictures of the incident, as documented in the report [61].

Figure 10-1
Damage from thermal runaway [62]

In the great rush to reduce battery costs and expand commercial offerings, battery safety can be easily overlooked. As the incident described above demonstrates, batteries contain an enormous amount of energy. When working properly, this energy is carefully charged and discharged by the battery controls. However, an uncontrolled release of this energy can easily turn deadly. Li-ion batteries have inherent safety risks that must be properly designed around and mitigated wherever possible. Since the McMicken incident, and several others like it, fire safety has been at the forefront of battery cell and system design, with more stringent certification, testing, and permitting standards becoming the norm. Industry standards for safety are continually evolving as we learn more about the risks around BESS operations.

This chapter covers the basics of battery safety, including risks to human health and to our environment. We first review the key risks associated with Li-ion technology, and how those risks can be mitigated through the design, construction, and operational phases. We cover all the key standards which govern battery projects, both in the US and around the world. Section 10.4 covers emergency response to battery incidents, and how that response should be planned for, coordinated, and executed.

The second half of the chapter analyzes the environmental concerns of batteries. While often considered a key part of the decarbonization of the power grid, batteries contain toxic substances that require stringent guidelines on their construction, operation, and disposal. Extraction of the raw materials used in Li-ion cells can cause widespread contamination, and some of the mining is happening in conflict zones with weak regulatory structure. We discuss the most common environmental hazards involved in the construction, operation, and disposal of battery cells, along with the related socio-political concerns associated with their manufacture.

While this section is of critical importance, we would also caution that battery safety is a rapidly changing field, with new standards, codes, and products cropping up nearly every month. All the information listed here is current as of this writing, but we urge the reader to verify the latest code provisions applicable to their project whenever this book is used as a reference.

10.1 Dangers of BESS operations

Li-ion battery systems are large masses of electrochemical materials that are designed to store huge amounts of energy. In normal operations, their currents flow in a controlled manner and they can charge and discharge according to the controls structure of the system.

However, there are a few pathways in which a battery can become unable to control, releasing the massive amounts of energy they have stored. If the battery no longer can limit the electrical current flowing to and from each cell, it creates a short circuit and quickly discharges the stored energy, heating it rapidly. If the cell temperature exceeds a certain threshold, this phenomenon is called thermal runaway. It is an uncontrollable increase in temperature (as high as 752°F or 400°C) within the battery cell. This can cause release of flammable gases such as hydrogen fluoride, hydrogen chloride, hydrogen cyanide, and carbon monoxide [63]. Moreover, this can spread as a chain reaction, increasing the temperature of nearby cells, modules, or racks to the point that they cannot be cooled off externally. Smoke and then fire will ensue. The resulting gas build-up, if not properly vented, can result in an acute explosion risk, as was the case in the McMicken incident described above.

Figure 10-2 shows the most common methods that thermal runaway is initiated within battery cells and/or modules:

- Mechanical abuse, in which the cell is physically deformed in a way that causes a short circuit, either by crushing or perforation
- Electrical abuse, where a short develops within the cell. This may happen under normal operational conditions, or if the cell is over-charged or over-discharged

- Thermal abuse, in which the cell is heated to extreme temperatures

Any one of these conditions can set off the chain reaction which can lead to the disastrous effects of thermal runaway, which in turn can induce runaway in other parts of the battery (usually through the heating they experience by being adjacent to the fire). The organic solvent in the cell is the flammable part, being very similar to diesel fuel. And lithium salt creates the toxic smog that occurs it vaporizes due to thermal runawalf

Analyzing battery related safety concerns reveals that non battery components are more likely to cause safety incidents than the battery itself. According to Electric Power Research Institute (EPRI) findings published in May 2024 from their BESS incident database, over 65% of the incidents could be attributed to integration aspects of BESS containerized systems compared to 11% caused by battery cells and modules themselves. Integrating elements such as sensors as part of fire suppression and piping for liquid-cooled thermal management is more manual, has more failure points, and not subject to as strict quality controls as the manufacturing of the cells and modules. These risks should be mitigated as more lessons learned from incidents get incorporated into product design: the rate of BESS incidents has already fallen by 97% between 2018 and 2023 (based on data reported to the EPRI database).

While less common, manufacturing-related issues related to battery cells and modules do persist, such as during electrode material mixing, winding of cells, and cell interconnection welding at module-level. As OEMs ramp up production faster, manufacturing lines are prone to quality issues despite high levels of automation. Factory audits and quality assurance checks are a critical aspect of ensuring safe operations of BESS. Figure 10-2 depicts these causes of battery fires, with the inner ring showing battery-related causes and the outer ring showing non-battery causes.

There are safety codes and standards mandated at a product and project site-level that are designed to make BESS products compliant and ready to operate safely – these are described further below. Most current codes assume that the entire BESS container will be affected by thermal runaway of individual cells if it occurs. Therefore, the best safety practices recommend letting the container burn for extended periods without making an effort to enter or save internal equipment, which prioritizes the safety of first responders and site personnel. Emphasis is placed on preventing propagation of thermal runaway from one container to the next and buildup of explosive gases. The former requirement is often included in the container design and demonstrated via destructive testing by product manufacturers. The latter is enabled by venting mechanisms in the container and/or incorporating deflagration panels that open to depressurize the container.

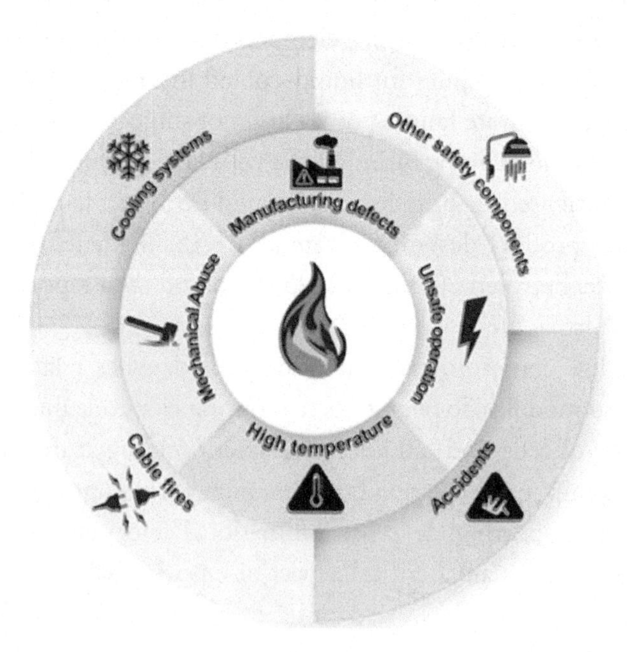

Figure 10-2
BESS fire causes related to battery and BOS [66]

10.2 Mitigating safety risks

The approach to battery safety involves three distinct phases: prevention, warning, and containment.

Prevention focuses on avoiding adverse conditions that could lead to a safety incident. This includes incorporating factors of safety and redundancy in the design, at the cell, module, rack, and container levels. It also includes imposing layers of monitoring and automatic control which keep the BESS within its desired window of operation. Wherever possible, preventing an equipment failure is preferred, since it heads off the problem before it has a chance to become more serious.

Despite the best prevention methods, there will still likely be failures that lead to more serious health and safety incidents, if not properly contained. There is nearly always a brief window between when the first signs of malfunction occur and when the situation becomes irreversible. This period allows for appropriate warning to be given, the second stage of addressing battery safety risks. This can include monitoring cell voltages and temperatures, ensuring smoke and gas detectors are highly sensitive and have redundancies, alerting fire departments of potential incidents, and shutting off power in the case of abnormal cell activity.

Lastly, should the warnings and mitigation taken not be sufficient to prevent a safety incident, the third phase of Containment is used. In this stage, steps are taken to avoid having the fire affect nearby people and the environment, as well as steps to avoid having the incident spread to other parts of the battery system. If fire cannot be controlled early enough, the approach taken by first responders is often to let the container in question burn while spraying adjacent containers to keep them from heating up.

These three methods are summarized in Figure 10-4, which shows the typical time between each phase.

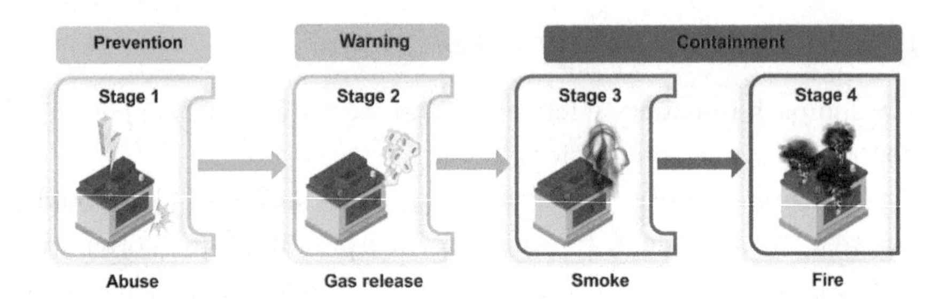

Figure 10-3
Stages of battery fire development

There are always inherent risks of equipment defects, failures, or operator error. Taking the appropriate steps of prevention, warning, and containment can reduce the costs associated with these failures and help protect first responders if an incident does occur. This section discusses how products can be designed, projects can be engineered, and systems can be built and maintained in ways that minimize the risks.

A note on the financial effect of these precautions: battery safety will always have some adverse effect on the cost and performance of the system. Safe design and planning, by their nature, cost time and money and limit the performance of a project. However, when safety incidents occur, their resulting cost to human health, property, and the environment are enormous, making safety-first an approach that makes sense from an ethical as well as a financial perspective.

Lastly, safety is of critical importance to the industry. If battery projects are risky and dangerous, or perceived as such by communities, regulators, and financiers, fewer BESS projects will be approved and connected to the grid. The goal of appropriate safety design is to strike a balance that limits the risk to life, property, and the environment in ways that make the batteries a viable financial prospect.

Product Design

The industry-leading battery integrators and integrators have taken BESS safety as a serious consideration into the design of their products. There is growing awareness and understanding of the standards the products need to adhere to, especially in the US, as elaborated in the next sub-section. Safety compliance requirements are constantly evolving as we learn more about large-scale battery systems and codes iterate.

There seems to be a range of design philosophies in the safety features being integrated into products, going from minimalist and practical to multiple safety measures. In all products, the BMS acts as a first layer of defense by detecting and reporting abnormalities in voltage, current, and temperature at the cell level.

Beyond BMS-level monitoring and control, some OEMs inject water or a clean agent directly into the module that is experiencing an overtemperature event and adjacent modules to cool them down, especially in the case of NMC or NCA cells. Overhead sprinkler systems can also be used but are less preferred since flooding the container with water causes permanent damage to all equipment and results in total loss. Others accept the total loss of a container as near-certainty if a single cell or module experiences a thermal runaway event, so they focus their efforts on ensuring that a thermal runaway event does not lead to build-up of explosive gases and propagate to adjacent containers. Modules may also be physically separated or have a thermal barrier so that propagation between modules is very unlikely. Both approaches are generally accepted by AHJs across the US, if the products comply with applicable codes, as elaborated in the next sub-section.

Some products have additional measures that are designed to address electrical fires using conventional methods, like a pressurized aerosol. Most products have smoke and gas sensors, built-in fire alarms that alert first responders using horns and strobe lights and control panels that communicate critical information to a central fire command station to prepare them before entering the project site.

There are advantages to having multiple layers and forms of protection in the containers as opposed to minimal safety features. But this needs to be balanced with additional time/effort for installation and maintenance, as well as the potential need to satisfy more codes that may ensue (for example, using stored water to spray on heated modules may call upon ASME boiler code standards for the water container, and the use of clean agents may be problematic as they are considered hazardous chemicals).

Project Design

A combination of factors will determine the measures required to ensure the relevant safety compliance requirements are met: local fire code, permitting requirements, and AHJs, such as the local fire department that need to approve the project's design. These measures range from designing the appropriate clearance between containers (sometimes more than manufacturer recommendations) to having fire walls to prevent fire propagation beyond the site. Sites are also required to have access roads that provide navigability within the site with specified width and turn-around radius – typically 20 ft (6.1 m) and 26 ft (1.8 m), respectively – for ease of fire trucks to navigate when responding to an incident.

Further, there are requirements to enable central monitoring of critical safety-related parameters of the whole site such that the source of the issue in the case of an event can be known at one central location, typically via a fire command center (FCC) installed at the entry to the site, similar to building monitoring. This must follow National Fire Protection Association (NFPA) 72 code for alarm and monitoring systems.

Finally, there are typically mandates to install generators or similar backup power sources to keep critical safety features like sensors, alarms, and vents continually powered even when the plant may be offline. Water tanks with specified capacity and responsibility to source the water needed for firefighting may be put on the project developer. This is all dependent on the local AHJ. Some states in the US, such as New York, California, and Arizona, are known to have more stringent requirements and inspections from local fire department authorities because of proximity to urban areas, vegetation, population, and more critically, prior experience with safety incidents pertaining to BESS. It is critical to align with the AHJ in early stages to discuss these details of the design and understand their requirements, as design changes/ enhancements may need to be made before it is too late in the project schedule. There needs to be adequate time to work these out with the battery integrator and EPC.

Construction and Operation

During the construction phase of the project, safety precautions that are mandated by the National Electric Code (NEC) and typical of electrical system construction sites must be adhered to. In addition, large-scale battery systems typically require some integration of components within the container on site such as wiring of individual modules, piping for water, etc. Manufacturers' guidelines must always be followed. Installation and commissioning of the battery systems are usually performed by technicians from the manufacturer who know the product best. Further, many local AHJs require a fire-safety inspection by the local fire-department and/or a certified fire-protection technician. This needs to be accounted for in the schedule as an essential milestone at a time agreed upon with the AHJ.

On-site operating personnel must also have dedicated BESS-specific training both for regular maintenance operations and safety incidents. An Emergency Response Plan specific to the energy storage technology at the project must be developed and shared in advance with the local fire department. There are public templates for this important document, for example the Draft Emergency Response Plan provided by American Clean Power [67].

Several hazards are present during construction and operation, associated with energy storage systems, such as shock or arcing hazard, resisting high pressure water intrusion, toxic gases catching fire, and explosion hazard from contact with air. The key aspect of addressing a thermal runaway event is not to exacerbate flammable conditions within the containers – one of the key learnings from incidents such as the McMicken fire is that in the case of gaseous build up within the container, it must not be opened but let to vent and/or burn itself out. Adjacent containers are often cooled to ensure temperatures do not climb beyond target levels.

10.3 Codes and standards

A set of codes and standards have been developed as safety compliance requirements related to stationary battery systems, including for utility-scale BESS. These are drawn from pre-existing electrical safety standards, as well as from experience gained from BESS-related fire events in recent years. These codes and requirements for BESS projects and products are continually being refined as experience is gained, projects expand in size and complexity, and technology evolves. The compliance requirements are being enforced to varying degrees across the county – depending on the state's fire code as well as requirements mandated by local AHJs and site conditions. These also evolve with time as awareness about BESS grows and lessons learned are shared across the industry. It is best practice to design systems to be compliant with all the major codes listed below. As mentioned, the local AHJ has the final say in safety compliance and may have requirements above and beyond the codes listed below. Contacting the local AHJ and coming to an understanding of the project and safety requirements early is critical to ensuring smooth execution of the project.

The following certifications apply to stationary BESS: UL 1973 for battery modules and packs, UL 991 for BMS, and UL 1741 for PCS, etc. The major compliance codes for the project to be compliant with are explained below. This is not a comprehensive list – as mentioned, it is best practice to check for specific requirements with the local AHJ. Notably, UL 1642 is omitted here, as it applies to battery cells but is primarily applicable for cells in consumer electronics.

NFPA 70 / National Electric Code

NFPA code 70, otherwise called the National Electric Code, is a US standard for the safe installation and maintenance of electrical wiring and equipment on any kind of electrical system. Most US states and their respective jurisdictions adopt and enforce the NEC through state laws. Some states use different versions of the code, and others have amendments that relate to their specific jurisdictional requirements. For example, in California, there are special provisions around residential PV installations to coincide with the state's renewable energy standards. Some countries even base their own codes on the NEC: According to the NFPA, Mexico, Guatemala, Nicaragua, El Salvador, Honduras, Costa Rica, Panama, the Dominican Republic, Colombia, Ecuador, Peru, and Venezuela rely, in some form, on the NEC.

Compliance with the NEC is typically enforced by the AHJ for a given project. Typically plan sets are required to be approved by the AHJ. The plans are reviewed, and any issues that the reviewer finds are flagged and addressed. Although the NEC is a comprehensive code, given its many nuances, it is subject to interpretation. Many approvals by the AHJ are iterative discussions in which issues are dealt with progressively with each successive round of comments.

The NEC goes hand-in-hand with NFPA 70E, the Standard for Electrical Safety in the Workplace, which focuses on electrical safety standards. NFPA 70E includes guidelines for safe design and construction such as lockout/tagout procedures to prevent unexpected energization of equipment and protection boundaries against arc-flash incidents. It also mandates the formulation and implementation of an Electrical Safety Program that addresses training, PPE, and safe practices for workers on site. This includes determinations of what level of protection must be worn when accessing or servicing energy storage systems. Lastly, NFPA 70E provides guidance on maintenance manuals of electrical equipment and guidelines for upkeep of equipment for safe operations.

NFPA 855

NFPA 855 is the Standard for the Installation of Stationary Energy Storage Systems. It covers fire safety aspects for the design, construction, installation, commissioning, operation, maintenance, and decommissioning of BESS. It is a comprehensive standard that includes aspects of other NFPA standards, such as NFPA 69 for explosion prevention, NFPA 68 for explosion prevention by deflagration venting, NFPA 72 for alarm and monitoring systems, NFPA 13 for sprinkler systems (if installed), and UL9540 for the battery system itself. The underlying philosophy of the standard is that thermal runaway will spread within a container, and that the most effective way for it to be tackled is to let it burn while preventing build-up of explosive gases. Hence, emphasis is placed on explosion prevention and off-gassing of flammable gases.

During the permitting and construction phase, the code typically requires submission of detailed planning documents for fire safety approval. This includes location, a layout of the site depicting the BESS, manufacturer specifications and ratings, details on fire suppression and smoke detection, thermal management and deflagration, and a commissioning plan. Installation and commissioning requirements largely defer to the manufacturer's guidelines, but additional constraints can be applied.

Given its overarching nature and acceptance at the national-level, NFPA 855 can be considered a comprehensive baseline for global BESS fire-safety design.

NFPA 68 and 69

One of the predominant causes of fire and explosion inside BESS enclosures is the release and ignition of combustible vapors in flammable concentrations when cells go into thermal runaway. Because of this, BESS explosion prevention is one of the crucial aspects of safe design. NFPA 855 requires explosion prevention by installing protective systems as prescribed by either NFPA 68 – the Standard for Explosion Prevention by Deflagration Venting or NFPA 69 – Standard for Explosion Prevention Systems. NFPA 69 is a more generic standard that applies to explosions of any type, while NFPA 68 is narrower, as it addresses one method of explosion mitigation – the design of deflagration venting. Specifically, this means the venting of explosive gases, and/or the inclusion of specialized panels intended to cede in the case of an explosion. These deflagration panels, as they are known, are intentionally designed as the weakest portion of the container, so that they will give way if an explosion should occur. Figure 10-4 shows a rendering of deflagration panels in a BESS container giving way in the instant of an explosion.

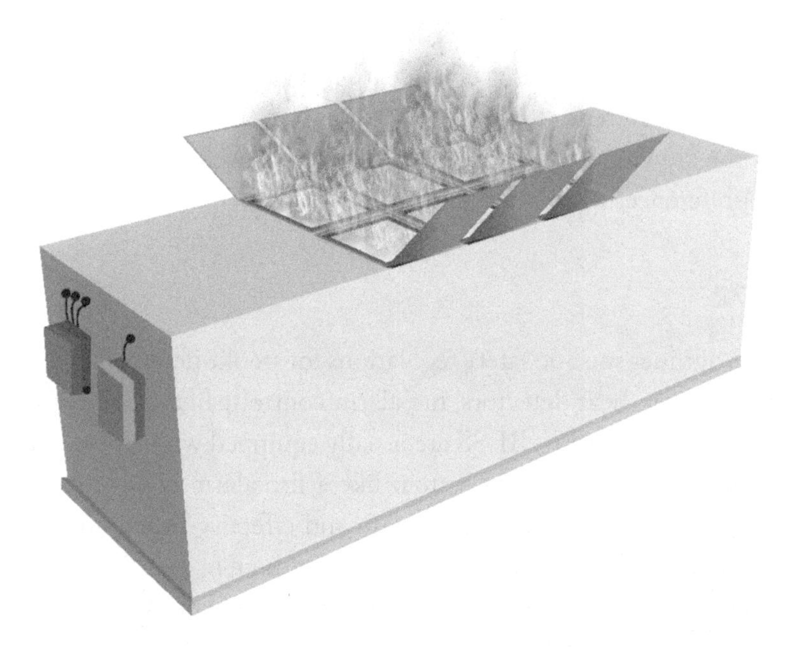

Figure 10-4
Illustration of deflagration panel function in BESS container

Not all battery enclosures have deflagration panels – some designers have opted to employ other methods allowed by NFPA 68 or 69. Others offer deflagration panels upon request at an added charge.

The NFPA 68/69 guidelines collectively include requirements for gas detection, ventilation, and explosion mitigation systems to mitigate the risks associated with BESS operations. NFPA 69 requires the combustible concentration within the enclosure to be maintained at or below 25% of the lower flammable limit (LFL), which is the lowest concentration at which a certain gas will combust, for a given temperature and pressure. The standard calls for more active measures including (but not limited to) gas sensors, smoke detectors, and exhaust ventilation to be designed and installed in the enclosures. NFPA 68 prescribes "passive" deflagration vents to be designed in/added to the container that will be forced open when there is increase in pressure caused by build-up of gases, and expelling the gases before they reach flammable limits.

At the time of writing this book, explosion prevention using measures that comply with either NFPA 69 or 68 is acceptable as per NFPA 855. However, this is evolving with learnings from various incidents and the preference for active explosion prevention per NFPA 69 seems to be growing. It is always best practice to check the local AHJs specific preferences and requirements.

NFPA 72

NFPA 72 outlines specific safety regulations for smoke detectors, alarm signaling devices, heat detectors, fire alarm control panels, and design redundancy requirements. BESS are usually equipped with smoke and gas detectors, and a reporting system like a fire alarm control panel. NFPA 72 sets requirements for the safe and effective functioning of the alarms and back-up power requirement for them in case the system is down. In the context of utility-scale BESS, where typically there are several containers spread around a large site of many acres, this code mandates a central fire command center to which all the individual fire alarm panels must be connected. This is intended to serve as the point of information for first responders in the case of an event. The BESS operator needs to have continuous remote monitoring of the system to alert the local fire department in the case of an event. Third-party monitoring companies can also perform this function, using an auto-dialer that is to the fire command center.

UL 9540

Developed by Underwriters Laboratories (UL) for energy storage systems first in 2016, the UL 9540 "Standard for Safety of Energy Storage Systems and Equipment" provides requirements and tests for the safety of energy storage systems as a whole, including batteries and their associated equipment such as the PCS; see Figure 10-5 for an overview of UL 9540 tests. It sets requirements for design, construction, testing, and operation of storage systems, covering various aspects such as electrical, mechanical, thermal, and environmental considerations to prevent fire, electric shock, and other hazards associated with storage systems. This calls for compliance with other standards, such as UL 1741 for power conversion systems that most major manufacturers readily meet, and UL 1973 which is a standard for battery systems at the component level.

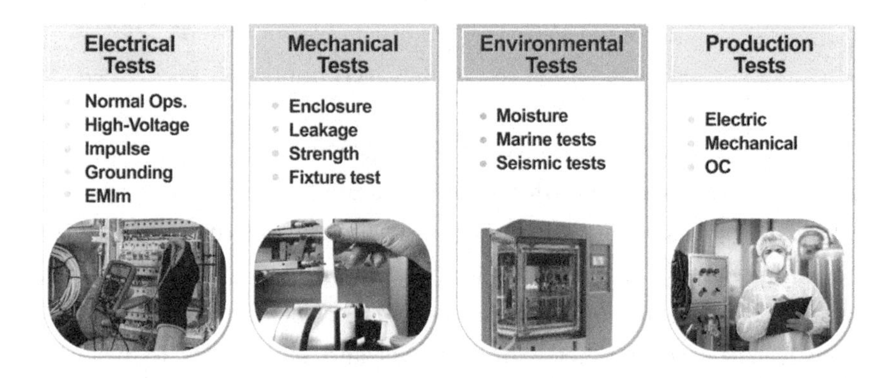

Figure 10-5
Overview of UL 9540 tests

The UL 9540 standard can be met as a product-listing by product manufacturers and integrators at the factory, which is increasingly being preferred by authorities in several parts of the US. If not, a field evaluation must be performed at the project site by a third-party certifying authority. Meeting this standard is a requirement per the 2023 version of NFPA 855, as well as state-level fire codes in states like California and New York.

UL 9540A

The UL 9540A is a test method for evaluating thermal runaway fire propagation in BESS. It is a series of progressively larger fire tests, going from the cell-level to module-, rack- and installation-level testing, as shown in Figure 10-6.

Figure 10-6
Picture from Installation-level test [68]

It is used to check the level of propagation, if any, of thermal runaway and its associated effects onto the adjacent unit. The method establishes criteria for when progressive testing at the next level is not required, that is, for example if thermal runaway is shown to be contained at the module-level, rack-level testing may not be required. However, it is best practice for OEMs and integrators to perform large-scale fire testing at the container- or installation-level, as this is what some AHJs would particularly require. See Figure 10-6 for an example of an installation level test. There is no pass or fail criteria in this test, instead the collected data is intended to be provided to the local authorities for evaluation of the product for the site in question. Performing the UL 9540A or an equivalent fire test is a mandatory requirement for most state fire-codes as well as the NFPA 855. This test is typically required to qualify for exemptions from clearance requirements set for energy storage systems as per NFPA 855 or local fire codes.

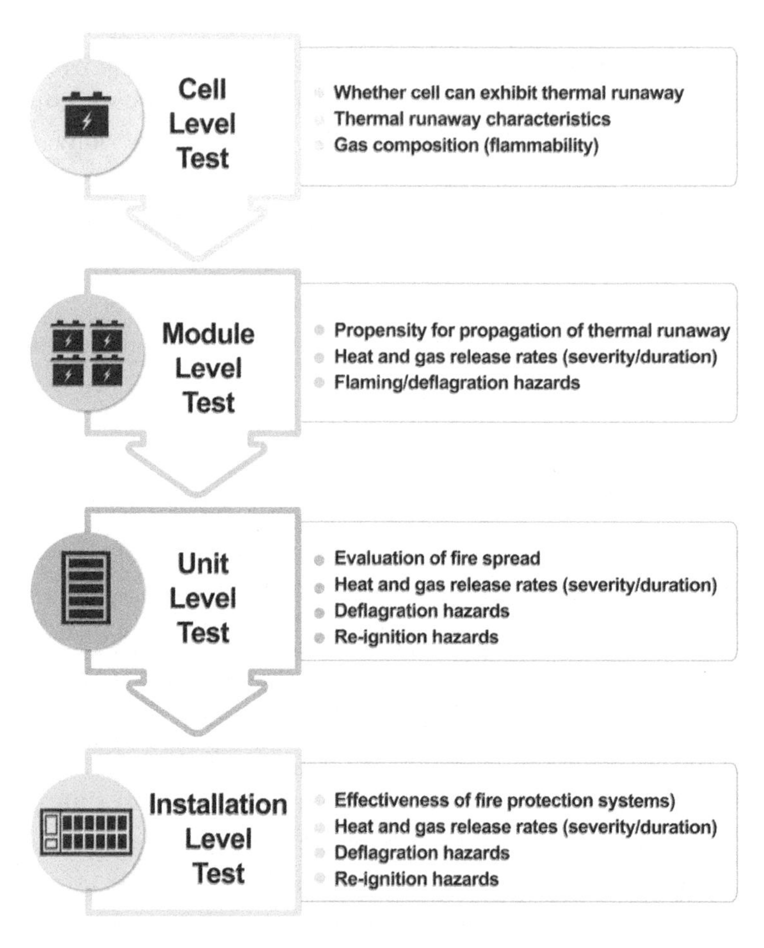

Figure 10-7
Sequence of 9540A testing

UL 1973

UL 1973 covers product design, construction, safety, and production tests for battery systems. It lays out a series of construction parameters, including requirements for nonmetallic materials, metallic parts resisting corrosion, enclosures, wiring and terminals, electrical spacing and separation of circuits, insulation and protective grounding, protective circuits and controls, cooling/thermal management, electrolyte containment, battery cell construction, and system safety analyses. It also outlines a series of safety performance tests for energy storage solutions, including electrical tests such as an overcharge test, short circuit test, over-discharge protection test, temperature and operating limits check test, imbalanced charging test, dielectric voltage test, continuity test, failure of cooling/thermal stability system test, and working voltage measurements [69]. It includes mechanical tests and safety tests – the latter involves an internal and external fire test. The battery system must be able to demonstrate a lack of propagation of fire between modules internally, as well as to other equipment and/or containers externally.

UL 1741

UL 1741 is the industry standard for inverter safety. It mandates inverters stay connected to the grid, even in times of instability, and meet rigorous requirements to support the grid. The five primary grid support functions that are expected with UL 1741 are to a) provide predictable and controllable ramp rates of power production when connecting to the grid, b) supply a consistent specified power factor, c) supply reactive power to control grid voltage, d) supply real power to control grid voltage, and e) supply real power to control grid frequency [70]. The inverter must be able to do this in interactive mode - meaning without instruction from the grid operator, by sensing local conditions.

IFC

The International Fire Code (IFC) is one of 15 International Codes (I-codes) published by the International Code Council (ICC), and is available for free at https://codes.iccsafe.org/. The IFC is intended to set an international standard for fire safety protocols, and has been adopted by some nations, often with amendments that make the code more appropriate for a certain country. The code covers the minimum fire safety requirements for new and existing facilities, storage, and processes. The fire-related sections of the IFC include fire prevention, fire protection, safe storage, and use of hazardous materials.

The IFC provides a comprehensive approach to controlling hazards for a variety of buildings and sites, regardless of the hazard being indoors or outdoors. The battery-specific safety approach and preventative measure requirements have been harmonized with NFPA 855, so the requirements are similar. This safety standard is used as the industry benchmark for storage safety in projects across the world, though additional local requirements may apply.

The IFC is issued every 3 years – the current edition is 2024, with a new version to be released in 2027. Note that some jurisdictions may not adopt the latest version, so it is important to check the law to see what version is referenced. Section 1207 of the IFC, Electrical Energy Storage Systems (ESS), is an 18-page guide of safety and fire requirements for energy storage systems. The section on Li-ion batteries is applicable for all systems over 20 kWh.

Some of the key elements of IFC's Section 1207 are summarized here:

- Li-ion battery systems greater than 600 kWh must be located more than 100 ft (30.5 m) from buildings, lot lines, public ways, stored materials, hazardous materials, high-piled stock, and other hazards.
- IFC requires a Hazard Mitigation Analysis (HMA) which can be a Failure Mode and Effects Analysis (FMEA) or similar of

all the potential risks, modes of failure, and documenting ways to address them. This must be conducted on a project level, although many aspects will be similar for all installations using any given battery product. This can be waived by the AHJ in some cases (see final bullet for more).

- For remote locations, the 600 kWh sizing can only be exceeded with appropriate results from UL9540A. All ESS requires some form of fire suppression, unless the AHJ has allowed it to be omitted. All units need smoke and automatic fire detection, and appropriate clearance to exposures.
- The IFC code gives extensive power to AHJ to interpret the code, such as in the inclusion of a fire suppression system, explosion control requirement, the HMA, the 100-ft buffer, or the requirement to transmit signals to a central station.

IEC 62619

The International Electrotechnical Commission (IEC) standard 62619 is a 39-page code devoted to battery system construction and testing. It is analogous to UL 1973. It is a system-level regulation and defers to other standards (such as IEC 62133 for cell-level requirements). Its guidelines are focused on making batteries designed for the application they are used for. The standard defines several failure scenario tests. It has special guidelines for large-format Li-ion batteries (which are the most common in utility-scale BESS products). IEC 62619 includes a fire propagation test in which one cell is forced into thermal runaway (similar to the UL9540A testing).

IEC 62933

IEC 62933 Electrical energy storage (EES) Systems is a large and comprehensive code. IEC codes are more common in Europe and countries with European geographical or historical links. Although the US-based 9540A fire testing standard is used in most of the world, many countries opt to use IEC codes for technical compliance. This code is over 400 pages, covering all aspects of the design, installation, and operation of Electric Energy Storage Systems. In many ways it is analogous to the requirements of UL 1973, but it has some distinct provisions

10.4 Emergency Response

At McMicken, first responders were injured because of a lack of knowledge and training on how to respond to safety events at a BESS site. Since then, there has been growing awareness on the type of actions first responders must take to address such emergency events. NFPA 855 mandates that every BESS project must have an Emergency Response Plan that documents site-specific risks, emergency mitigation measures recommended by the battery supplier, and site monitoring and communication measures that have been put in place. This needs to be discussed with the local fire department officials for smooth addressal of issues.

The NFPA lays out some guidelines for management of BESS events in its BESS Emergencies Quick Reference Guide [71]. It recommends an "identify, shut-down, watch-out" approach. The battery containers should always be considered energized, as they retain energy and remain charged even if they are turned off/shutdown. In addition, the American Clean Power association has put out a guideline document for response to safety events called "First responders guide to lithium-ion battery energy storage incidents" [63]. It recommends that once an incident has been reported, the first responders should assess the site for issues along with site personnel and when possible, sites should be remotely monitored. It further lays out that "full firefighter protective gear should be worn where there is any possibility of fire or explosion, including proper use of self-contained breathing apparatus (SCBA). If there is no risk of fire or explosion per the project incident command, protective clothing for arc-flash and shock hazards should be worn by anyone operating within the arc-flash boundary." In the case of Li-ion battery systems, an observable fire is positive because it ensures that flammable gases will be consumed as they are released, and an explosion is unlikely. Hence the "let it burn" approach – which means to allow the enclosure to burn in a controlled manner, so that all fuel is consumed and the possibility of reignition is minimized – is considered the safest outcome for such an event. Containers adjacent to the one in fault must be monitored and may be sprayed with water defensively to prevent propagation of heat.

If a thermal runaway has been detected and reported by the BMS and EMS, but no flames are observed, then the system is at risk for a potential explosion as thermal runaway is likely to have generated flammable gases that may build up to explosive levels if not ventilated. The container should be observed from a safe distance ideally using BMS data, or using thermal scanning if the BMS is damaged. All BESS must be equipped with measures compliant with NFPA 69 or 68, as elaborated above, to mitigate these explosion hazards.

10.5 Environmental

Beyond safety concerns, batteries have extensive effects on their environment. These effects can be divided into three stages: manufacturing, operation, and decommissioning.

This section deals with the environmental impacts around the manufacture of Li-ion batteries. For discussion of the environmental compliance of individual battery projects, see Section 6.4.

Environmental Impacts of Battery Manufacturing

The production of Li-ion batteries, like any industrial process, has environmental impacts. One of the key environmental concerns associated with the manufacturing of Li-ion batteries is around extraction of raw materials; the mining of lithium, cobalt, nickel, and other minerals required for battery production can have adverse environmental impacts. It can involve deforestation, habitat destruction, soil degradation, water pollution, and greenhouse gas emissions. Lithium mining requires vast amounts of water. Especially, the brine extraction method involves pumping large volumes of water– estimated between 500,000 to two million gallons of water per ton of lithium – to extract lithium salt from underground brine pools. This is particularly challenging in the planet's most Lithium-rich places like the Salar de Atacama in Chile or the Bolivian salt-flats, which are extremely dry. This process can deplete and/or contaminate local water supplies, impacting ecosystems and communities. If not handled correctly, toxic chemicals and heavy metals are released into the environment [72].

The energy consumption of manufacturing Li-ion batteries is also significant. Extracting and treating the raw material extraction is energy intensive. The fabrication of cells, assembly of modules and racks, and transportation from factory to site collectively consume considerable energy. Energy consumption and sustainability depends on the country of manufacturing. Although there are OEMs such as Tesla that have pledged to power their factories exclusively with renewable energy [73], most manufacturing that happens in Asian countries is predominantly powered by fossil-fueled generation.

Researchers are exploring sustainable alternatives for battery components, including aluminum, sodium-based salts, iron, silicon, and even hemp [74]. Technologies such as flow batteries and gravity-based storage are gaining traction, especially with the desire for longer duration batteries (> 6 hours) for which they are better suited. There is hope that energy storage will become more sustainable in the coming decade, though lithium is expected to remain the battery leader for many years.

Recycling

At the end of their life cycle, Li-ion batteries need to be properly managed to minimize further environmental impact. If not handled correctly, they can again release toxic chemicals and heavy metals into the environment. While most of the materials in Li-ion batteries are recyclable, they contain hazardous materials such as lithium salts and transition metals, which can pose environmental and health risks if they enter the environment or the food chain.

Current recycling processes include pyrometallurgy, which involves a high-temperature treatment, and hydrometallurgy, which is extraction of metals using aqueous solutions or water-based solutions. The recycling process typically involves the following steps: collection and sorting of used batteries, mechanical shredding to break them down into smaller pieces, including the battery cells and components, leaching to dissolve the metals from the battery components, filtration and separation, purification to separate and concentrate specific metals and recover desired metals.

These processes are energy-intensive and often result in the loss of valuable materials. Extraction of lithium from old batteries is typically more expensive than mined lithium and cannot be done profitably [75]. The other challenge is that the state of health and composition of the batteries at third-party facilities can vary significantly, which can make the sorting and processing inefficient. Nonetheless, lithium extraction from Li-ion batteries has been demonstrated in small setups by various entities as well as in production scale [75]. This facet of the industry is expected to experience an imminent boom. As a result, Li-ion battery recycling has already gained momentum in the US, with companies like Redwood Materials, Ascend Elements, and Li-Cycle planning recycling plants in the US [76].

An alternative to recycling is reusing Li-ion batteries in second-life applications. For example, EV batteries eventually lose their efficacy for transportation but still have enough life for stationary

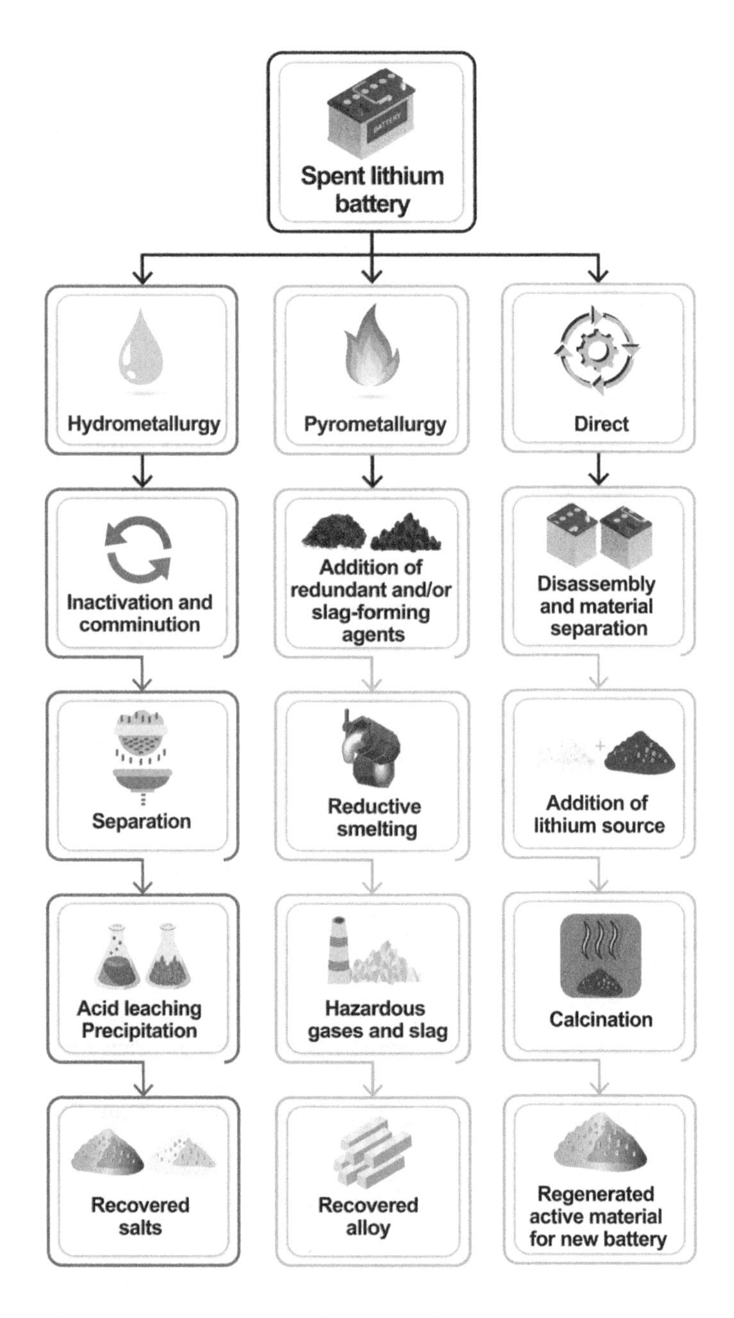

Figure 10-8
Recycling process types

purposes. Due to age, these batteries have a higher risk of developing dendrites that can lead to accelerated aging and heightened risk of thermal runaway. Therefore second-life applications are still in the research and demonstration phases and require further testing before widescale adoption.

Is Storage Renewable?

As we have discussed throughout the book, BESS plants can provide a wide variety of services to the grid, from frequency regulation to transmission deferral, to energy arbitrage. However, one of the key uses of storage is to store energy caused by intermittency of solar and wind resources. In this use case, batteries absorb excess energy during peaks and release energy to fill troughs to smooth generation and meet grid demand as shown in Figure 10-9 from the EU Smart Cities Information System.

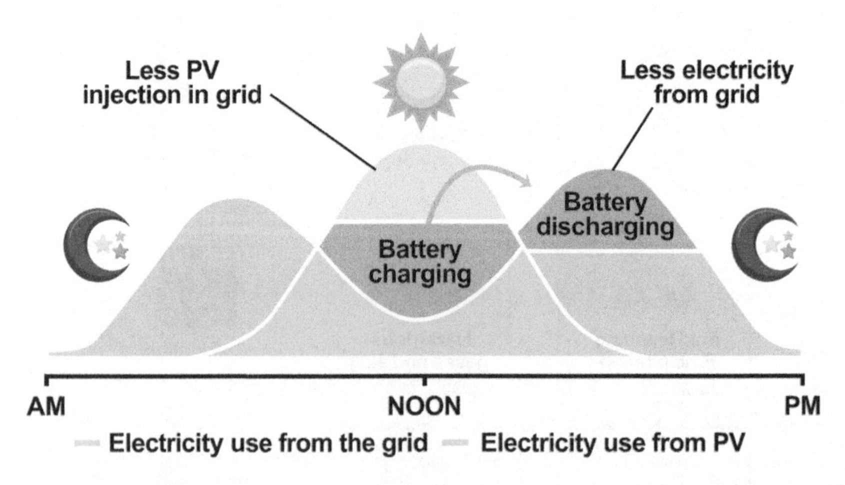

Figure 10-9
Illustrative typical profile of PV+Storage [77]

This shift is frequently driven by market mechanisms – since high amounts of solar on the grid typically drive prices down, PV+Storage plants are incentivized to store the PV energy in the daytime (when energy is cheap) and discharge it after the sun has gone down (when energy prices are higher). In this role batteries play a strong role in enabling more renewables to come on to the grid. In its 2023 Energy Transition Outlook, DNV stated that "In 2050, solar PV will be in unassailable position as the cheapest source of new electricity globally [4]." In this position, energy storage clearly appears to be an enabling technology for the decarbonization of the power grid, a cornerstone of efforts to combat climate change.

However, as we have seen, Li-ion batteries are typically not economical for storage beyond four hours, meaning that, coupled with solar, they will not allow for full decarbonization of the grid. This will likely be accomplished with other long-duration energy storage (LDES) technologies. These are discussed further in Section 11.2. Additionally, batteries come with a host of other environmental and social concerns. They are also a non-renewable resource, since they depend on finite minerals that come from the ground.

Deciding whether batteries are green is a nuanced decision which, by necessity, includes assigning value to intangible things such as the value of our environment, and social / political effects of batteries. One method for evaluating the worth of batteries is via their carbon footprint. According to environmental group 8 billion Trees [78], a modern EV battery of 200 kWh has emissions of 8 to 40 tons of CO_2 equivalent. Using the low end of this range results in 40,000 tons of CO_2 in a 5 MWh battery container!

Although these numbers involve many assumptions, they demonstrate that there is a significant amount of carbon involved in the manufacture of Li-ion batteries. In order for batteries to have a net-negative carbon impact, they must show that over their life they are able to justify the carbon invested in their creation. Calculating the net carbon effect of a BESS requires a complex process of evaluating the expected charge and discharge of the grid, and what type of power is offset. For example, if a certain market incentivizes a battery to charge when the grid is heavily dependent on fossil fuels, and to discharge when renewables represent a large amount of the power mix, the net footprint of the BESS may be positive, meaning the project increases the amount of carbon.

Several consulting firms have arisen to track this for a given battery project, since this varies based on the market, the technology, and the way the battery is operated. Overall, batteries built in competitive energy markets with heavy renewables contributions will tend to be carbon negative, since they are incentivized to charge when renewables are plentiful and discharge when the grid is more dependent on non-renewable sources.

10.6 Socio-political Issues

Beyond environmental issues, batteries also have some significant socio-political impacts which should be considered. While most energy storage projects are seen as positive developments as part of the decarbonization process, their manufacturing, installation, and operation come with several detrimental side effects. This section covers some of the more salient concerns in this area, and how the industry has attempted to mitigate them as it grows.

Cobalt in Batteries

Research in the lab has considered hundreds of different materials before discovering the elements that create the most efficient chemistries used in Li-ion cells. This research typically does not consider the source of the materials, but some of these elements are quite rare and can only be sourced in certain areas.

Cobalt is at the forefront of this concern. Cobalt was discovered to be an ideal ingredient in high-power energy cells. It is found in two of the leading chemistries: lithium nickel manganese cobalt oxides (commonly referred to as NMC or NCM) and lithium nickel cobalt aluminum oxides (commonly known as NCA). The initial Tesla vehicles that were released in the 2010s used Panasonic NCA cells, and they were also used in electronic devices by manufacturers such as Sony and Samsung. Prior to approximately 2020, many stationary storage batteries used NCM cells. In 2021 and later, the stationary storage market became dominated by lithium iron phosphate (LFP) cells (these chemistries are covered in depth in Section 3.5).

NCM remains a popular chemistry in many EVs due to its high-power applications and high energy mass density (Wh / kg). Cobalt is a rare commodity, and 73% of it comes from Democratic Republic of The Congo (DRC), according to the Cobalt Institute [79]. DRC is the second-largest country in Africa at 2.3M km^2, and its government does not exert control over large parts of the country or its economy. Much of the cobalt production is from artisanal mining that often involves dangerous working conditions like extraction by hand, unsupported tunnels, and a lack of protective equipment. Many children are forced to work in unsafe conditions, and cobalt ore from these artisanal mines is often mixed with that coming from more regulated mining operations, making it difficult to track.

Because of these issues, many firms in the BESS industry have made efforts at tracking the provenance of Cobalt in their cells. Others have focused on R&D of cells that use less Cobalt. Since the stationary industry has largely switched to LFP chemistry (primarily due to improved safety performance and lower cost), Cobalt has become less of a concern; however, it is still a hot topic in energy storage, especially around EVs.

Mining of Limited Resources

The global lithium reserves are estimated to be around 98 million tons, with most of the accessible resources being in Chile and Australia [80]. At the current rate of extraction, it might seem sufficient for the foreseeable future, but with the growing demand for EVs, personal devices, and stationary energy storage, the demand for lithium could outpace supply. According to some industry sources, the demand for lithium is expected to grow to over two million tons per year by 2030, growing by about four times of current estimates [81]. A related challenge is to build up the infrastructure to harness lithium as fast as we need it as battery manufacturing can and is being grown much faster than lithium extraction.

Similarly, cobalt and nickel are finite resources and can face similar challenges. As demand for batteries soars, depletion of resources may be a bottleneck, and the use of any limited resource is by definition not sustainable.

The BESS industry has sought to mitigate these concerns in several ways. One is the development of battery technology that uses more abundant minerals – two chemistries currently under development are iron-air and sodium-ion, both of which utilize minerals that, while also finite, are present in much more abundant quantities. These types of new chemistry are covered in depth in Section 11.2.

Other strategies have been to pursue novel sources of rare elements. One of the more prominent strategies is to harvest "nodules," from the sea floor, where elements common to batteries have accumulated over millions of years. Nodules have been found to have high concentrations of cobalt, copper, iron, manganese, and nickel, which some have referred to as a "battery in a rock" [82]. While this strategy appears to sidestep many of the social concerns around mining in places like the DRC, it comes with a host of potential issues – disturbing a little-understood, precious ecological zone, difficulty in regulation due to their remote location, and private companies muscling out claims from nations. Despite these concerns, seabed mining and exploration has been ramping up, and is a growing source for minerals that will be essential to energy storage.

Political tensions

Although batteries are a global enterprise, there are many political tensions in the industry. The biggest is between the US and China, international rivals who are both key players in energy storage. China manufactures most of the cells used in energy storage products and is the biggest single deployer of battery technology. The US has developed some of the most competitive energy storage markets, some of the software used in BESS products, and is a leading installer of BESS. Over the past 20 years, China has made heavy investments into the mining of raw materials, the refining of those materials, and the manufacture of energy cells, making their control of the industry a strategic risk for the US The US has tried to counter these advantages with the 2022 Inflation Reduction Act (IRA) which includes tax advantages for battery cell manufacturers who build factories in the US, and special tax credits for projects which are sourced primarily from US materials. Similar to the raw materials chain for PV modules, the US has put in place protectionist policies that have attempted to combat China's lower production price.

The Uyghur Region in northwest China has been one of the main hubs for manufacture of materials involved in EV and solar supply [83]. In a July 2022 report to the U.N.'s Human Rights Council, the UN Special Rapporteur on Contemporary Forms of Slavery stated that he "regards it as reasonable to conclude that forced labor among Uighur, Kazakh and other ethnic minorities in sectors such as agriculture and manufacturing has been occurring in the Xinjiang Uighur Autonomous Region of China." [84] In 2021 the US passed the Uyghur Forced Labor Prevention Act (H.R. 6256) imposed sanctions on anyone who "knowingly engages" in business with goods produced in Xinjiang using forced labor. The bill also widened sanctions that curtailed visas and blocked assets for foreign entities and individuals responsible for abuses of human rights in connection with forced labor. In an August 2022 letter, China's Permanent Mission to the UN in Geneva stated that "Anti-China forces in the US and the West ... have tried to obstruct and undermine the progress of human rights in Xinjiang under the pretext of protecting human rights."

In addition to human rights concerns, the US has cybersecurity concerns around Chinese infrastructure. In February 2024, Duke Energy decided to remove a Chinese-made battery system from US Marine Corps Base Camp Lejeune, less than one year after the system had been commissioned, in favor of a "domestic or allied nation supplier" [85].

Figure 10-10
Camp Lejeune BESS decommissioned by Duke Energy in 2024 [85]

As of publication of this book, most utility-scale BESS projects continue to be built with Chinese cells, but this may be an ongoing concern as the US and China continue their rivalry.

10.7 Usual Suspects

Safety Advisors

Designing energy storage systems that are compliant with increasing safety requirements mandated by standards and AHJs is critical to project success. Since safety mandates are evolving across the country, it is highly beneficial to engage safety advisors who are involved in the development of safety standards and have experience working with local fire departments to understand the exact requirements pertaining to the project. Some of the well-renowned names offering this type of advisory are listed below. Renowned traditional engineering consulting companies are increasingly growing this area of expertise in-house and helpful as well.

DNV: In addition to being a leading IE and OE advisory firm, Norway-based DNV offers specialized safety evaluation and advisory services. They regularly publish public papers on risk appetite for financiers and insurance providers owing to independent engineering expertise in the renewables industry. Their annual Battery Performance Scorecard also includes some safety insights.

Energy Safety Response Group (ESRG): Apart from design, permitting, engineering support and training, ESRG specializes in testing of BESS equipment to UL standards and other destructive testing methods and product certification

Fire and Risk Alliance: A firm specializing in solutions for fire protection, risk, and life safety projects. Fire & Risk Alliance provides services including fire protection consulting, hazard analysis, risk analysis, incident investigation, and applied research services

Jensen and Hughes, Inc.: Offering specialized fire-safety consulting, risk and hazard analysis, process safety, forensic investigations, and emergency management since 1939.

The Hiller Companies (formerly American Fire Technologies Inc): Provide a full range of offerings including design, stakeholder management, and installation, service and maintenance of fire protection systems

NRTLs

Nationally Recognized Testing Laboratories (NRTLs) are accredited as approved certifiers of specific industry standards. The Occupational Safety and Health Administration (OSHA) accreditation as an NRTL recognizes private sector organizations to perform certification for certain products to ensure that they meet the requirements of both the construction and general industry OSHA electrical standards. The NRTLs who work in BESS product and site certification are:

CSA Group: Widely recognized by numerous North American and international organizations for providing testing and certification services for a broad range of product categories including electrical, gas-fired, plumbing, personal protective equipment, and others.

Intertek: Provides assurance, testing, inspection, and certification services to a wide range of industries.

TUV Rheinland: A renowned independent provider of inspection and testing services, boasting over 150 years of service and 22,000 experts worldwide.

UL: A globally recognized name in standardization, safety science, and reliability testing of a vast range of equipment including BESS. As one can gather from the above sections, several BESS-related standards set by UL are viewed as industry benchmarks. UL performs a lot of in-house testing to award certifications of standard compliance across the industry.

Battery Testing Authorities

Beyond the NRTLs, there are lab testing facilities where testing of cells and modules can be done for performance assessment, cycling life evaluation, and safety testing. A couple of noteworthy ones are:

ESRG: As mentioned above, ESRG also offers battery fire testing services during product development and pre-certification stages so that results can be incorporated into the developing product design. They offer real-world experience as well as guidance from academic experts to guide the product development process. They are also qualified to perform safety-testing for UL-certification purposes.

NYBEST: The New York Battery and Energy Storage Technology (BEST) Test & Commercialization Center (BEST Test Center) in Rochester, New York, provides an extensive range of testing services covering performance and safety for batteries, controllers, and complete energy storage systems and has conducted over 7.5 million hours of battery cycle testing. This Lab was formerly run by DNV until it was taken over by the Rochester Institute of Technology (RIT) in 2024.

Battery Recyclers

As touched-upon previously, some of the reputed names in battery recycling in the US include [86]:

Ascend Elements: A closed-loop recycling firm founded in 2015 that has been operating out of pilot plants in Massachusetts and opened a commercial recycling plant in Covington, Georgia in early 2023. It has secured $300 million in equity and debt financing as of late 2022.

American Battery Technology Company: Founded in 2011, is a startup that programs off-the-shelf manufacturing machines to disassemble instead of assemble battery modules. It received a $10 million grant from the US Department of Energy in late 2022.

Li-Cycle: This firm has a facility in Arizona that shreds batteries while submerged in a solution to tamp down fire risk, then use hydrometallurgy to isolate valuable minerals at a planned facility in Rochester, NY, for which it recently obtained a conditional loan of $375 million from the US Department of Energy.

Moment Energy: British Columbia-based Moment Energy is manufacturing a stationary energy storage product manufactured from repurposed Li-ion EV battery modules. They are the first to be certified to UL 1974 (the standard on repurposed batteries) and have developed a product focused on the C&I market.

Redwood Materials: Led by Tesla's co-founder and Chief Technology Officer, JB Straubel, based in Carson City, Nevada. It has acquired a $2 billion conditional loan in May 2023 from the US Department of Energy for development of the recycling facility.

11. THE FUTURE OF ENERGY STORAGE

As of the end of 2023, there was approximately 90 GW / 230 GWh of battery energy storage connected to the world's power grids, according to BNEF – and around 42 GW of this was added in 2023 alone [87]. While this is steep growth, DNV predicts that stationary battery capacity will rise to over 20,000 GWh (20 TWh) by 2050 [4]. Although this may include several other battery technologies beyond Li-ion, one thing is certain: energy storage will be one of the key technologies propelling the decarbonization of the global economy.

Imagine traveling forward in time to see the grid of 2050. What would an interconnected energy system with 20 TWh of batteries look like? Will it be dominated by Li-ion technology as it is today, or will some other chemistry be king? Will there still be centralized power plants, or will our entire grid be composed of bi-directional flows of energy between our houses, vehicles, and devices? Will electro-chemical batteries still be popular, or will advances in hydrogen, compressed air, or mechanical storage prove to be lower-cost options?

These are all open questions, and everyone in the industry seems to have a different opinion on what the future holds. New battery breakthroughs seem to be announced every day, promising to be game changers for the industry. For as much innovation as there is in the sector, there is also a lot of hype, driven by those promoting a certain technology, stock, or market. This chapter attempts to cut through the talk by focusing on the numbers. We cover the current state of the battery industry and explore some of the most promising technologies being deployed and their potential for growth. Lastly, the chapter discusses the current and future state of the energy storage regulatory environment, one of the most important drivers of energy storage deployments.

11.1 Growth of the Battery Sector

As of this writing, utility-scale batteries are one of the world's fastest-growing infrastructure sectors. Growth has been exponential from 2016, when utility-scale projects began coming on the grid and is predicted to maintain a CAGR of over 9% through 2030 [88]. Projects are also getting much bigger: the largest US projects in 2016 were around 40 MW, while as of 2024 there is an 860 MW battery under construction and several projects announced over 500 MW (see Section 13.1 for more on the 860 MW project). The costs of battery systems have continued to decline – as of this writing, battery packs recently crossed the $100 / kWh threshold, with fully integrated 4-hour systems that can now be built for under $300 / kWh. Figure 11-1 shows NREL's prediction for the future of battery pricing through 2050.

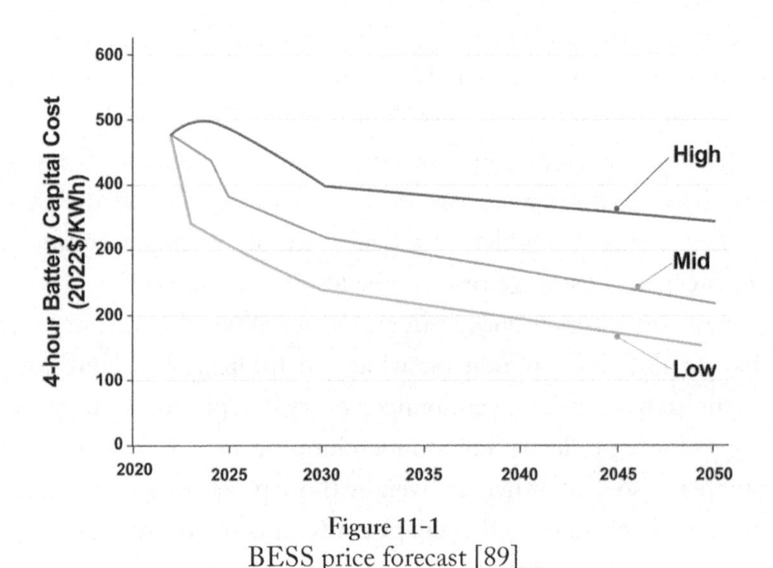

Figure 11-1
BESS price forecast [89]

In 2022, the price of raw material lithium carbonate (Li_2Co_3), the key input for nearly all lithium-based chemistries, skyrocketed. This spike, driven by high demand, limited supply, and supply-chain issues, left many integrators fearful that the rising cost of their battery cells would leave them unable to fulfill existing sales orders. To compensate for the spike, many cell producers, and in turn integrators, began including a pricing mechanism which indexed the price of their Li-ion cells to the price of the underlying commodity. This mechanism was disliked by developers and investors, as it meant that costs could unexpectedly rise, potentially reducing or eliminating the returns on proposed projects. As of 2024, the price of Li_2Co_3 has come down, making this no longer a critical input in the price, though some manufacturers have maintained indexing in their prices.

The financing of BESS projects has evolved alongside battery technology. In 2016 Li-ion battery projects were seen by banks as fringe, expensive technology with unknown risks and unclear revenue streams. In 2024, batteries have received a wave of funding, with nearly $5 billion in debt in the US alone [90]. Finance has come from some of the largest banks and funds in the world – hedge funds, private equity groups, and asset managers have all made large investments in energy storage technologies, developers, and projects. Projects are routinely financed on a non-recourse basis (as discussed in Chapter 6), a vote of confidence in the value of projects, which are now routinely seeing useful lives of 25 years – on par with the finance of traditional power plants. Current trends project that investment will continue growing along with the battery industry. Energy storage has come to be seen as a reliable and stable infrastructure investment, similar to solar PV plants, wind farms, and natural gas plants.

Regionally, batteries have been installed in a few leading markets: the US, China, Germany, the U.K., Canada, Australia, and South Korea account for the largest current markets for storage, as shown in Figure 11-2. In some markets, batteries can generate revenues without specific advantages, such as energy arbitrage or DC-coupled

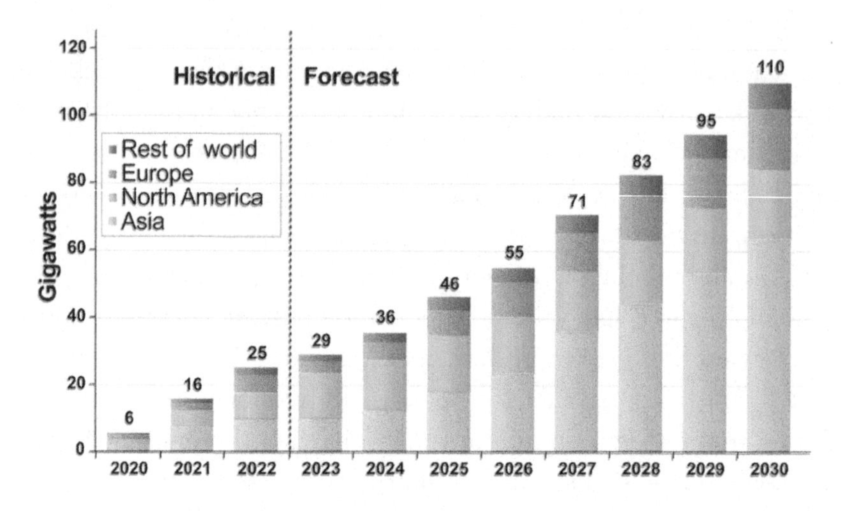

Figure 11-2
BESS construction by region [91]

PV+Storage projects. However, most fast-growing energy markets are driven by three external factors:

- **Government incentives** offer cash or tax credits to developers to offset the CapEx or OpEx of a project. The largest of these is the US Investment Tax Credit (ITC), which was extended to BESS projects in the 2023 Inflation Reduction Act (IRA)
- **Energy storage procurements** are run by utilities and grid operators specifically targeting energy storage, such as the 2023 Long-Term Request for Qualifications (LT-1 RFQ) in Ontario, run by the grid operator IESO, which solicited bids for over 2.5 GW of energy storage projects
- **BESS-specific markets** around the globe have recently been redesigned to call for batteries, or to require services that only BESS projects can provide. An example of this type of market is the Firm Frequency Response (FFR) market in the U.K., which attracted large battery investments despite few available tax incentives.

As of this writing, several new markets are opening, such as Chile, Greece, Japan, Italy, India, Brazil, Spain, and South Africa. Although each market is different, nearly all are driven by some combination of the three investment drivers listed above. As costs continue to fall, more markets will repeat the path that the pioneering markets have taken, devising markets and incentives to encourage energy storage projects to participate in their grids.

11.2 Evolution of Battery Technology

As with any nascent technology, batteries are going through progressive changes as the industry matures. Since the discovery of the Li-ion battery cell in 1991, many different chemistries, form factors, and integrated battery products have been researched, developed, and built. When utility-scale batteries began to come on to the power grid in the early 2010s, most systems were air-cooled batteries with NMC chemistry, in retrofitted shipping containers, with each 20-foot container housing less than 1 MWh of energy storage. In 2024, most utility-scale BESS projects are liquid-cooled LFP batteries in purpose-built enclosures, in which the 20-foot containers house 5 MWh or more. These changes represent shifts in the industry that will continue to evolve as technology improves, and more and more projects result in advancements in battery technology.

This section details six areas in which batteries have evolved, and where they may be heading: Chemistry, enclosure, safety, AC vs. DC-block, energy density, and degradation.

Chemistry

Today, most utility-scale stationary storage projects are currently using some type of lithium iron phosphate (LFP, or $LiFePO_4$) chemistry. The shift toward this chemistry has been driven by a variety of factors – cobalt has serious detrimental environmental and social impacts, in the case of battery fires, NMC burns hotter and faster than LFP, and LFP is more economical to produce. However, even as LFP cells are currently dominant in utility-scale production, several new chemistries claim to offer superior performance, safety, and cost factors, and hope to dethrone LFP, just as LFP did to NMC. This section covers some of the current contenders as alternative chemistries for stationary storage.

One of the current focuses of research is on cells chemistry based on **sodium-ion cells**, rather than Li-ion. Sodium (Na) is the third-lightest metal (after lithium and beryllium) and is more plentiful and safer than chemistries based on lithium. However, sodium-ion cells typically have a shorter lifespan than Li-ion cells. As of this writing, CATL's first-generation Na-ion cells, announced in 2021, had a density 160 Wh/kg, on par with similar values for LFP, [92] but significantly less than NMC at 255 Wh/kg. As of this writing, there have been no major announcements of updated energy density from Na-ion cells. BNEF claims that they expect the technology to improve dramatically by 2025. The relative densities are shown in Figure 11-3, an estimate and prediction from May 2023 showing the BNEF estimates of past and future energy densities [93].

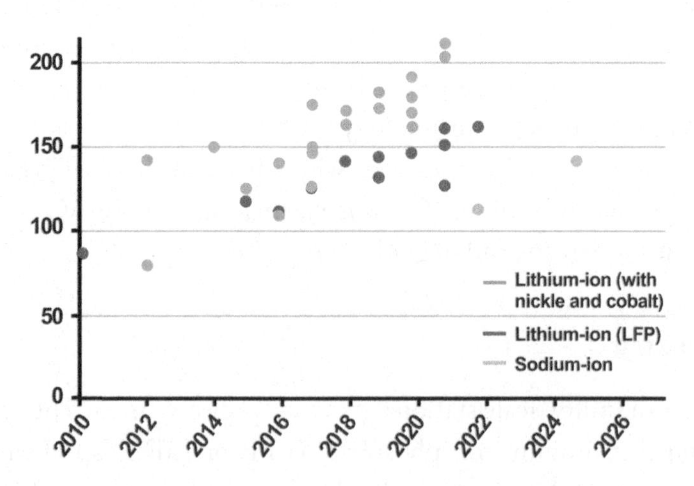

Figure 11-3
Battery pack energy Density (Wh / kG) [93]

Metal-air batteries have had a strong spurt of growth in recent years. This chemistry uses a metal (typically iron, zinc, or aluminum) as the anode, a water-based electrolyte, and a stream of air that serves as the cathode [94]. These batteries have high theoretical energy densities, but in practice they have issues such as consumption of the electrode and low round-trip efficiencies.

One of the more promising chemistries in this class is iron-air, also known as a "reversible rust" reaction since it is based on the oxidation of the iron anode. These batteries are leading candidates for long duration energy storage (LDES) – in some cases up to 240 hours, or 10 days of energy storage. The most prominent company using this chemistry is Form Energy, a startup begun in 2017 and funded by Breakthrough Energy, an investment fund focused on climate technology led by Bill Gates. Form claims that their battery can exceed 100 hours of duration. Additional benefits are that the chemistry uses plentiful materials (such as iron), has a relatively low cost, and has strong safety performance. Downsides include slow recharge and relatively low round-trip efficiency. [95] In December 2023, Form received a $30m grant from the California Energy Commission (CEC) to build a 5 MW / 500 MWh battery as part of the PG&E electrical grid. While there is abundant interest in this chemistry, the deployments as of this writing are miniscule, on a power and energy basis, compared to the operational LFP and NMC projects.

Solid-state battery cells are often touted as the future of energy storage. These cells use a solid electrolyte, rather than the liquid that is commonly used in current prismatic cells. This technology has been under research since the 1960s, but has been limited by high costs, or requirements of high temperatures for operation. In The 2010s, a renewed interest in solid-state Li-ion based cells was ignited by publication of research led by Noriaki Kamaya [96]. Several major automakers have announced their intention to pursue solid-state chemistries – notably, Toyota's 2023 Battery Technology Roadmap targeted use of solid-state Li-ion batteries starting in 2027, anticipating a driving range over 1,000 km with fast-charge time of 10 minutes

[97]. However, despite the speculation around solid-state, it has not yet proven commercially viable in either EV or stationary applications.

According to BNEF [98], "*Solid-state batteries progress, with new announcements potentially adding more than 40GWh. Solid-state batteries have become the most promising technology for pushing cell-level energy density up to 500 watt-hours per kilogram and driving battery prices down in the second half of the decade. Several leading battery manufacturers, like LG Energy Solution, CATL, and SK, as well as startups like Solid Power, Prologium, and Quantumscape, have laid out clear roadmaps to commercialize solid-state batteries within this decade.*"

Flow batteries, discussed in Section 3.7, have shown promise as potential long-duration energy storage technology. Several firms have made advancements in their technology, with the two leading chemistries: Vanadium Redux flow and iron flow.

In October 2021 prominent Oregon-based flow battery manufacturer ESS, Inc. signed a battery supply agreement for 2 GWh of flow batteries with SBEnergy, a Japanese renewable energy developer building projects throughout North America. In 2023, they signed a deal to build a 50 MW / 500 MWh system in Germany. Such deals signal that the industry may be overcoming some of the technical issues and mitigating risks that have impeded the financing of large-scale projects in the past. We expect to see flow batteries continue to evolve and they are one of the technologies that could become a strong competitor with the currently dominant Li-ion cell. In particular, the long-duration ability of flow batteries makes them a strong contender for increasing penetration of renewables into the energy mix.

Technology advances and breakthroughs may be enough to overcome the size and cost issues that flow is experiencing, and flow batteries the only viable solution for certain projects. As of this writing, the investment and deployment in flow batteries are a small fraction of that into Li-ion batteries.

Lastly, **nickel-hydrogen (NiH)** batteries have shown some modest potential as a competitive stationary battery cell technology. This technology uses a pressure vessel which stores energy in a gaseous state, some of which may exceed 1000 psi. Traditionally this technology has been used to store energy for satellites and space probes, but providers such as EnerVenue have begun to promote this technology in a stationary BESS application. This chemistry is known for its long life, with cells being able to perform over 20,000 cycles, maintaining 85% round-trip efficiency. However, the downsides of this technology are the relatively high cost and low energy density, as compared with Li-ion.

Enclosure Capacity

One area of innovation and change in batteries has been in their enclosures, also known as containers. As discussed in Chapter 4, early stationary batteries (2014-16) were installed in retrofitted shipping containers. As the industry evolved, most battery manufacturers moved toward custom-built enclosures that had no walk-through access. This shift was primarily driven by safety concerns about having a container that could fill with flammable gases. This is because custom-built enclosures are designed with many features that mitigate some issues with the shipping containers, such as elimination of empty space, incorporation of dedicated venting channels for better airflow, and the incorporation of features such as deflagration panels.

As the designs evolved, most enclosures kept the same 2.4 m (8 foot) width of shipping containers, but length varied from a 2.4 m square to extra-long 16.2 m containers. However, as of this writing most products appear to be coalescing around a 6.1 m (20 foot) shipping container enclosure. This dimension allows for shipping that doesn't require special trucking or lifting capabilities, while maintaining a small enough sizing that allows the project to isolate individual containers in the case of a fire that consumes a battery enclosure.

There are notable outliers, such as Tesla, whose Megapack product line is closer to a full 12.2 m (40 foot) container, or Wartsila with smaller units from their QuantumHE battery product. As of this writing, many leading manufacturers such as CATL, Sungrow, Hithium, REPT, BYD and Narada have all moved toward a 6.2 x 2.4 m (20 x 8-foot) footprint.

Duration

When the BESS industry was stepping into commercialization around 2016, the duration (roughly the MWh to MW ratio) was generally between 15 minutes to 1 hour. Fast forward 8 years, at the time of writing this book, the most common BESS duration configuration is 4 hours. This has become possible thanks to BESS prices consistently declining – plus a 4-hour BESS is generally more economical than lesser duration on a $/kWh basis. It is also well known that the C-rate of a 4-hour system, 0.25, is much better for Li-ion battery performance and longevity. Lastly, utilities who contract BESS projects for grid needs have become more comfortable and have more use-cases for 4-hour systems, most common being providing energy in peak windows that last at least 2-3 hours instead of natural gas-based peaker plants.

The industry is transitioning toward long duration energy storage (LDES) which is the term applied for 8-hour and greater duration. As more renewables come online, there will be the need to shave longer peaks of excess renewable production and supply longer hours of need in a reliable manner. This can result in the complete elimination of gas-fired peaker plants which are still needed for reliability and ramp-up purposes. Seasonal shifts can be enabled, such as from summer to winter, providing benefits in managing seasonal variations in energy supply and demand. LDES will also become critical in the wake of increasing extreme natural events like snowstorms and wildfires which are often mired with days-long grid outages.

There is consensus that Li-ion chemistry is best suited for durations up to 6 hours – beyond that they lose economic and performance advantages. The contenders for LDES are:

- **Pumped Hydro** is a technology that has been around for decades, and accounts for 95% of utility-scale energy storage by capacity simply because this can only be executed in GW scale. In terms of energy storage, it involves pumping water uphill to store it in a reservoir. When additional generation is needed, the water is released downhill to generate electricity. While this is a well-proven technology and a good fit for LDES, it comes with added siting and development complexities.

- **Compressed Air Energy Storage (CAES)** plants compress air using electricity during off-peak hours and store it in underground reservoirs. The compressed air is then combusted together with natural gas in a turbine, increasing the turbine's overall efficiency. The system can be scaled to provide as long duration as needed. Drawbacks of this technology are low overall fuel-efficiency and limitations of sites, as most viable sites are underground caverns or defunct mines.

- **Flow Batteries**, discussed in depth in Section 3.7, flow batteries circulate liquid electrolytes to charge or discharge electrons via redox reactions. They store energy in large tanks of liquid electrolytes, enabling scalability by simply increasing the size of the tanks and thus making them very good candidates for long duration storage requirements. The attributes of scalability, long service life and high operational safety position them as the current closest economically viable alternative to Li-ion batteries. In December 2023, Form Energy, an iron-air flow battery developer, was awarded $30M to build a 5 MW, 100 hour duration storage system in Pacific Gas & Electric Company's service area [99].

AC vs. DC-block

Most battery systems are sold as a DC product – typically 1500 VDC nominal battery racks, which are wired in parallel to form containers ranging from 1.5 to 6 MWh. As described in Chapter 4, the designer of the system would be able to add an PCS of their choosing to the DC battery product to form an integrated BESS project. This allowed the designer some versatility to pick different PCSs, or to size the system with a higher or lower power (while staying within the technical requirements of the battery).

In 2019 Tesla launched their Megapack product which embraced a different design philosophy – instead of having an external PCS, the battery container was sold with the DC batteries and PCS together as a single product. This had the drawback of not allowing the designer to pick a specific size or model of PCS, since Tesla only offers the product in a 2-hour or 4-hour version, ranging from 1.3 to 3.9 MWh per enclosure. However, it had the advantage of simplifying design, reducing balance of plant construction, and the large advantage of including the PCS within Tesla's performance guarantee (although some guarantees require the owner to cover the cost of PCS failure). Tesla supplies its own PCS to the Megapack, which is integrated directly into the block of battery modules. Each individual enclosure outputs low-voltage AC (typically 480 to 770V three-phase). This arrangement is known as an AC-block, to distinguish it from the DC-block used in most battery products without an PCS.

Other manufacturers have begun to mimic this design – for example, the Sungrow PowerTitan 2.0 was announced in June 2023 with a capacity of approximately 5 MWh and up to 2.5 MW of power, using an integrated PCS. As power electronics evolve and the BESS industry becomes more standardized, it is likely that there will be more products which offer an PCS that is integrated into the battery enclosure. EnergyVault offers an AC-block version of their product B-Vault with a capacity of 2.98 MWh and 0.745 or 1.49 MW power rating. Others may have similar products planned in the pipeline depending on market demand.

Energy density

Energy density can be measured in two ways: volumetrically, typically expressed in Wh / L, or by mass, most often seen as Wh / kg. Mass density has traditionally been more important to EVs, since the battery must accelerate with the vehicle. In contrast, stationary storage is more concerned with volume, since weight is irrelevant, but a higher Wh / L means that more energy can be packed in a single container. As enclosures have coalesced around the dimensions of a 20-foot ISO shipping container (6.2 x 2.4 x 2.6m), it is common to compare the density against what the 20-foot container can hold. Table 11-1 shows the relative energy densities being advertised by some leading ESS producers as of March 2024. A picture of these products can be found in Figure 4-12 and Figure 4-13.

Table 11-1
BESS enclosure energy densities, as of March 2024

Manufacturer	**Product name**	**Energy density Mwh per 160 ft^2 (14.9 m^2)**	**Notes**
AESI	Terastor	3.8	Product is 7.3 MWhDC in an 8.2 x 3.8 m enclosure
BYD	MC Cube T	6.4	
Canadian Solar	SolBank	4.8	
CATL	TENER	6.25	Announced in April 2024. Claims to offer 5-year "zero degradation."
Clou	Aqua C-1	3.7	
EnergyVault	B-Vault	3.2	The EnergyVault B-Vault BV2.2 offers 4.3 MWh of storage in a 9.8 x 2.4 m (31 x 8 ft) container. Note that this product contains both the PCS and 50 kVA transformer. Density is corrected to reflect the inclusion of these components.

Manufacturer	Product name	Energy density Mwh per 160 ft^2 (14.9 m^2)	Notes
Fluence	GridStack Pro	5.0	
Gotion	ESD1126	2.7	
Hithium	Block	5.0	
LGChem	JF1 DC-Link	2.4	
Narada	NESP Series	1.9	
Powin	Powin Pod	5.0	
REPT	Wending	6.9	
Risen / SYL	SU Series	3.4	
Samsung	SBB	3.8	
Sungrow	PowerTitan 2.0	5.0	The Sungrow PowerTitan 2.0 includes PCSs.
Sunwoda	Noah X	4.2	
Tesla	Megapack 2XL	4.0	Tesla Megapack 2XL offers up to 3.9 MWh of storage, but the enclosure is 14.5 m^2 (156 ft^2). This product also includes PCSs.
Trina Storage	Elementa	2.5	The Trina Storage Elementa comes as a 7.8 x 1.7 m (25.6 x 5.6 ft) enclosure, with a 2.2 MWh capacity. Since this is smaller than the standard 20-ft container.
Wartsila	Quantum2	4.1	

Given that the densities have been rising rapidly, it is likely that the energy in a 20-foot container will continue to rise. Most energy storage chemistries have a limit to the theoretical density, based on the atomic structure of the energy storage molecules. While this is a physical limit, as variations of each chemistry have evolved, this value has risen, as has the amount of Li-ion cells that can be packed into a container. As research continues, the amount of energy that can be packed into a 20-foot container will continue to rise, although it is expected to converge upon a value for each chemistry that is used, based on physical limitations.

Degradation Improvements

As they are cycled, all batteries have some slight loss of their capacity. As described in depth in Chapter 3, this is known as their degradation, and it is a property of the physical structure of the batteries. As of this writing, most utility-scale batteries can offer approximately 70% capacity retention, while being cycled over 20 years at one cycle per day. In practice this means that a battery that has 1 MWh of capacity at beginning-of-life (BOL) will have at least 800 kWh after being cycled a total of 7,300 times over the 20 years. One of the main goals of research and development in batteries is to offer batteries that can resist a higher number of cycles with less degradation.

All major cell manufacturers are chasing aggressive targets for these numbers. At the 2023 Re+ conference in Las Vegas, manufacturer Ampace unveiled a cell that can withstand 15,000 cycles [100], which would equate to a cycle per day for 40 years. Hithium stated in October 2023 that their 300 Ah cell could achieve a 12,000-cycle life [101] – however, it is important to note that these lifetimes are usually predicted, since cycling a cell that many times would take over 10 years. CATL, the world's leading cell manufacturer, stated in December 2023 that their current batteries had reached 12,000 cycles, and their goal is to increase cycle life to 18,000 [102].

In April 2024, CATL announced that their newest battery product will have "zero degradation" for the first 5 years. Based on previous technology, this would have been impossible, since all prior LFP cells have declined in capacity immediately after beginning to be put into use. CATL has claimed that the innovation powering this is known as pre-lithiation, in which the anode and/or cathode are treated with an organic lithium salt, such as squarate ($Li_2C_4O_4$). The net result is a period in which the energy capacity increases after the cell is cycled. The research behind this phenomenon was published in 2022 by Gomez-Martin et. al. [103] as shown in Figure 11-4.

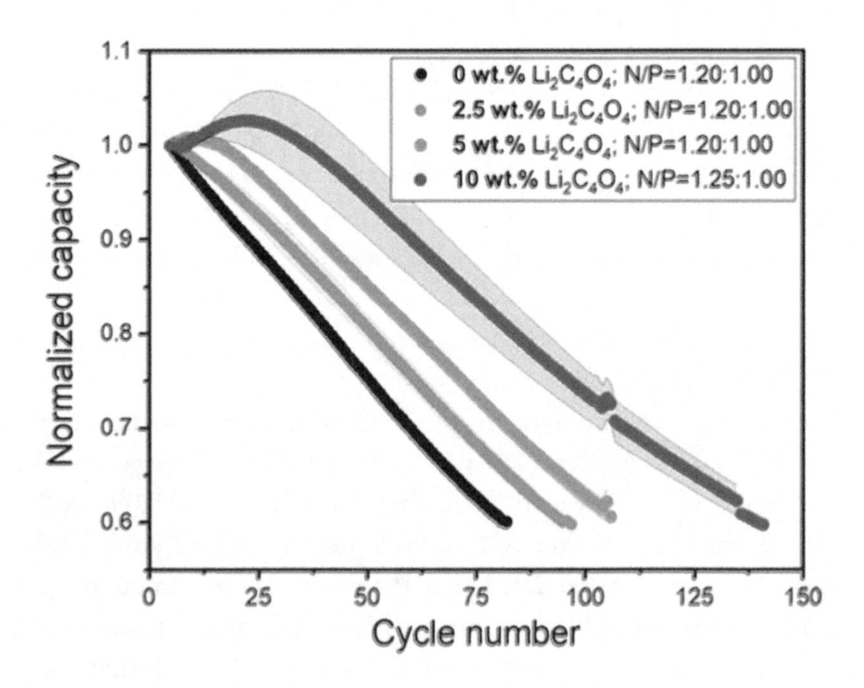

Figure 11-4
Effect of pre-lithiation from 2022 Advanced Science paper by Gomez-Martin, et.al. [103]

The quest for low degradation is not limited to LFP – Nickel-hydrogen battery manufacturer, Enervenue, has offered a 20,000 cycle warranty, guaranteeing 88% maximum degradation [104]. In December 2023, Volkswagen announced [105] that their testing of QuantumScape solid state battery had achieved 1000 cycles with 95% SOC, which experts predict could lead to a vehicle that has no noticeable loss in range after up to 500,000 km.

As the technology race continues to accelerate, improvements in cycle life and degradation are certainly going to continue, both within Li-ion technologies and with new competing chemistries. These will result in vehicles that have very long lives with little degradation, and stationary batteries with stronger warranties, allowing the cost of their services to continue to decline. As with energy density, there may be some limit to how many cycles a cell can achieve, but at present there are still large gains being made across the industry.

11.3 Vehicle-to-grid (V2G)

Stationary energy storage systems, the topic of this book, are an exciting and growing field, and have seen a huge surge in innovation and investment. However, the total energy deployed in stationary systems is a small fraction of the total energy of Li-ion cells deployed in the world. By far, the biggest application of Li-ion has been in EVs. The numbers paint this picture – in 2022 alone, 550 GWh of cells were produced for EVs, according to the IEA [106], compared with 35 GWh for ESS, according to Bloomberg [1]. This disparity begs the question: Is there a way that the massive store of energy, roaming our streets in the form of EVs, could be connected to the grid to provide services as if they were stationary storage?

The answer is an open question, but one that many experts believe will play a key role in expanding storage. The concept of having batteries back-feed power into the grid is known as Vehicle-to-Grid (V2G). The V2G concept is similar to the VPP concept, but with one key difference – EVs are only connected to the grid a fraction of their lifetime, when they are charging.

As with VPPs, in V2G the vehicle owner agrees to have a portion of their battery be used via an external controller. The controller monitors the SOC, location, and available power and energy for all vehicles in the fleet. By charging or discharging these assets simultaneously, the fleet of vehicles can act as a single large battery. With enough vehicles performing this task, the fleet can act identically to a BESS, providing a distributed power source that can do anything a BESS can do – charging, discharging, frequency regulation, injecting capacity.

The advantages of V2G are vast:

- Large storage capacity that can be connected without having to go through a costly and time-consuming high voltage interconnection
- A rotating pool of battery ownership that allows the battery to be perpetually replenished
- Insulation against outages or technical problems, since the system is comprised of thousands of individual contributors, rather than a single centralized plant.
- Providing a value stream to vehicles that have daily or seasonal downtime, such as school bus fleets

V2G is still in its infancy – according to industry group V2G Hub, there are 131 V2G pilot projects in 27 countries, as of January 2024, including 24 in the US Although these pilots are beginning to show promise, it is a long road to developing a cohesive plan that works to deliver power from the grid.

Technical challenges

The complex task of coordinating thousands of vehicles to work in concert to feed power into the grid has several technical challenges.

Firstly, most charging infrastructure has been designed to be one-way, meaning that the charger is designed to feed the vehicle and not the other way around, for safety purposes. While there is no physical limitation that would prevent power flowing from vehicle to the grid, there are software controls designed to prevent this that would have to be circumvented. As of January 2024, there are only a few EVs which are capable of bi-directional charging, such as the Nissan Leaf, Hyundai Ioniq, Volkswagen ID models, and the Ford F-150 Lightning. Tesla has stated that all Tesla vehicles will support bi-directional charging by 2025 [107]. In addition to a vehicle that supports it, a special bidirectional EV charger must also be used, requiring an additional investment on behalf of users.

V2G has a natural limitation of customer participation – as with VPPs, the owner of the battery purchased it for the services it could provide, and so they typically do not want to sell all their asset's capabilities onto the grid. Typically, the driver sets a reserve level, which is the minimum charge that the battery can ever be allowed to have. In practice, this is the minimum level of range that the driver would allow – for example, if the vehicle had a 200-mile range, and the driver set the vehicle.

Another challenge is that the organizer of the VPP must know how many EVs will be available to charge or discharge at any point in time. This presents difficulties in guaranteeing to an offtaker how much power will be available at any given time; this difficulty can be remedied by having a very large program, making the available demand more predictable.

While it is still in its infancy, many believe that V2G could revolutionize the power grid – if even 10% of the EVs produced in 2022 were to connect to the power grid, this would provide a collective 50 GWh of energy storage capacity. This technology will continue to evolve and may become a key part of the grids' storage capacity.

Figure 11-5
Rendering of a residential home backup and EV-charging battery [108]

11.4 Regulatory Outlook

As discussed throughout the book, energy storage deployment is largely a function of policy. Where markets, incentives, and trade rules are favorable to energy storage, projects tend to pop up soon after; conversely, any markets lacking these drivers is unlikely to attract high volumes of storage projects. Based on current trends, the cost and technical benefits of Li-ion batteries may prove so widespread that all markets will provide them with access, or risk being left behind. However, this is a long process – incumbent technologies are entrenched, utilities are slow to change, and financiers can be risk adverse. For the near future, batteries are highly dependent on the policies set in the municipalities, states/provinces, and countries of each potential market.

Given the importance of regulatory drivers, energy storage developers are constantly monitoring the latest policies, tenders, and announcements from governments, grids, and grid operators to find new potential markets. Some of the most well-established developers have succeeded because they have established an early presence in markets which expand to become market leaders.

This section details some of the key regulatory and policy drivers of energy storage, what markets have been some of the leaders in storage deployment, and what markets are expected to be central to growth in coming years. The focus of this chapter is on North American energy storage, but the final section covers global markets, and which are expected to emerge in the coming years.

Markets

The ability of efficient markets is one of the key factors that make energy storage projects viable. As described in Section 6.1, the main drivers of energy storage projects are either contracted or merchant markets. Contracted revenues typically take the form of an offtake contract that the project signs with the utility. Some utilities have put out tenders specifically aimed at attracting energy storage – for example, Hawaiian Electric has issued tenders for Renewable Dispatchable Generation (RDG), which provide a strong incentive for PV+Storage projects to come online. Other offtakers, such as NVEnergy in Nevada, have sought to sign PPAs which compensate offtakers for time-of-use production. Under this arrangement, the offtaker pays a higher price for energy provided during peak demand times, which for most utilities is from approximately 6:00 – 9:00 pm. This structure also creates a strong market for offtake agreements that incorporate energy storage (especially in the case of PV+Storage).

Beyond contractual revenues, the establishment of merchant markets in which batteries can participate is critical to unlocking revenue for batteries which make them attractive. The mechanics of these markets are detailed in Section 6.1. They are largely driven by either laws or directives which either create new energy-storage specific markets or modify the rules for existing markets so that energy storage can participate. Frequency regulation, spinning reserve, and black-start services are examples of markets which energy storage can participate in, which have proved to be key revenue drivers – in some cases, payments from merchant market participation may represent 80% or more of a project's total revenue.

Many of the key regulations in the US come from the Federal Energy Regulatory Commission (FERC), who issue rules that bind all the Independent Service Operators (ISOs) and Regional Transmission Operators (RTOs). These federal rules do not apply to the Texas grid regulator, ERCOT, since its grid is located entirely within the state of Texas. Some of the key FERC rules related to energy storage have been:

- FERC Order 841, issued in 2018, which amended FERC's regulations to allow the participation of energy storage projects in the capacity, energy, and ancillary service markets operated by the ISOs and RTOs, outside of ERCOT
- Three FERC Orders in 2022 (RM22-4, RM22-12, and RD22-5) which addressed inverter-based resources (IBRs, which include Solar PV, wind, and battery projects), aimed at increasing the reliability of the grid by imposing new rules on these generators. This was in response to several grid reliability issues.
- FERC Order 2023, issued in 2023, reformed the processes that transmission systems used to admit new generators onto the grid. This was in response to the large backlogs in interconnection queues (over 10,000 requests representing over 2,000 GW of projects, as of the end of 2022 [109])

Even prior to Order 841, several ISOs had begun opening their markets to energy storage. PJM, covering Pennsylvania, New Jersey, Maryland, and Chicago, was an early adopter, and was one of the hot markets with some of the first energy storage projects in 2014-2016. The main revenue driver was PJM's Regulation-D market, which allowed batteries to participate in frequency regulation, which became quite lucrative for those projects. Although in 2018 a controversial change to cycling caused many of these projects to earn less money, the market proved that batteries could compete as valuable resources.

In California, the energy storage market rose swiftly in the wake of an environmental disaster. In October 2015 a natural gas leak was discovered at the Aliso Canyon underground natural gas storage facility. The leak released over 100,000 combined tons of methane and ethane into the environment, in one of the worst environmental disasters in US history. In the wake of this disaster, California moved quickly to find alternatives to the peaker plants that used natural gas. Batteries were one of the main beneficiaries, via a program called Resource Adequacy, which was modified to become Flexible Resource Adequacy in 2015, providing a strong incentive for energy storage assets to participate in this market. Furthermore, California Executive Order B-55-18 committed the state to achieving a carbon-neutral grid by 2045, increasing the need for both PV+Storage and standalone energy storage projects. As of the writing of this book, California is the leading state for deployment of energy storage, with over 7 GW of batteries deployed through the end of 2023, and more projects continuing to come online. They also host some of the world's largest projects – including the world's largest project (as of April 2024), Edwards and Sanborn with a total storage capacity of 3.3 GWh.

The Texas market, run by the Electricity Reliability Council of Texas (ERCOT), is unique as a market in that it is based only on the energy produced, rather than other markets which make payments based on both energy and capacity. Since the Texas grid is not interconnected outside of the borders of the state, it does not have to abide by the rules of FERC as the other ISOs are RTOs do. This, along with a rapid growth in solar and wind assets, presents challenges to the grid which ultimately create revenue opportunities for dispatchable resources like BESS. The effect of these developments is highly volatile energy prices, especially in the real-time market, that BESS can be designed to capitalize on in a dynamic and quick-responding manner.

Several markets in the US are continuing to grow their attractiveness for energy storage, treading in the wake of the leaders like California and Texas. Some of these markets are:

- New York
- Midwestern Interconnection (MISO)
- ISO-New England (ISO-NE)
- Southwest Power Pool (SPP)

Incentives

The main driver of wind and solar in the US had been the investment tax credit (ITC) and production tax credit (PTC), which are discussed in detail in Chapter 6. These incentives provide either a portion of the CapEx of the project (ITC) or a set payment per unit energy discharged by the project (PTC). Rather than making these payments directly, the US government has opted instead to pay these as tax credits that can be transferred to holding entities.

Up until 2023, these incentives could not be applied to storage since the law was intended exclusively for wind and solar projects. The 2023 ITC was a boon to energy storage, since for the first time it allowed batteries to claim the 30% tax credit, with adders that can result in a 50% credit. This provided a massive boost to the finance of the projects, resulting in an enormous rush in investment in the US energy storage market, especially opening it up for stand-alone BESS projects as the need for pairing with renewables is no longer applicable. These credits will continue to be offered until 2032 or may be extended depending on the level of greenhouse gas levels [110]. In March 2023, Canada passed legislation establishing an ITC similar to the American incentive, with provisions for up to a 30% credit to stimulate investment.

Many individual states and municipalities have incentives for energy storage. For example, in New York State, the New York State Energy Research and Development Authority (NYSERDA) offers several different types of incentives which can offset the costs of battery projects.

Some countries, states, and municipalities have put up barriers to energy storage rather than incentives. These come in many forms, but some of the more common are:

- Many countries have raised taxes on the import of batteries, either to raise revenue or to incentivize domestic production
- Some US counties have instituted bans on battery projects due to safety concerns
- Some procurements are designed to give an edge to fossil fuel generation, such as natural gas plants.

Hopefully these barriers will fall along with the price of batteries, but many are driven by larger political or economic conditions in each state.

The goal of most incentives is to accelerate the energy transition by easing the economics of battery storage. Although some of these subsidies have persisted for decades, the stated intention of incentives is for them to expire at a certain point once technology has matured and can compete without economic assistance. As has played out with wind and solar PV over the past 15-20 years, the incentives are intended to lead to economies of scale, in which larger battery factories, battery projects, and financing avenues lead to reduced unit costs. Additionally, there is a reduction based on learning-by-doing, in which the ecosystem of developers, EPCs, manufacturers, operators, and financiers becomes more and more skilled in energy storage products. The increase in projects provides a boost to the industry workforce, continuing to accelerate construction of projects.

Global markets

Outside of North America there are many active and growing energy storage markets. This section details some of the largest markets to date, along with growth projections.

China: The Chinese battery market is driven by government requirements to pair energy storage with solar and wind projects. China's cumulative energy storage was 59.4 GW in 2022, according to APCO Worldwide [111], and was anticipated to grow at a CAGR of 18.9% from 2023 to 2032. China's market is dominated by Chinese companies and by state-owned enterprises, given that they are the leader in cell manufacture. China's market is notable in that it has sought the installation of other storage technologies besides Li-ion batteries, such as compressed air, flywheels, and supercapacitors.

South Korea was an early adopter of energy storage and features some of the most well-established Li-ion battery manufacturers in the world. From roughly $69 million in 2013, the market grew to over $1.8 billion in 2018. [112] Energy storage was at the center of the "Renewable Energy 3020 Plan," which was the government's effort to increase renewable penetration to 20% by 2030. Although the market slowed based on some safety incidents in 2019-2020, it has seen renewed growth after the government set a goal to capture 35% of the global ESS market.

Australia has embraced energy storage heavily in their power grid. One of the most prominent early energy storage projects, the 100 MW / 129 MWh Hornsdale Power Reserve near Adelaide, was built in 2017 using Tesla batteries by French developer Neoen. The driver of that project and many others is the volatility of the Australian grid. Outages of coal plants, outdated transmission lines, and natural disasters such as brush fires have driven spot prices to commonly show large spreads between various locations. The National Electricity Market (NEM), which serves the population centers of Queensland, New South Wales, Victoria, Tasmania, and South Australia, has had several bilateral agreements to install energy storage, as well as running a competitive process to install battery projects. As of January 2024, Australia had a combined pipeline of over 40 GW of energy storage projects, according to Rated Power [113]. The world's most powerful planned battery, an 850 MW 2-hour Waratah system, which has a planned COD in late 2024, is located on the Central Coast of New South Wales. The Waratah is being developed by Akaysha Energy with battery integrator Powin.

The European Union (EU) has been a growing market for storage, although most projects have been smaller in scale than those built in other markets and concentrated in a few countries. In March 2023 the European Commission listed energy storage as a key driver of the energy transition, predicting that energy storage capacity would grow to over 270 GW by 2026. As of 2022, Spain, Germany, and Italy were the leading markets for the pipeline of projects. The adjacent United Kingdom has installed more energy storage than any country in Europe, primarily driven by the Fast Frequency Response (FFR) program, which is an ideal market for energy storage to be compensated for providing frequency regulation services.

Latin America has been slow to adopt energy storage policies, but Chile has been far and away the leader, having 12 projects with a total capacity of 1.3 GW having reached COD as of October 2023, with 85 projects, totaling 6.4 GW, in various stages of development. Other countries are working to get involved, with Colombia, Panama, Dominican Republic, and Puerto Rico all launching energy storage tenders in 2023. Brazil is the region's largest energy market but has a relatively small market for energy storage with approximately 250 MWh of installed capacity at the end of 2023.

Japan boasts a robust wholesale market in the Japan Electric Power Exchange (JEPX), a spot market which manages approximately 40% of the total electricity demand. This makes an ideal market for batteries, and there have been many new projects announced in 2023. The Japanese government has provided investment incentives for battery projects, and a new low-carbon capacity market has created a large demand 3-hour capacity batteries to be built. However, high capital costs, land constraints, and unstable markets have tempered interest. In December 2023 Gurin Energy, a Singapore-based developer, announced a 2 GWh BESS in either the Fukushima or Tochigi prefectures, to begin construction in 2026.

12. THE GROWING BATTERY WORKFORCE

According to consulting firm E2, the energy storage workforce in the US grew by 6.2% in 2022 to over 85,000 [114], making it one of the fastest growing segments in the economy. From engineers to financiers, electricians to manufacturers, the sector will require new entrants to build this critical part of the world's energy infrastructure. Some may be transferring from other technical fields, while others may be joining the field as the start of their career. This section details the key roles in the industry and how anyone can get involved.

12.1 Manufacturing

The manufacturing of any given battery product is a complex global process that employs thousands of people across engineering, procurement, integration, delivery, installation, and service. Most LFP battery cells are currently manufactured in China, but the 2022 Inflation Reduction Act (IRA) provided strong incentives for manufacturers to build plants in the US. These plants will generate thousands of jobs from construction, operations, logistics, and service. Most cell manufacturers house large research and development (R&D) laboratories as well as testing facilities in house. These jobs range in skills from mass manufacturing to electrical engineering, sales, and research scientist. There is a role for software developers, as most cell, module, and rack products involve some level of controls, such as in the BMS. Some jobs require only a high school degree, while others require degrees ranging to Ph.D.-level for the more advanced research posts. Students of business are sought after as the battery development process requires advanced finance, contracting, and planning to adapt to an ever-changing market. Since many cell manufacturing facilities are geared toward cells for EV applications, those with experience in the automotive industry have an advantage to pivot into the battery field. Since the US manufacturing industry has been in years of decline, as of 2024 there was ample opportunity for early- and mid-career professionals to enter the manufacturing space.

Moving up the supply chain, battery integrators require a different set of skills. Given that most integrators purchase cells, modules, or racks from separate manufacturers, integrators are more focused on engineering of the interconnected battery system and software to ensure the components function together. Hardware designers, electrical engineers, and software developers are some of the most sought-after professionals for this segment. Both in the case of battery integrators and integrators, customer-facing application and sales engineers are required to serve as technical leads during contract negotiations, as well as to enhance product development by relaying customers' requests. Given the large role of safety, battery safety engineering has become a growing niche in this sector. This role is typically heavy on laboratory experience, given that much of the design process is geared toward passing lab criteria.

Since integrators often provide sales as well as service contracts, customer service and field engineering support are required. Many traveling positions are available with integrators to support installation, commissioning, and operational service calls for batteries. These roles often require extensive electrician training since they are involved with high-voltage AC and DC circuits. However, since much of the battery controls system is based on software, computer knowledge is essential for this type of work.

Throughout the manufacturing process there is a strong drive for legal expertise involved in the contracting, negotiation, sales, and M&A process that is inherent to battery products. Many large law firms have added an energy storage section to their existing renewable energy practices to support this growing segment of the industry.

12.2 Development

Developers of energy storage projects are one of the largest growing segments of the industry. Chapter 6 details the Development process, describing the lengthy and complex process involved in moving an energy storage project from initial conception to finance, construction, and eventually operation.

Development firms vary widely in their workforce depending on their business model, but most firms encompass five key areas: development, engineering, execution, finance, and legal.

Initial development requires a diverse set of skills: business and finance to perform the modeling, regulatory expertise to navigate the markets and incentives, and on-the-ground staff to negotiate land agreements, attend community meetings, and comply with local rules and filings. Many early-career professionals are attracted to development as it can entail extensive travel to site projects and attend meetings.

Many developers have in-house engineering staff that perform sizing of components, CAD layouts, and provide an interface with the integrators and manufacturers used on each project. While most developers use an outside engineering firm to prepare the detailed design drawings, many have professional engineers (P.E.s) on staff. Engineers support the coordination between the various contractors on a project, from procurement to BoP to controls. Many development engineers have experience from wind, solar, or fossil fuel design, pivoting to energy storage, while others have entered the field focused on energy storage and have been able to carve out a niche in this growing field. Engineers play a critical role in supporting operational projects, reviewing operational data, supporting degradation/augmentation of projects, and assisting in the decommissioning of project installations.

Project execution is a critical area for developers, since keeping projects on time and on budget is the biggest driver for keeping a developer profitable. Execution staff often come from the construction world and are specialized in project budgeting, scheduling, and coordination between parties. Most developers hire out the field work on a BESS project to one or several BoP contractors, but keeping these parties on track and dealing with the inevitable unexpected conditions requires the developer to have diligent support dedicated to execution. The lenders and equity investors are often closely connected to the project execution, often requiring monthly progress reports, independent engineer certificates, and loan disbursements. The professional working on execution of large BESS projects must be well-versed in field work, contracting, and finance.

As discussed in Chapter 7, project finance is key to the development of any BESS project. The skillset of a successful battery financier includes basic business concepts, project modeling, negotiation of key contractual terms, and the more delicate aspects of structuring a deal with the equity and debt partners. The larger and more complex deals have included financial instruments such as hedge agreements, intricate liquidated damage provisions, and other instruments intended to limit risk to developers. Given the complexity of the incentives and tax implications of the investment, most developers have on-staff experts devoted to these topics. The CFO of the developer is often the role overseeing the team devoted to financing, and this involves building long-term relationships with lenders, investors, and funds involved in the large-scale finance of infrastructure. Project finance is not all numbers – the lawyers and businesspeople involved in the transactions often work with engineers to understand the technical aspects of projects and ensure that the projects are designed to limit risk in accordance with the evolving battery technology.

Finally, the legal department of developers is key to their success. Whether this is done in-house or using outside counsel, nearly every aspect of an energy storage project is governed by contracts. Whether cell supply, battery supply, long-term service agreement, employment engagement, land lease / option, offtake agreements and EPC contracts, the successful negotiation of these documents is what differentiates successful developers from those who run into legal or financial trouble. As the industry matures there has become more standardization of these documents, but there is no substitute for having sharp legal expertise to advise on the negotiation of battery agreements and limit risk to a project. Particularly in the growing world of mergers and acquisitions (M&A) of battery projects, portfolios, and developers, attorneys knowledgeable in energy storage are in high demand.

12.3 Construction

Construction of BESS projects has increasingly become a specialized workforce. While most large construction contractors have renewable energy units that specialize in wind or solar installations, more recently many have formed teams that specialize in the construction of energy storage projects. Since most construction contracts do not include supply of the battery units (see Section 8.4 for more on this), these units are typically involved in the construction of the BoP, which means everything other than the battery units themselves (and sometimes the PCS also). Everything else is the contractor's responsibility – everything from civil grading, fencing and foundations, to the substation, wiring between components, and supporting the commissioning process. Although battery sites are typically smaller than wind or solar plants of the same power, they are still large-scale construction sites that require a large amount of specialized labor.

Some of the skills involved in battery construction are generic to the construction industry, such as the civil work, while other skills are more specific to energy storage, such as making terminations in battery cabinets, or supporting the commissioning process.

Electricians are invaluable on a battery project. Since most BoP contracts stipulate that the contractor is responsible for connecting all components, there are thousands of DC, AC, and communications wires to terminate, and all of them must be done precisely. Electrical workers must connect medium-voltage (MV) lines in trenches, wire up MV transformers, and connect communications systems that allow the components to talk to one another. Some larger projects may have over 100 electrical workers at a given time working on a project.

Construction includes many management and higher-level positions such as project manager, construction manager, scheduler, accounting, and safety personnel. Although the jobs may require workers to travel to far-flung locations, the pay is typically quite high and allows for excellent job security as the industry grows.

In the US, the 2002 IRA included a provision which required all laborers on a job, whether prime contractors or subcontractors, be paid wages equal to or greater than prevailing wages, as determined by the Department of Labor. While this increased costs for many projects, the tax credits issued to the project more than compensated for the increase. The net result is that laborers on the hundreds of renewable projects that received IRA incentives have all been granted a pay raise that will continue in the long term, making these jobs even more attractive.

12.4 Operations and Asset Management

BESS projects are mostly operated remotely, by observing the digital outputs of battery voltages, currents, temperatures, and other indicators. While some very large projects have on-site staff, many are unstaffed, except when preventative or corrective maintenance is being performed. However, that does not mean there are not opportunities in this field – the O&M of battery plants include many different careers which are also essential to BESS operations. These include:

- Staff from the BESS and PCS manufacturers who offer ongoing preventative and corrective maintenance to their products
- Commercial electrical contractors who typically provide the ongoing maintenance for the remainder of the system
- Operators, usually working in command centers who continuously monitor a portfolio of projects for safety, performance, and revenue
- Safety and performance monitoring services, often provided from a third-party which provide the ongoing data analysis to identify safety and performance issues in the data sets produced by the project

Besides the physical management of a battery plant, it must be managed from a financial point of view. Asset managers do just this – they manage a portfolio of projects and aim to keep costs low and revenues high. This involves managing the O&M timeline to balance the safety and performance needs of a site with the cost to maintain it. Asset managers keep an eye out for trends that may indicate a project has issues, such as elevated operational temperatures, or above-average service calls.

A growing field of BESS asset management is the use of data-driven analytics and Artificial Intelligence (AI) for predicting the behavior of market signals and achieving higher BESS revenues. By constantly analyzing data and integrating with the EMS, AI facilitates effective decision-making on when to charge and discharge the BESS in real-time, leading to more efficient and lucrative energy storage operations. Further, BESS operators are beginning to use sophisticated predictive analytics to monitor and improve BESS safety, performance, and asset lifetime. Machine learning and AI-based approaches will be required to rapidly process large quantities of field data that operators can use to minimize unplanned maintenance, mitigate risks, and increase system availability.

Although there are fewer jobs in this class than others, they are critical to the successful operation of a project and will become an increasingly specialized discipline as more BESS projects are deployed.

12.5 Finance

As described in Chapter 6, no battery projects are built without being financed. From modelers to bankers to underwriters, the BESS finance infrastructure encompasses dozens of professionals who work to make these projects happen. Given the large size of these assets (some transactions have exceeded $500 million), they require meticulous planning of every aspect of how the project will be executed, what the risks are, and expectations of the revenue and profit the battery will earn as an investment. CFOs of battery developers are often some of the most well-compensated professionals in the industry, as are those associated with the transaction, including:

- Financial analysts working for equity investors
- Analysts working for the lenders supporting these projects
- Attorneys who support the financing, mergers, and acquisitions associated with battery investments
- Insurance company staff who underwrite battery projects
- Independent engineers who provide technical expertise to the finance and M&A of projects

While perhaps less glamorous or visible than other parts of a battery project, firms specialized in BESS finance will continue to grow as the pipeline of BESS investments continues to grow.

12.6 How Can I Get Involved?

After reading this book, we hope the reader recognizes that the energy storage industry is a fascinating and dynamic field. If already in the industry, we hope the reader takes away a deeper knowledge, and if eager to join, there are many paths to getting involved.

Here we cover three of the more conventional paths to joining: academia, pivoting from another field in renewables, and joining energy storage from a separate career.

New Entrants

For those just starting their career, this is a great time to get involved in energy storage. While there are few energy storage specific programs, most electrical engineering departments have a concentration in renewable energy which includes coursework on energy storage. Chemical engineering has traditionally offered programs focused on cell design, research, and materials, alongside materials science programs. Most advanced work on battery-specific topics is at the graduate level in master's or doctoral programs. Depending on which area of the industry being targeted, it may not be necessary to get graduate-level training, but many professionals in battery research, finance, or business have a master's degree or doctorate in a relevant field.

A professional engineering license (P.E. in the US, P.Eng. in Canada, and varying across different countries) can be a useful addition to any engineering focused career in the energy storage industry. Requirements vary by country, state, and province, but in the US the path to become a P.E. typically involves the following:

- Pass the Fundamentals of Engineering (FE) exam
- Accumulate a combination of education and/or years of experience. This varies but the most common path is four years at

an undergraduate engineering school followed by four years of work experience under a professional engineer.

- Pass the P.E. exam, an 8-hour exam in a chosen discipline of (for battery engineers the most common discipline is Electrical and Computer Engineering, with a specialization in Power Engineering)
- Submit the P.E. license application, together with all recommendation letters, experience documentation, and test results to the state engineering certification board.

While the P.E. license is state-specific, a national system exists in the US under the National Council of Examiners for Engineering and Surveying (NCEES). If an engineer wishes to obtain her or his license in multiple states, applying through NCEES can expedite this process.

For anyone interested in the trades, training as an electrician is an excellent path to doing so. As with P.E. licensure, the requirements vary from state to state, but the most common path is to start as an apprentice and work to become a journeyman and eventually a fully licensed electrician.

Licensed electricians looking to specialize in battery energy storage often work for large commercial firms performing electrical work on renewable projects. Although work on BESS projects may not be available immediately, by working on wind and solar projects, an electrician can gain the necessary field experience to work on storage projects when they come available.

One important resource for training in energy storage is NABCEP, the North American Board of Certified Energy Practitioners. This non-profit organization runs trainings and provides accreditations for workers in the renewable energy field. Their focus is on solar PV projects, and they run courses and tests for the design, procurement, installation, and O&M of these systems. However, as batteries have become more present on residential and commercial installations, NABCEP also offers courses around energy storage. Although the size of residential and commercial projects is quite different from that of utility-scale, the principles are the same, and the barrier for entry lower. Many professionals currently working in the energy storage industry got their start taking NABCEP courses.

Renewables to Storage

In the early days of utility-scale storage, circa 2015, one of the first large-scale applications of energy storage was in a coupled arrangement together with solar PV, driven by the batteries' ability to capture clipping losses. As costs came down, solar and BESS projects became the norm, and by 2022 a large portion of new solar projects in the US began to come with storage as an integral component. With this shift in technology came a shift in labor practices – workers who previously had focused on solar were now working on energy storage projects. Similarly, many developers who formerly focused on solar have pivoted to focus on energy storage, given the rapid growth, limited workforce, and similarities of the installations.

While wind isn't frequently co-located with storage, the finance and contracting structure shares many similarities – both are focused on a single high-value item that forms the majority of the capital cost, interconnect to the grid at high voltages, and require BoP contractors who are commonly hired to perform all the installation outside of the expensive component itself (either turbines or BESS units). Many professionals in the design, construction, and operation of wind farms have moved into the battery field as their employers have begun working on battery projects. The skillsets are different, as are the large lifts, complex mechanical components, and power converters used on today's wind farms which are quite different than batteries. However, the similarities are enough to make a wind to storage transition an easy one in most fields.

For readers that are working in solar and wind, there are many pathways to move into the storage industry. The easiest way to pivot is to volunteer for roles on storage-specific projects when available. Whether this is building models, completing designs, or actively being involved on a jobsite, being close to storage projects is a great way to understand the similarities and differences between solar, wind, and batteries. Some people choose to study, either through informal or organized courses and research, while others learn their skills on the job.

For those working in wind or solar whose firms are not pursuing energy storage projects, it is typically a short leap to move into energy storage. At the time of this writing, many employers are flexible on work experience, given the high demand for energy storage workers. Some may offer entry into the field at a reduced rate, but these opportunities may be a chance for paid training, allowing the workers to specialize in energy storage and transfer into a more senior role in a relatively short time.

Career Pivot

Lastly, those in unrelated fields to energy storage may be interested in pivoting into the field, given its rapid growth and expanding workforce. This transition is often done by people working in the business side of energy storage – jobs from upper management to sales to finance often are easily able to move their similar concepts from other industries into energy storage. Many have pivoted from real estate, given the similarity in the development and finance processes.

Pivoting on the technical side is more challenging, but not impossible. Many engineers from disciplines outside of electrical or chemical engineering (such as civil, mechanical, or industrial) have been able to transition into technical battery careers. For those looking to obtain P.E. licensure, the rules vary by state, but many locations allow an engineer to practice in any field provided they have sufficient experience to do so.

Others have gone back to the drawing board, returning to the classroom to study, often part-time, to earn degrees that can qualify them for careers in energy storage. For example, many master's degree programs in the US are focused on renewable energy, and some are even energy-storage specific. Some of these programs are designed for mid-career professionals, allowing workers to earn a degree at night or on a part-time basis, so that they may finish their careers in one field before pivoting into energy storage.

Some jobs, like a master electrician working on the job site, must be physically present at the project, for obvious reasons! However, although many jobs offer hybrid or even fully remote positions, we would encourage anyone transitioning to spend as much in-person time as possible at their new workplace. The in-person interactions and proximity to those in the field often gives workers in the office or on the jobsite a strong advantage over those who are working remotely.

Regardless of the path, the costs of switching careers must be weighed with the benefits. That said, many professionals have been pleasantly surprised with the benefits that come with this switch. Although there may be short-term issues such as reduced pay, costs for training or education, and delays in career advancement, the rapid growth of the industry often ends up benefiting those who switch in the long run.

Having a mentor can be a great way to consider a change of career. Ideally, the mentor will be in the energy storage field, but even someone in an adjacent field can provide valuable advice as a sounding board. Many large companies may have opportunities for establishing mentor relationships, although this may not be an appropriate path for those considering leaving their employer. Professional organizations such as IEEE, the Energy Storage Association (ESA), or similar organizations can be a great place to start studying the options available for a transition into energy storage.

Whatever the path, we wish success in the journey to get involved in energy storage. As we hope the reader has come to appreciate through these pages, storage is a tremendously dynamic field, ranging across a wide variety of specialties. From the chemical engineers designing the newest cell chemistries, to manufacturing specialists maximizing cell production, to finance professionals figuring out the implications of the latest round of policy changes to their battery project – all of these are facets of the energy storage resolution that is moving ahead at breakneck pace. For those who want to get involved, where there is a will, there is a way, and it is rarely too late to join the field.

The remainder of the book has some valuable sections of reference material, including case studies which dive into the specifics of three different types of energy storage projects. On a closing note, the authors welcome any questions, comments, or feedback. All the best journeying into the world of energy storage!

13. CASE STUDIES

13.1 Utility-Scale Standalone Storage: The Waratah Super Battery

Figure 13-1
Digital rendering of the Waratah Super Battery [115]

Australia is a highly active market for energy storage – the combination of an island grid with several clustered population centers, along with aggressive renewable energy targets has resulted in the development of many large-scale batteries. Akaysha Energy, which was acquired by Blackrock in 2022, has been a key player in this market as a developer of standalone energy storage projects. Akaysha's flagship project, the Waratah Super Battery (WSB) began construction in May 2023 and plans to achieve COD by March 2025. At the time of this writing, WSB has the highest power rating of any battery in the world, 850 MW, with an energy storage capacity of 1,680 MWh. The WSB forms part of the Sydney Ring Project, identified as a priority effort in the Australian Energy Market Operator (AEMO) Integrated System Plan (ISP) 2022, which maps the investment required to achieve a targeted 82% renewables by 2030.

The main offtaker of the project is the System Integrity Protection Scheme (SIPS) run by the Energy Corporation of New South Wales (NSW). The goals of SIPS are to release latent capacity in the transmission system, enable renewable energy integration, and boost grid reliability. NSW plans to close the Eraring coal-fired power plant (2880 MW) in 2025 as part of its plan to retire four of its five coal-powered energy stations in the next 11 years. Following the closure of Eraring, the grid will require modifications to ensure meeting the SIPS requirements. Rather than adding thermal generation or building new transmission lines, NSW will use the WSB to meet these requirements. In the event of a failure of a transmission line or generation plant, WSB is available to charge or discharge 700 MW of its power onto the grid, serving as a shock absorber for the NSW grid.

The additional power and energy of the system is used as a buffer for future degradation and will earn additional merchant revenue for the Project in the NSW markets.

The plant was financed by Blackrock through a raise of Australian banks along with global institutional and sovereign co-investors. Australia's green bank, the Clean Energy Finance Corporation (CEFC), committed A$100 million to the project. Other investors include NGS Super, a not-for-profit fund aiming to achieve a carbon-neutral investment portfolio by 2030.

The project site, approximately 60 miles north of Sydney, is 13.8 ha (34.1 acres) or approximately 25 National Football League (NFL) fields (or more than 8 Australian Football League (AFL) fields). The project interconnects at the Munmorah Substation at 330 kV, which was the former interconnection for the former 1400 MW Munmorah Coal Station, which was retired in 2012. A layout drawing of the project is shown in Figure 13-3.

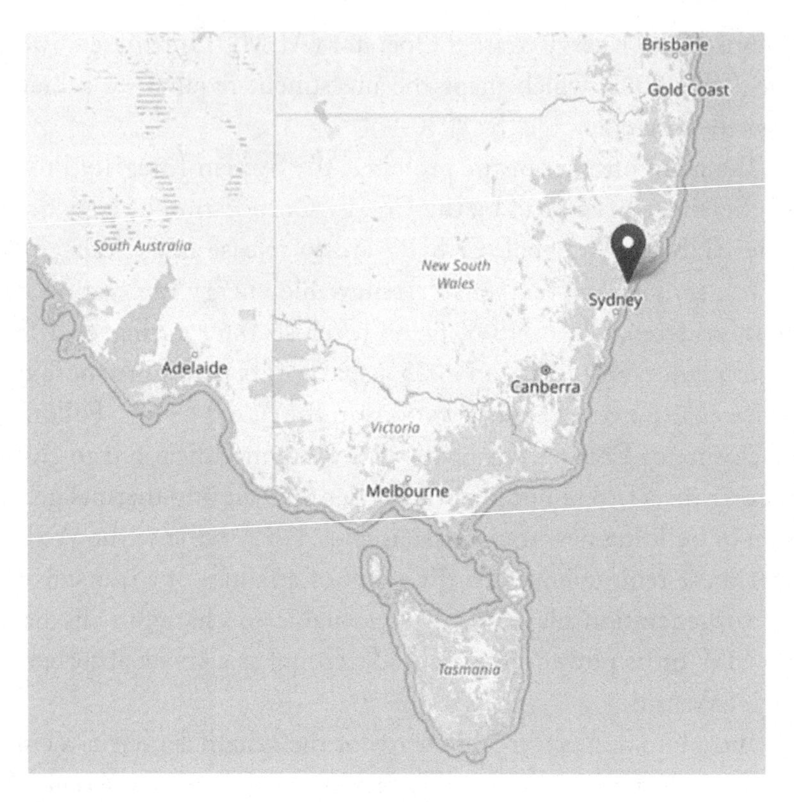

Figure 13-2
WSB project site [116]

The EPC on the project is Consolidated Power Projects Australia Ltd (CPP), a subsidiary of Quanta Services, a Texas-based infrastructure company.

The key equipment used:

- Powin Centipede S750 BESS, 288 lineups which each consist of:
    - 1 Collection Segment
    - 9 Energy Segments (750 kWh each)
    - Powin BMS, EMS, and Stack OS control system
    - A total of 2,592 Stack750 units for a total rated DC energy capacity of 1,901 MWh

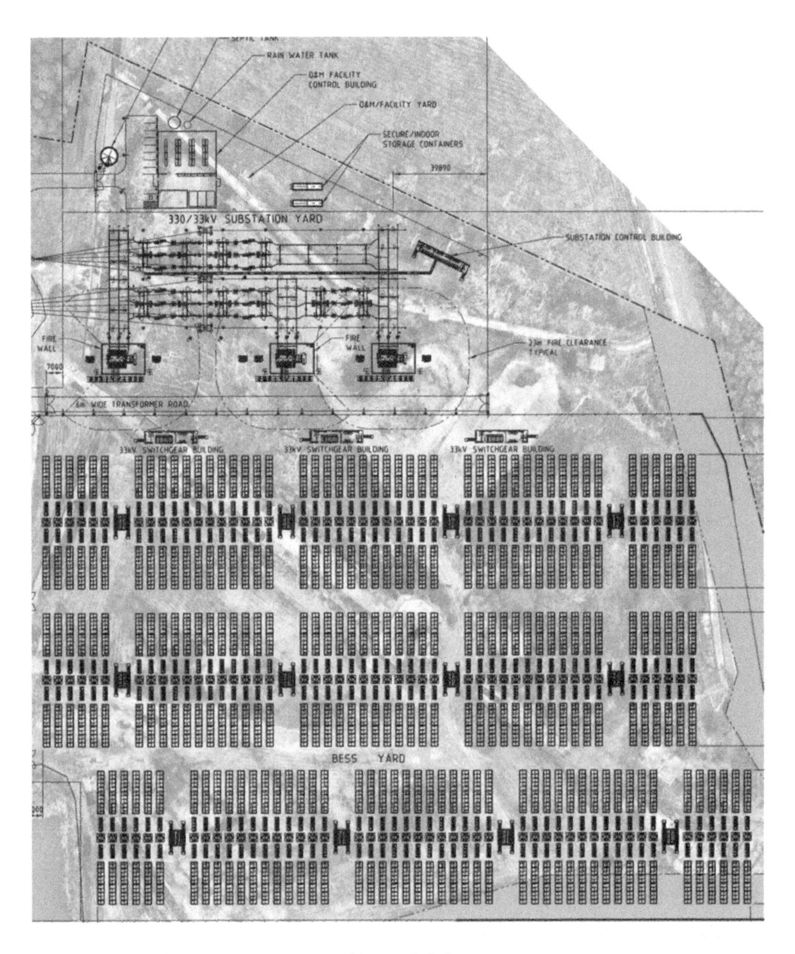

Figure 13-3
Layout of Waratah Super Battery [53]

- 288 eks PCS units
- 3 – 350 MVA main power transformers (MPT) supplied by Wilson Transformers
- 2 – 330 kV feeder bays that connect the project substation to the existing Munmorah Substation

It is notable that the power output at the battery terminals is 950.5 MW (1901 MWh / 2), which is 12% over the 850 MW rated output of the BESS on the grid. This overhead is necessary to account for losses incurred between the battery terminals and the grid (typically 4-5%). Additionally, the discharge energy must supply the auxiliary loads used by the battery units, primarily to cool the racks as they charge and discharge.

To comply with requirements of the NSW Planning Secretary, the project developed several Key Management Plans, which are publicly available:

- Environmental Impact Plan
- Remediation Action Plan
- Biodiversity Plan
- Site Layout Plan
- Environmental Impact Plan
- Infrastructure Approval
- Traffic Management Plan

The WSB project's main use case is rare among large batteries in that it comprises a virtual transmission solution. This use case functions as a result of the SIPS grid monitoring system, which immediately signals the battery in the case of sudden faults. A schematic of the WSB controls system is shown in Figure 13-4.

Figure 13-5 is a drone image of the construction progress as of January 2024. The infrastructure for the substation is seen in the front left of the photo: the aerial support infrastructure has been installed, along with the three foundations for the main power transformers. In the back right of the photo are the foundations for the battery lineups and PCSs. A single PCS was installed early to test alignment between a PCS and medium-voltage transformer (MVT).

Figure 13-6 shows the journey of one of the three 350 MVA 330/33/33 kV MPTs from the Wilson Transformer factory outside

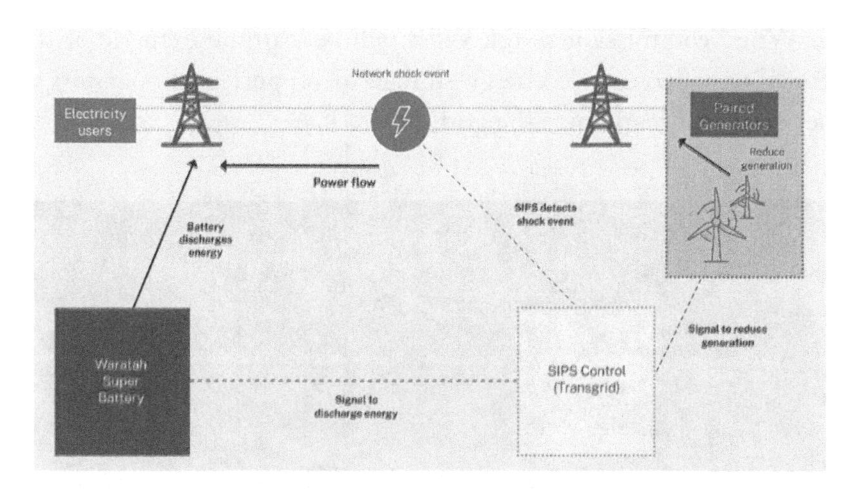

Figure 13-4
Waratah transmission grid diagram [117]

Figure 13-5
Waratah construction January 2024 – Substation shown front left, BESS shown top right [115]

Melbourne. Once installed on site, two radiators will be attached on each side of the transformer, making it almost three times the size.

When commissioned, the WSB will be a prime example of the versatility of large-scale energy storage to support grid operators as they work to decarbonize the grid.

Figure 13-6
Transport of a main power transformer for WSB [108]

13.2 Solar PV+Storage: Fifth Standard Project

Energy storage coupled with solar PV plants has been one of the earliest applications of large-scale battery projects. The trend started with DC-coupled storage, but most utility-scale plants are now built with AC-coupled storage. The Fifth Standard project is an example of large-scale AC-coupled PV+Storage project in California, one of the largest and most active markets for both solar PV and battery systems.

Fifth Standard is situated on approximately 1,600 acres about 70 km (45 mi) southwest of Fresno in Huron, CA, and is owned and operated by RWE Clean Energy. The BESS component has a nameplate capacity of 137 MWac and 548 MWh (4-hour duration) and is AC-coupled with a 150 MWac PV plant [118]. It is noteworthy that the BESS occupies only approximately 5 acres, or around 0.4% of the total project site. The project began construction in 2022 and came online in June 2023. The revenue streams are both from the CAISO merchant market and from a contractual offtake agreement with a private entity. Specifics could not be disclosed, but ballpark estimates from public information issued by the California Public Utility Commission put the offtake price in the range of $6-8/kW-month [119].

Figure 13-7
Fifth Standard Project site [116]

The total project cost, based on PV module and battery cost estimates at the time they were procured for the project, is in the range of $350-400 million which was financed with a combination of equity and tax equity funding from a leading bank.

Figure 13-8
Fifth Standard PV+Storage Project

Fifth Standard was designed and approved for construction before the IRA was passed in 2022. At this time, the law required that the BESS was exclusively able to charge from the grid – meaning it could draw no power from the grid. Since this requirement was lifted by the IRA, the plant was able to operate at higher-than-forecast revenue, improving the financial performance of the plant. The BESS currently is participating in day-ahead energy arbitrage and ancillary services – this will evolve as market conditions change over time.

The BESS is composed of 240 battery DC-blocks provided by LG Energy Solution (LGES's "DC LINK") utilizing an air-cooled NMC chemistry-based cells. The project uses 40 SMA Sunny Central 3950 PCS series as the BESS PCSs. The PV and BESS share a common interconnection point at Pacific Gas & Electric's Gates substation. Blattner Inc. served as the EPC on this project, having provided complete engineering, procurement (excluding major equipment) and construction services for the plant. RWE Clean Energy serves as the

Figure 13-9
Fifth Standard site layout

O&M provider, operator, and qualified scheduling entity for operating the project in the CAISO system.

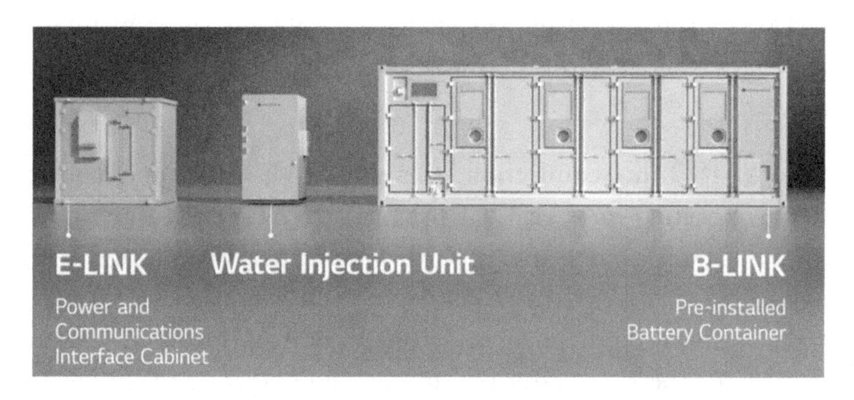

Figure 13-10
DC LINK Product from LG Energy Solution

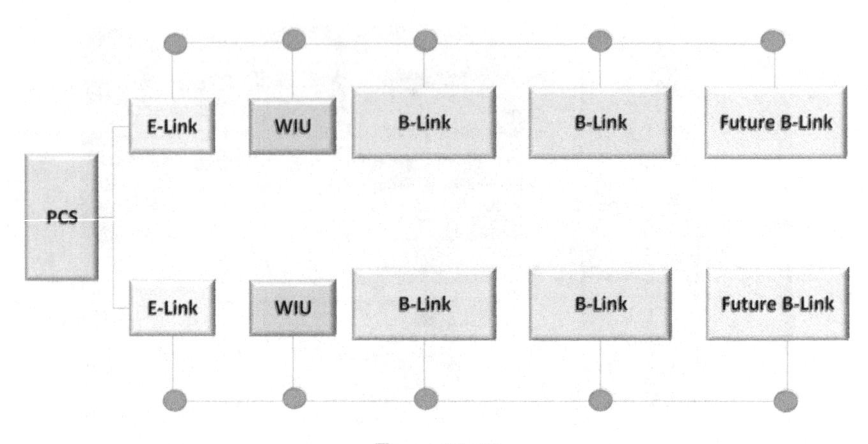

Figure 13-11
Illustrative BESS AC-block

The project's development, like several PV plants, spanned many years. The project secured an offtake agreement in February 2022, after which the execution of the project began, encompassing detailed design development and construction. Commissioning was initiated in November 2022 and took roughly six months to complete because of challenges on both the PV and BESS components of the project. Energization of the combined system was achieved in April 2023. Testing as required by CAISO for qualification as a resource was carried out in May. COD was declared on June 2, 2023.

Every utility-scale project comes with its fair share of unique challenges. Some of the noteworthy challenges this project overcame are:

- **CAISO interconnection**: This was RWE's first project interconnecting to CAISO at the time of completion (note: since this project came online, RWE Clean Energy was formed via a merger of RWE Renewables and Con Edison's Clean Energy Business, the latter of which had assets operating in CAISO area). The ISO's asset registration and operating requirements, response times, and testing protocols for resource adequacy took more time than budgeted during the project's commissioning phase.

- **Use-case switching**: Different signal mechanisms are used by CAISO to control DA energy commitments and ancillary services at different time-intervals. Controls had to be configured to manage both signal modes for the BESS simultaneously.
- **Commissioning**: This step is tricky and under-estimated for most battery projects because of the evolving nature of battery products. It is also important to have technicians from the suppliers of the BESS and PCS on site at the same time to support the commissioning activities and troubleshoot any challenges. Suppliers and supporting staff being in international locations and differing time zones add to the challenge of completing the activities on schedule. Commissioning for a PV+Storage plant as one unified system can be especially more challenging because of increased complexity and interdependency between the PV and BESS components. This was experienced in this project when supply-chain related delays on PV module installation added to delays in overall commissioning. Utility connection for auxiliary power and substation energization were not available in time for commissioning, so temporary generators needed to be used to provide auxiliary power during cold commissioning, which is needed for all internal communication and control checks.
- **New product deployment**: This project was one of the first deployments of LGES's containerized battery solution, with the bulk of their prior experience being focused on module and rack products. This resulted in challenges of the project team discovering product nuances on the fly and adjusting the scope of the EPC specific to the product. This is to be expected of any product that is new to the owner and the EPC and is a challenge that is diminishing quickly as the industry matures,

containerized products become more standardized and EPCs gain experience with a variety of BESS products.

- **Capacity test:** This had to be done more than once because of procedural kinks and iterative coordination that was required between the battery, PCS, and EMS providers (the latter two were provided by SMA in this case) which was the first time for the suppliers at this project. Once these kinks were resolved, the project was able to successfully pass all required tests and move toward COD.
- **Telemetry issues:** CAISO requires dedicated data circuits and internet connectivity for the metering equipment. These were not ordered ahead enough, leading to delays in CAISO qualification.

Fifth Standard has been in successful operation for over a year as of publication and is producing healthy financial returns for its owner and operator, while being able to inject its solar production both during the daytime and after dark thanks to the BESS. While the project presented challenges like most large-scale PV+Storage projects to the developer, EPC, and battery integrator, it represents a successful early project in the California market. Since being built, RWECE, Blattner, and LGES have gone on to play roles in many GWh of additional projects, building on lessons learned in the Fifth Standard project.

13.3 Commercial / Industrial Storage

This book is focused on utility-scale solar, but we have included many passing references to both smaller installations and to microgrids. Given the popularity and interest in these types of systems, we present here a 800 kWh commercial BESS installation that is typical of grid-tied smaller systems which can also be "islanded" in a microgrid application. The Miller Community Center is a pilot project built by the utility Seattle City Light (SCL). Located in the Capitol Hill neighborhood of Seattle, the Miller facility is home to programs, activities, and events for the community, including sports activities, camps, after-school care, and senior programming. The Center also houses shower and restroom facilities for Seattle Public School students and for residents in need.

The project's goals were three-fold:

- To use solar energy to provide a more sustainable energy source for the facility, reducing the electric bill paid by the center over the year.
- To increase the resilience of the facility and allow it to become a community hub that could be islanded from the main grid in the case of unplanned emergency events.
- To serve as an example project that can prove the ability of a city-owned facility to provide resiliency and energy independence, serving as a template for future microgrid and battery projects.

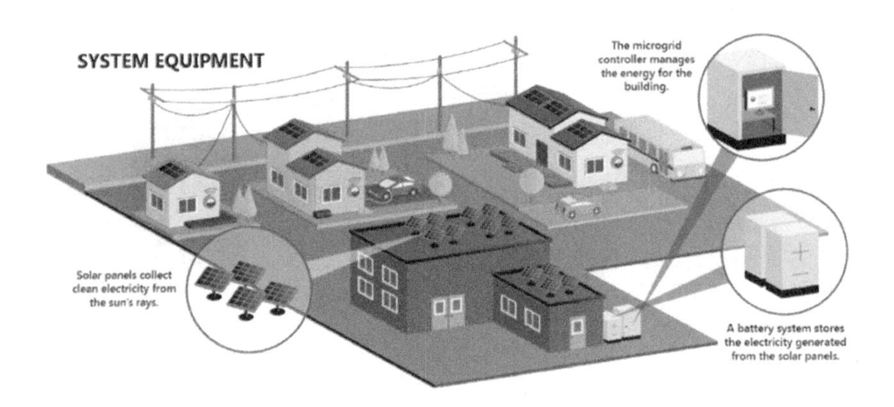

Figure 13-12
Miller Microgrid concept, as provided by SCL [120]

The project was funded by a $1.5 million grant from Washington State. The City of Seattle partnered with the University of Washington to analyze and study the microgrid's benefits. The site for the microgrid was provided by Seattle Parks and Recreation, and it was built in partnership with SCL and the Seattle Office of Arts & Culture.

The project was built between January and June 2021. The main contractor for the facility was Worley Parsons Group, Inc., who designed, built, tested, and commissioned the facility, and technical advisory services to SCL were provided by DNV.

The main technical components of the Miller system are:

- A 200 kW / 800 kWh BESS by manufacturer NEC, which can power operations of the center as a microgrid for approximately 24-hours
- 200 kW of PCS capacity powered by WSTech inverters
- A 43 kW rooftop PV array, provided by Puget Sound Solar
- A monitoring and control system by Trimark Associates
- The budget for the project was $3.3 million, according to the Capitol Hill Seattle Blog [121].
- The electrical layout of the system is shown in Figure 13-13.

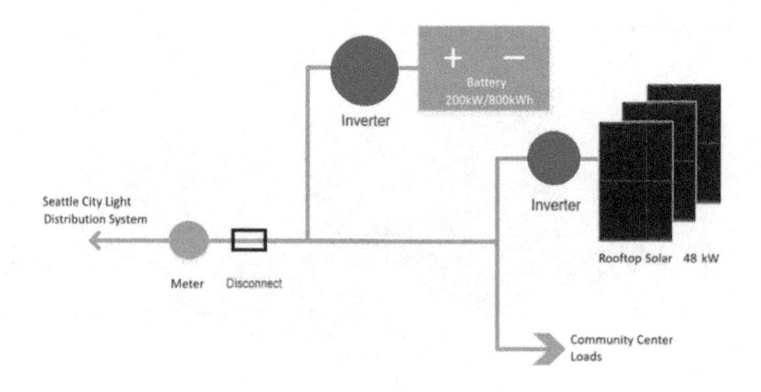

Figure 13-13
Miller Microgrid SLD [120]

As configured, the battery can provide several different services. Since the battery is behind the facility meter, it is a BTM system. This means that it can be charged or discharged to aid in the load management of the system. If the facility has Time-of-Use (TOU) energy billing with the utility, the PV energy can be sent into the battery to be discharged later. Alternatively, when the demand from the facility is highest, the battery can inject power into the grid, effectively lowering the amount of power drawn from the grid – this is known as peak shaving. All these functions are available because the output of the BESS is synchronized to the grid via the PCS, which is in grid-following mode.

When there is a power outage, the facility will lose power for a moment – this is because the PCS will have no grid to follow. At that point the disconnect located adjacent to the facility meter will open, to isolate the system from the grid. This is a safety function, since if not isolated, the system could be injecting power into the local lines, posing a safety hazard. Within a few seconds of operation, the grid will come back online when the PCS switches to "grid-forming" mode – this means that rather than searching for a grid frequency, they will form a grid on their own, without reference to the main grid. Once the grid is formed by the PCSs, the loads can be re-energized, at which point the facility will come back online. The 200 kW maximum output of the PCSs is sufficient to fully power the facility – for reference, smaller homes typically have approximately a 24 kW maximum input from the grid, while larger ones may have 48 kW.

It is notable that the ability to perform grid-forming is not a standard feature for the PCS on utility-scale projects. This is because most large projects operate, and earn their revenue, exclusively while connected to the grid. Most major PCS manufacturers offer a version of their products with grid-forming mode, but there may usually be an additional cost.

With exclusively the batteries, the facility would have access to the 800 kWh of energy, assuming the system is fully charged. A typical facility like the Miller center might use around 200 kWh per day, meaning the facility could run off-grid for approximately 4 days. However, this system benefits from having an added source of energy in the form of the solar PV system. On sunny days, the 43 kW rooftop array may provide up to 200 kWh of energy – meaning that for every fully sunny day, the system would be able to extend its energy use beyond the initial 4-day period. In off-grid systems, this time is known as the period of autonomy. Seattle being a somewhat rainy city, there is a risk that the PV array may not contribute significant energy during the autonomy period, but it is likely that over a period of several days, the PV will at a minimum extend the period by around 1 day. In a scenario where the power is out for a several day period, the facility would likely limit its energy consumption, focusing on essential energy uses only – this can extend the autonomy period even further. In this scenario, the Miller center is intended for use as a community gathering point, allowing residents to charge their devices and congregate in a warm public space until the grid is restored.

Figure 13-14 and Figure 13-15 show the physical infrastructure of the system. The pad-mounted unit on the right in the top photo is the WSTech PCS, while the unit on the left is the NEC battery. Behind the two units in the foreground are an identical set of units. This system is comprised of two independent 4-hour systems, each with 400 MWh of energy storage capacity – this has the advantage of allowing one system to be taken down for maintenance while the other is still functional.

The lower photo shows the external location of the system. It is adjacent to the miller facility and is protected by a gated security fence.

Figure 13-16 shows the technical drawing with the battery layout, and the location of each component. As with the larger utility-scale system, the same components are present – battery enclosures, PCSs (inverters), and transformers, as well as the switchgear and controls system.

Figure 13-14
Batteries and PCS at the Miller Microgrid [120]

Figure 13-15
Miller Microgrid Project [120]

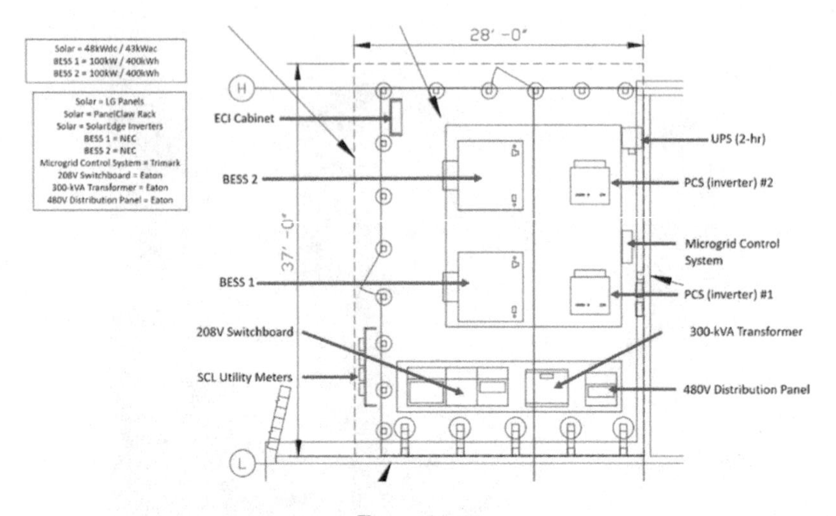

Figure 13-16
Miller Microgrid components [120]

The are several differences between a system of this type and of a utility-scale system:

- There is no medium-voltage on this site, since the output of the PCSs (480 V three-phase) can be immediately stepped down to the internal voltage of the facility (208 V three-phase)
- The PCS (PCS) in this system can provide a grid-forming function, while most utility-scale PCSs cannot
- This system has a microgrid control system, in addition to the standard communications infrastructure
- The scale is smaller – each enclosure in this system has 400 kWh capacity, while most utility-scale system enclosures usually have 2 to 5 MWh per enclosure (which are most commonly 20-foot containers)

After commissioning, the project was formally inaugurated on April 22, 2022 (Earth Day), which was attended by Washington State Governor Jay Inslee, a major promoter of climate change awareness.

Inslee called the project the "beginning of a revolution," stating that the state intended to replicate the project across the state. Since the beginning of its operation, the microgrid has been able to reduce the energy bills of the Miller center, as well as serving as a backup power source in the case of power outages.

REFERENCES

[1] BNEF, "2023 Energy Storage Market Outlook," Bloomberg New Energy Finance (BNEF), 21 March 2023. [Online]. Available: https://about.bnef.com/blog/1h-2023-energy-storage-market-outlook/. [Accessed 13 January 2024].

[2] Bloomberg New Energy Finance, "Energy Storage Investments Boom As Battery Costs Halve in the Next Decade," BNEF, [Online]. Available: https://about.bnef.com/blog/2h-2023-energy-storage-market-outlook/. [Accessed 03 January 2024].

[3] EIA, "World Energy Investment 2023: Overview and key findings," EIA, 1 May 2023. [Online]. Available: https://www.iea.org/reports/world-energy-investment-2023/overview-and-key-findings. [Accessed 22 April 2024].

[4] DNV, "Energy Transition Outlook 2023," DNV, Hovik, Norway, 2023.

[5] G. S. E. J. S. v. d. R. A. a. L. M. Jarbratt, "Enabling renewable energy with battery energy storage systems," McKinsey & Company, 2 August 2023. [Online]. Available: https://www.mckinsey.com/industries/automotive-and-assembly/our-insights/enabling-renewable-energy-with-battery-energy-storage-systems. [Accessed 6 January 2024].

[6] U.S. Department of Energy (DOE), "United States Energy & Employment Report 2023," U.S. Department of Energy, Washington, DC, 2023.

[7] J. T. V. M. M. T. T. E. B. a. S. A. H. C. S. Johnson, "Layered Lithium-Manganese Oxide Electrodes Derived from Rock-Salt LixMnyOz (x+y=z) Precursors," in *194th Meeting of the Electrochemical Society*, Boston, MA, 1998.

[8] J. Landers, "Large Battery Energy Storage System Now Operating in Hawaii," ASCE, 9 March 2024. [Online]. Available: https://www.asce.org/publications-and-news/civil-engineering-source/civil-engineering-magazine/article/2024/03/large-battery-energy-storage-system-now-operating-in-hawaii. [Accessed 28 May 2024].

[9] N. Maluf, "The Inner Sanctum of a Battery & Fast Charging," Qnovo, 12 June 2015. [Online]. Available: https://www.qnovo.com/blogs/inner-sanctum-of-a-battery-fast-charging. [Accessed 2 June 2024].

[10] B. Chapman, "How does a lithium-Ion battery work?," let's talk science, 23 September 2019. [Online]. Available: https://letstalkscience.ca/educational-resources/stem-in-context/how-does-a-lithium-ion-battery-work. [Accessed 3 June 2024].

[11] D.Abraham,"https://www.azom.com/article.aspx?ArticleID=21260," Argonna National Laboratory, 5 January 2022. [Online]. [Accessed 2 June 2024].

[12] A. Neverman, "Everything You Need to Know About the 18650 Battery," Common Sense Home, 12 November 2022. [Online]. Available: https://commonsensehome.com/18650-battery/. [Accessed 6 January 2024].

[13] "LiFePO4 Voltage Chart (3.2V, 12V, 24V & 48V)," Battery Finds, 6 April 2023. [Online]. Available: https://batteryfinds.com/lifepo4-voltage-chart-3-2v-12v-24v-48v/. [Accessed 2 6 2024].

[14] B. Wang, "EV Battery Glut Drives Prices Down to $70-75 Per kWh," Next Big Future, 10 October 2023. [Online]. Available: https://www.nextbigfuture.com/2023/10/ev-battery-glut-drives-prices-down-to-70-75-per-kwh.html#:~:text=Sources%20are%20reporting%20that%20Chinese,price%20average%20of%20below%20%24100.. [Accessed 7 May 2024].

[15] DNV, "2024 Battery Scorecard," DNV, Katy, TX, 2024.

[16] Wood Mackenzie, "LFP to overtake NMC as dominant stationary storage chemistry by 2030," Wood Mackenzie, 17 August 2020. [Online]. Available: https://www.woodmac.com/press-releases/lfp-to-overtake-nmc-as-dominant-stationary-storage-chemistry-by-2030/. [Accessed 23 April 2024].

[17] A. Colthorpe, "Tesla Megapack on fire in 'minor incident' at battery storage site in Australia," Energy Storage News, 27 September 2023. [Online]. Available: https://www.energy-storage.news/tesla-megapack-on-fire-in-minor-incident-at-battery-storage-site-in-australia/. [Accessed 3 November 2023].

[18] B. University. [Online]. Available: https://batteryuniversity.com/img/content/BU-205_chart-2-web.jpg.

[19] T. e. a. Igogo, "Supply Chain of Raw Materials Used in the Manufacturing of Light-duty Vehicle Lithium-ion Batteries," Clean Energy Manufacturing Analysis Center / NREL, Denver, 2022.

[20] A. O. e. a. Aydin, "Lithium-Ion Battery Manufacturing: Industrial View on Processing Challenges, Possible Solutions and Recent Advances," *Batteries,* vol. 9, no. 11, p. 555, 2023.

[21] R. N. Charette, "Can Flow Batteries Finally Beat Lithium?," IEEE Spectrum, 23 December 2023. [Online]. Available: https://spectrum.ieee.org/flow-battery-2666672335. [Accessed 6 January 2024].

[22] Fuel Cell and Hydrogen Energy Association, "Fuel Cell & Hydrogen Energy Basics," FCHEA, 1 August 2019. [Online]. Available: https://www.fchea.org/h2-day-2019-events-activities/2019/8/1/fuel-cell-amp-hydrogen-energy-basics. [Accessed 4 June 2024].

[23] V. Sprenkle, "Energy Storage Grand Challenge Cost and Performance Assessment 2020," U.S. Department of Energy, Portland, 2020.

[24] A. a. S. S. Headley, "U.S. DOE Energy Storage Handbook, Chapter 11 Hydrogen Energy Storage," U.S. Department of Energy, Washington, 2020.

[25] DNV, "2022 Battery Scorecard," 2022. [Online]. Available: https://www.dnv.com/power-renewables/energy-storage/2022-battery-scorecard.html.

[26] A. Colthorpe, "LFP to dominate 3Twh global lithium ion battery market by 2030," Energy Storage News, 22 March 2022. [Online]. Available: https://www.energy-storage.news/lfp-to-dominate-3twh-global-lithium-ion-battery-market-by-2030/. [Accessed 25 May 2024].

[27] A. Colthorpe, "CATL partners with integrator FlexGen on 220MWh of Texas battery storage projects," Energy Stoarge News, 6 January 2021. [Online]. Available: https://www.energy-storage.news/catl-partners-with-integrator-flexgen-on-220mwh-of-texas-battery-storage-projects/. [Accessed 2 June 2024].

[28] "Powin Signs Agreement With Top Tier BatteryCell Provider to Meet Customer Demand," Powin, 11 August 2021. [Online]. Available: https://powin.com/powin-signs-agreement-with-top-tier-battery-cell-provider-to-meet-customer-demand/. [Accessed 2 June 2024].

[29] A. Colthorpe, "Energy storage product launches in Europe from major Chinese solar companies Trina, Sungrow," Energy Storage News , 12 October 2021. [Online]. Available: https://www.energy-storage.news/energy-storage-product-launches-in-europe-from-major-chinese-solar-companies-trina-sungrow/. [Accessed 2 June 2024].

[30] S&C Electric Company, "Are You Looking After Your Renewable Substation?," S&C Electric Company, 14 November 2017. [Online]. Available: https://www.sandc.com/en/gridtalk/2017/november/14/are-you-looking-after-your-renewable-substation/. [Accessed 13 May 2024].

[31] J. Parnell, "Huawei confirms US solar market exit," PV Tech, 19 July 2019. [Online]. Available: https://www.pv-tech.org/huawei-confirms-us-solar-market-exit/. [Accessed 23 April 2024].

[32] Fisher Engineering, Inc. / Energy Safety Response Group, "Victorian Big Battery Fire: Report of Technical Findings," Fisher Engineering, Inc. / Energy Safety Response Group, Johns Creek, 2022.

[33] GSES, "Implementing Virtual Power Plants (VPP) in Australia," GSES, 25 November 2020. [Online]. Available: https://www.gses.com.au/implementing-virtual-power-plants-vpp-in-australia/. [Accessed 6 June 2024].

[34] "Join the Tesla Virtual Power Plant," Tesla, [Online]. Available: https://www.tesla.com/support/energy/tesla-virtual-power-plant-sce#2022-performance. [Accessed 6 July 2023].

[35] R. Walton, "Texas regulators look to distributed resources, additional coal reserves, to boost reliability," Utility Dive, 22 April 2022. [Online]. Available: https://www.utilitydive.com/news/texas-regulators-look-to-distributed-resources-additional-coal-reserves-t/622525/. [Accessed 10 March 2024].

[36] IESO, "Price Overview: Global Adjustment," IESO, 4 October 2022. [Online]. Available: https://www.ieso.ca/en/Power-Data/Price-Overview/Global-Adjustment. [Accessed 10 March 2024].

[37] R. Ma, "Sr. Machine Learning Specialist," LinkedIn, 1 October 2023. [Online]. Available: https://www.linkedin.com/posts/rohanma_sr-machine-learning-engineer-autobidder-activity-7107765356907171840-TjZC?utm_source=share&utm_medium=member_desktop. [Accessed 10 March 2024].

[38] "2022 ACP Annual Report," ACP, 1 May 2023. [Online]. Available: https://cleanpower.org/wp-content/uploads/2023/05/2022-ACP-Annual-Report_Public.pdf. [Accessed 17 April 2024].

[39] Monitoring Analytics, LLC, "State of the Market Report for PJM," Monitoring Analytics, LLC, 11 August 2022.

[40] Rocky Mountain Institute, "The Economics of Battery Energy Storage," RMI, 1 March 2017. [Online]. Available: https://rmi.org/wp-content/uploads/2017/03/RMI-TheEconomicsOfBatteryEnergyStorage-FullReport-FINAL.pdf. [Accessed 17 April 2024].

[41] California ISO, "California ISO Peak Load History 1998 through 2022," [Online]. Available: http://www.caiso.com/Documents/CaliforniaISOPeakLoadHistory.pdf. [Accessed 14 2 2023].

[42] ERCOT, "Real-time locational prices," ERCOT, 3 June 2024. [Online]. Available: https://www.ercot.com/content/cdr/contours/rtmLmp.html. [Accessed 6 June 2024].

[43] J. Hawkins, "Optimized Integration of PV with Battery Storage: A Real World Success Story," Sandia National Lab, August 2016. [Online]. Available: https://www.sandia.gov/files/ess/docs/NMRESGI/2016/2_Optimized_Integration_Of_PV_Hawkins.pdf. [Accessed 3 June 2024].

[44] GreenBiz, "Meet the power plant of the future: Solar and battery hybrids," GreenBiz, 1 June 2022. [Online]. Available: https://www.greenbiz.com/article/meet-power-plant-future-solar-and-battery-hybrids.

[45] Energy Storage News, "PJM's frequency regulation rule changes causing 'significant and detrimental harm'," Energy Storage News, 18 April 2018. [Online]. Available: https://www.energy-storage.news/pjms-frequency-regulation-rule-changes-causing-significant-and-detrimental-harm/. [Accessed 6 July 2023].

[46] D. Johnston, "Limiting Liability: Roman Law and the Civil Law Tradition," *Chicago-Kent Law Review,* vol. 70, no. 4, pp. 1515-1538, 1995.

[47] N. Cuthburt, "A Guide to Project Finance," Dentons, London, 2018.

[48] National Renewable Energy Lab, "Utility-Scale Battery Storage - Annual Technology Baseline," NREL, 15 July 2023. [Online]. Available: https://atb.nrel.gov/electricity/2023/utility-scale_battery_storage. [Accessed 14 April 2024].

[49] Berkeley Lab, "New study refocuses learning curve analysis on LCOE rather than up-front installed costs in order to provide a more-holistic view of technology advancement," Electricity Markets and Policy, 22 May 2022. [Online]. Available: https://emp.lbl.gov/news/new-study-refocuses-learning-curve-analysis. [Accessed 7 August 2023].

[50] U.S. Environmental Protection Agency, "Summary of Inflation Reduction Act provisions related to renewable energy," 1 June 2023. [Online]. Available: https://www.epa.gov/green-power-markets/summary-inflation-reduction-act-provisions-related-renewable-energy. [Accessed 7 August 2023].

[51] ACORE, "The Risk Profile of Renewable Energy Tax Equity Investments," ACORE, 1 December 2023. [Online]. Available: https://acore.org/wp-content/uploads/2023/12/ACORE-The-Risk-Profile-of-Renewable-Energy-Tax-Equity-Investments.pdf. [Accessed 17 February 2024].

[52] N. Weaver, "ERCOT: how much money did battery energy storage make in H1 2023?," Modo Energy, 26 September 2023. [Online]. Available: https://modoenergy.com/research/ercot-battery-energy-storage-systems-revenues-january-june-2023-leaderboard. [Accessed 24 May 2024].

[53] A. Kokkinis, Consollidated Power Projects, 16 August 2023. [Online]. Available: https://issuu.com/akayshaenergy/docs/wsb-site_layout_plan_final_?fr=xKAE9_zU1NQ. [Accessed 2 June 204].

[54] "Global lithium-ion battery capacity to rise five-fold by 2030," Wood Mackenzie, 22 March 2022. [Online]. Available: https://www.woodmac.com/press-releases/global-lithium-ion-battery-capacity-to-rise-five-fold-by-2030/. [Accessed 9 November 2023].

[55] Enovix, "Winning the Global Battery Arms Race," 24 August 2021. [Online]. Available: https://enovix.medium.com/winning-the-global-battery-arms-race-dc411ce2f649. [Accessed 24 June 2024].

[56] Benchmark Source, "One year on, Biden's IRA has changed the battery landscape," Benchmark Source, 15 August 2023. [Online]. Available: https://source.benchmarkminerals.com/article/one-year-on-the-ira-has-changed-the-battery-landscape-in-the-us . [Accessed 18 May 2024].

[57] C. Murray, "Lithium prices to remain elevated this year, battery packs to fall to US$100/kWh by 2025-27," Energy Storage News, 24 May 2023. [Online]. Available: https://www.energy-storage.news/lithium-prices-to-remain-elevated-this-year-battery-packs-to-fall-to-us100-kwh-by-2025-27/. [Accessed 3 June 2024].

[58] A. Soule, "In CT and New England Batteries Gain Juice for Emergency Power," Cadenza Innovation, 23 March 2023. [Online]. Available: https://cadenzainnovation.com/news/in-ct-and-new-england-batteries-gain-juice-for-emergency-power/. [Accessed 18 March 2024].

[59] LiuGong, "LiuGong Dressta Machinery Takes Customers' Needs as Top Priority in Uzbekistan," LiuGong, 15 September 2018. [Online]. Available: https://www.liugong.com/en/news/20180915/. [Accessed 3 June 2024].

[60] Trelic, "Bathtub curve – a useful tool to understand failure rates," Trelic, 20 December 2020. [Online]. Available: https://www.trelic.fi/bathtub-curve-a-useful-tool-to-understand-failure-rates/. [Accessed 1 April 2024].

[61] D. P. Hill, "McMicken Battery Energy Storage System Event Technical Analysis and Recommendations," DNVGL, Chalfont, 2020.

[62] APS, "McMicken investigation," 27 July 2020. [Online]. Available: https://www.aps.com/en/About/Our-Company/Newsroom/Articles/Equipment-failure-at-McMicken-Battery-Facility. [Accessed 2 June 2024].

[63] American Clean Power, "First Responders Guide to Lithium-Ion," American Clean Power, Washington, D.C., 2023.

[64] "BESS manufacturing quality: Lessons learned from more than 30GWh of factory inspections," Clean Energy Associates, 2 February 2024. [Online]. Available: https://www.cea3.com/cea-blog/bess-manufacturing-quality-lessons-learned.

[65] B. Wu, "Battery Thermal Runaway," Battery Design, 1 May 2023. [Online]. Available: https://www.batterydesign.net/thermal-runaway/. [Accessed 3 June 2024].

[66] "From Cause to Prevention: Decoding Lithium Ion Battery Fires," Twaice, 27 July 2023. [Online]. Available: https://www.twaice.com/article/decoding-battery-fires. [Accessed 2024 March 19].

[67] "Enegy Storage Draft Emergency Response Plan," American Clean Power, 10 June 2022. [Online]. Available: https://cleanpower.org/wp-content/uploads/2022/11/ACP_Energy_Storage_Emergency_Response_Plan_Template.pdf. [Accessed 20 March 2024].

[68] A. K. J. R. a. M. M. Adam Barowy, "Energy Storage System Installation Test Report Now Available," FSRI, 12 April 2021. [Online]. Available: https://fsri.org/research-update/energy-storage-system-installation-test-report-now-available. [Accessed 3 February 2024].

[69] "Overview of Lithium battery safety testing- UL 1973," EverExceed, 24 December 2021. [Online]. Available: https://www.everexceed.com/overview-of-lithium-battery-safety-testing-ul-1973_n433. [Accessed 3 February 2024].

[70] "UL 1741SA Standards for Renewable Energy Inverters," Windurance, 11 August 2022. [Online]. Available: https://blog.windurance.com/standards-for-renewable-energy-inverters-understanding-ul-1741sa. [Accessed 3 February 2024].

[71] National Fire Protection Agency, "Battery Energy Storage Systems (Bess) Emergencies: Quick Reference Guide," NFPA, Quincy, 2016.

[72] A. Katwala, "The Spiralling Environmental Cost of Our Lithium Battery Addiction," Wired Magazine, 5 August 2018. [Online]. Available: https://www.wired.com/story/lithium-batteries-environment-impact/. [Accessed 22 April 2024].

[73] G. Dabi, "Tesla Gigafactory: A Road to a Sustainable Future," ASME IIEST Shibpur Student Section, 9 October 2020. [Online]. Available: https://medium.com/the-treatise/tesla-gigafactory-a-road-to-a-sustainable-future-b95e137ac2c3. [Accessed 24 April 2024].

[74] B. Pilkington, "Sustainable Alternatives to Lithium Use in Batteries," AZO Cleantech, 16 May 2022. [Online]. Available: https://www.azocleantech.com/article.aspx?ArticleID=1538. [Accessed 24 April 2024].

[75] O. e. a. Dolotko, "Recycling of Batteries: 70 Percent of Lithium Recovered," Karlsruhe Institue of Technology, 28 March 2023. [Online]. Available: https://www.kit.edu/kit/english/pi_2023_015_recycling-of-batteries-70-percent-of-lithium-recovered.php#:~:text=Presently%2C%20mainly%20nickel%20and%20cobalt,is%20expensive%20and%20hardly%20profitable.. [Accessed 22 April 2024].

[76] CNBC, "Dead EV batteries turn to gold with U.S. incentives," Reuters, 21 July 2023. [Online]. Available: https://www-cnbc-com.cdn.ampproject.org/c/s/www.cnbc.com/amp/2023/07/21/dead-ev-batteries-turn-to-gold-with-us-incentives.html. [Accessed 22 April 2024].

[77] H. e. a. Vandevyvere, "Positive Energy Districts Solution Booklet," EU Smart Cities Information System, 1 November 2020. [Online]. Available: https://www.researchgate.net/publication/354424161_POSITIVE_ENERGY_DISTRICTS_SOLUTION_BOOKLET_EU_Smart_Cities_Information_System/download?_tp=eyJjb250ZXh0Ijp7ImZpcnN0UGFnZSI6I19kaXJ1Y3QiLCJwYWdlIjoiX2RpcmVjdCJ9fQ. [Accessed 22 April 2024].

[78] G. Kilgore, "Carbon Footprint of Lithium-Ion Battery Production (vs Gasoline, Lead-Acid)," 8 Billion Trees, 18 March 2024. [Online]. Available: https://8billiontrees.com/carbon-offsets-credits/carbon-footprint-of-lithium-ion-battery-production/. [Accessed 22 April 2024].

[79] Cobalt Institute, "Cobalt Mining," Cobalt Institue, 1 January 2024. [Online]. Available: https://www.cobaltinstitute.org/about-cobalt/cobalt-life-cycle/cobalt-mining/. [Accessed 22 April 2024].

[80] Natural Resources Canada, "Lithium facts," Government of Canada, 29 February 2024. [Online]. Available: https://natural-resources.canada.ca/our-natural-resources/minerals-mining/mining-data-statistics-and-analysis/minerals-metals-facts/lithium-facts/24009. [Accessed 9 April 2024].

[81] Statista, "Demand for lithium worldwide in 2020 and 2021 with a forecast from 2022 to 2035," Statista, 21 February 2024. [Online]. Available: https://www.statista.com/statistics/452025/projected-total-demand-for-lithium-globally/. [Accessed 22 April 2024].

[82] S. a. C. J. Imbler, "Deep-Sea Riches: Mining," The New York Times, 29 August 2022. [Online]. Available: https://www.nytimes.com/interactive/2022/08/29/world/deep-sea-riches-mining-nodules.html. [Accessed 22 April 2024].

[83] R. Whitlock, "Research warns Uyghur forced labour risks in renewable energy sector not adequately addressed," Renewable Energy Magazine, 23 January 2024. [Online]. Available: https://www.renewableenergymagazine.com/panorama/new-research-warns-that-governments-and-investors-20240123. [Accessed 22 April 2024].

[84] T. Obokata, "Contemporary forms of slavery affecting persons belonging," United Nations Human Rights Council, New York, 2022.

[85] M. Martina, "Exclusive: Duke Energy to remove Chinese battery giant CATL from Marine Corps Base," Reuters, 9 February 2024. [Online]. Available: https://www.reuters.com/business/energy/duke-energy-remove-chinese-battery-giant-catl-marine-corps-base-2024-02-09/. [Accessed 22 April 2024].

[86] J. Spector, "EV Battery Recycling is Costly. These 5 Startups Could Change That.," Canary Media, 13 June 2022. [Online]. Available: https://www.canarymedia.com/articles/electric-vehicles/ev-battery-recycling-is-costly-these-five-startups-could-change-that. [Accessed 22 April 2024].

[87] BNEF, "2H 2023 Energy Storage Market Outlook," Bloomberg New Energy Finance, 9 October 2024. [Online]. Available: https://about.bnef.com/blog/2h-2023-energy-storage-market-outlook/#:~:text=Three%20years%20into%20the%20decade,the%20rest%20of%20the%20decade.. [Accessed 2 April 2024].

[88] BNEF, "2H 2023 Energy Storage Market Outlook," Bloomberg New Energy Finance, 9 October 2023. [Online]. Available: https://about.bnef.com/blog/2h-2023-energy-storage-market-outlook/#:~:text=Global%20energy%20storage's%20record%20additions,times%20expected%202023%20gigawatt%20installations.. [Accessed 21 April 2024].

[89] W. a. K. A. Cole, "Cost Projections for Utility-Scale Battery Storage: 2023 Update," NREL, Colorado Springs, 2023.

[90] Enerdatics, "Debt raised for standalone BESS projects in the US surges 3X y/y to $4.6bn in 2023, eclipsing growth in activity for solar + storage system," Enerdatics, 26 January 2024. [Online]. Available: https://enerdatics.com/blog/debt-raised-for-standalone-bess-projects-in-the-us-surges-3x-y-y-to-46bn-in-2023-eclipsing-growth-in-activity-for-solar-storage-system/. [Accessed 2 April 2024].

[91] Energy Storage News, "Global BESS deployments to exceed 400GWh annually by 2030, says Rystad Energy," Rystad Energy, 15 June 2023. [Online]. Available: https://www.energy-storage.news/global-bess-deployments-to-exceed-400gw-annually-by-2030-says-rystad-energy/. [Accessed 3 March 2024].

[92] "Future Sodium Ion Batteries Could Be Ten Times Cheaper for Energy Storage," Next Big Future, 1 9 2023. [Online]. Available: https://www.nextbigfuture.com/2023/09/future-sodium-ion-batteries-could-be-ten-times-cheaper-for-energy-storage.html. [Accessed 27 12 2023].

[93] M. E. Initiative, "The Future of Energy Storage," 2022.

[94] F. T. Ahmadi, "Metal Air Battery," Simulation of Battery Systems, 2020. [Online]. Available: https://www.sciencedirect.com/topics/earth-and-planetary-sciences/metal-air-battery. [Accessed 2 January 2024].

[95] D. Orf, "Iron-air batteries are here. They may alter the future of energy.," Popular Mechanics, 17 1 2023. [Online]. Available: https://www.popularmechanics.com/science/energy/a42532492/iron-air-battery-energy-storage/. [Accessed 27 12 2023].

[96] N. e. a. Kamaya, "A lithium superionic conductor," *Nature Materials,* vol. 10, pp. 682-686, 2011.

[97] M. Duff, "Toyota lays out its EV Battery Road Map, Including a Solid-State Battery (Eventually)," Car and Driver, 26 11 2023. [Online]. Available: https://www.caranddriver.com/news/a45942785/toyota-future-ev-battery-plans/. [Accessed 27 12 2023].

[98] BloombergNEF, "Top 10 Energy Storage Trends in 2023," News, 11 1 2023. [Online]. Available: https://about.bnef.com/blog/top-10-energy-storage-trends-in-2023/#:~:text=Solid%2Dstate%20batteries%20progress%2C%20with,second%20half%20of%20the%20decade.. [Accessed 27 12 2023].

[99] K. Balaraman, "Form Energy $30m Grant is California's Largest Long Duration Energy Storage," Utility Dive, 18 December 2023. [Online]. Available: https://www.utilitydive.com/news/form-energy-30m-grant-california-largest-long-duration-energy-storage/702765/. [Accessed 2 April 2024].

[100] E. Bellini, "Ampace unveils battery tech with 15,000-cycle lifespan," PV Magazine, 21 September 2023. [Online]. Available: https://www.pv-magazine.com/2023/09/21/ampace-unveils-battery-tech-with-15000-cycle-lifespan/. [Accessed 13 January 2024].

[101] Hithium, "Hithium's 300 Ah stationary battery cell reaches 12,000 cycle life," ECD Online, 1 October 2023. [Online]. Available: https://www.ecdonline.com.au/content/efficiency-renewables/sponsored/hithium-s-300-ah-stationary-battery-cell-reaches-12-000-cycle-life-714399247. [Accessed 13 January 2024].

[102] M. Maisch, "CATL Staying on Top of the Battery Game," PV Magazine, 1 December 2023. [Online]. Available: https://www.pv-magazine.com/2023/12/01/catl-staying-on-top-of-the-battery-game/. [Accessed 13 January 2024].

[103] A. e. a. Gomez-Martin, "Opportunities and Challenges of Li2C4O4 as Pre-Lithiation Additive for the Positive Electrode in NMC622 Silicon/Graphite Lithium Ion Cells," *Advanced Science,* vol. 9, no. 4, pp. 2201742 (1-15), 2022.

[104] R. Selesky, "EnerVenue provides 20-year extended warranty for its nickel-hydrogen batteries," List Solar, 11 October 2022. [Online]. Available: https://list.solar/news/enervenue-provides/. [Accessed 13 January 2024].

[105] A. Groß, "PowerCo confirms results: QuantumScape's solid-state cell passes first endurance test," Volkswagen Group, 3 1 2024. [Online]. Available: https://www.volkswagen-group.com/en/press-releases/powerco-confirms-results-quantumscapes-solid-state-cell-passes-first-endurance-test-18031. [Accessed 13 1 2024].

[106] IEA, "Trends in Batteries," Global EV Outlook 2023, 2023. [Online]. Available: https://www.iea.org/reports/global-ev-outlook-2023/trends-in-batteries. [Accessed 13 January 2024].

[107] "Tesla Plans To Adopt Bi-Directional Charging By 2025," CleanTechnica, 19 August 2023. [Online]. Available: https://cleantechnica.com/2023/08/19/tesla-plans-to-adopt-bi-directional-charging-by-2025/. [Accessed 19 January 2024].

[108] P. Lau, "How to save money on your home electricity bill? Explore the advantages and working principle of solar photovoltaic power generation systems," LinkedIn, 29 May 2023. [Online]. Available: https://www.linkedin.com/pulse/how-save-money-your-home-electricity-bill-explore-advantages-paul-lau/. [Accessed 2 June 2024].

[109] FERC, "Explainer on the Interconnection Final Rule," FERC Office of Public Participation, 30 October 2023. [Online]. Available: https://www.ferc.gov/explainer-interconnection-final-rule. [Accessed 19 January 2024].

[110] McGuire Woods, "Inflation Reduction Act Creates New Tax Credit Opportunities for Energy Storage Projects," McGuire Woods, 27 December 2022. [Online]. Available: https://www.mcguirewoods.com/client-resources/alerts/2022/12/inflation-reduction-act-creates-new-tax-credit-opportunities-for-energy-storage-projects/. [Accessed 19 January 2024].

[111] X. Zhang, "China's Booming Energy Storage: A Policy-Driven and Highly Concentrated Market," APCO Worldwide, 14 November 2023. [Online]. Available: https://apcoworldwide.com/blog/chinas-booming-energy-storage-a-policy-driven-and-highly-concentrated-market/#:~:text=China's%20energy%20storage%20market%20size,storage%20systems%E2%80%9D%20(NTESS).. [Accessed 20 January 2024].

[112] I. Hwang, "Korea's Energy Storage System Development," World Bank Group, 1 January 2020. [Online]. Available: https://documents1.worldbank.org/curated/en/152501583149273660/pdf/Koreas-Energy-Storage-System-Development-The-Synergy-of-Public-Pull-and-Private-Push.pdf. [Accessed 20 January 2024].

[113] J. Vickerman, "The rise of BESS in Australia," Rated Power, 9 January 2024. [Online]. Available: https://www.woodmac.com/press-releases/australia-leads-global-market-for-battery-energy-storage-systems/. [Accessed 2 April 2024].

[114] E2, "2023 Clean Jobs America Report," E2, Washington DC, 2023.

[115] A. Energy, "Waratah Super Battery," Akaysha Energy, [Online]. Available: https://akayshaenergy.com/projects/waratah-super-battery. [Accessed 2 June 2024].

[116] "Open Street Map," [Online]. Available: https://wiki.openstreetmap.org/wiki/Import/Catalogue.

[117] NSW EnergyCo, "Waratah Super Battery," NSW Government, 2024. [Online]. Available: https://www.energyco.nsw.gov.au/projects/waratah-super-battery. [Accessed 7 February 2024].

[118] V. Bücker, "RWE connects its first utility-scale battery storage project to the California grid," RWE Clean Energy, 14 June 2023. [Online]. Available: https://www.rwe.com/en/press/rwe-clean-energy/2023-06-14-rwe-connects-its-first-utility-scale-battery-storage-project-to-the-california-grid/. [Accessed 2 June 2024].

[119] S. Cole, "2021 Resource Adequacy Report," California Public Utilities Commission, March 2023. [Online]. Available: https://www.cpuc.ca.gov/RA/ . [Accessed 2 June 2024].

[120] S. C. Light, *Provided by Seattle City Light*, 2024.

[121] Capitol Hill Seattle (CHS), "Governor says new solar microgrid for emergency energy at Capitol Hill community center first of 'hundreds' across state," CHS, 22 April 2022. [Online]. Available: https://www.capitolhillseattle.com/2022/04/governor-says-new-solar-microgrid-for-emergency-energy-at-capitol-hill-community-center-first-of-hundreds-across-state/. [Accessed 22 February 2024].

[122] Yahoo, "Yahoo," [Online]. Available: www.yahoo.com. [Accessed 13 7 2023].

[123] BloombergNEF, "Global Energy Storage Market set to hit one terawatt-hour by 2030," 2021.

[124] ACP, "Energy Storage Draft Emergency Response Plan," 2022.

[125] M. Merano, "Tesla California VPP is supplying the grid with 20MW+ in second event," 2022.

[126] T. E. ToolBox, "Energy Storage Density," [Online]. Available: https://www.engineeringtoolbox.com/energy-density-d_1362.html. [Accessed 2023].

[127] C. Crownhart, "How sodium could change the game for batteries," MIT Technology Review, 11 5 2023. [Online]. Available: https://www.technologyreview.com/2023/05/11/1072865/how-sodium-could-change-the-game-for-batteries/. [Accessed 27 12 2023].

[128] S. Jaffe, "$250 per kWh: The battery price that will herald the terawatt-hour age," E Source, 21 December 2022. [Online]. Available: https://www.esource.com/white-paper/437221l3ux/250-kwh-battery-price-will-herald-terawatt-hour-age. [Accessed 6 January 2024].

[129] D. Power, "Pivotal Virtual Power Plant Pilot Comes to Texas," Guidehouse Insights, 27 July 2023. [Online]. Available: https://guidehouseinsights.com/news-and-views/pivotal-virtual-power-plant-pilot-comes-to-texas. [Accessed 10 March 2024].

[130] San Miguel Global Power, "Leading the Philippine energy industry with BESS," San Miguel Global Power, 01 June 2022. [Online]. Available: https://www.smcglobalpower.com.ph/our-business-bess?p=1. [Accessed 17 March 2024].

[131] G. Weyl, "Key Criteria that Drive Large-scale Energy Storage Success," Power Magazine, 1 July 2023. [Online]. Available: https://www.powermag.com/partner-content/key-criteria-that-drive-large-scale-energy-storage-success/. [Accessed 17 March 2024].

[132] "First Responders Guide to Lithium-Ion Battery Storage System Incidents," Cleanpower, July 2023. [Online]. Available: https://cleanpower.org/wp-content/uploads/2023/07/ACP-ES-Product-7-First-Responders-Guide-to-BESS-Incidents-6.28.23.pdf. [Accessed 19 March 2024].

[133] U.S. Energy Information Administration, "Battery Storage in the United States: An Update on Market Trends," EIA, 24 July 2023. [Online]. Available: https://www.eia.gov/analysis/studies/electricity/batterystorage/. [Accessed 22 April 2024].

[134] Permanent Mission of the People's Republic of China to the United Nations, "Fight against Terrorism and Extremism in Xinjiang: Truth and Facts," Information Office of the People's Government of Xinjiang Uyghur Autonomous Region, Geneva, 2022.

[135] T. Egan, "The Problem with Current Lithium Extraction Methods," EnergyX, 23 February 2023. [Online]. Available: https://energyx.com/blog/the-problem-with-current-lithium-extraction-methods/. [Accessed 24 April 2024].

[136] I. Crawford, "How much CO2 is emitted by manufacturing batteries?," MIT Climate, 15 July 2022. [Online]. Available: https://climate.mit.edu/ask-mit/how-much-co2-emitted-manufacturing-batteries. [Accessed 22 April 2024].

[137] A. Colthorpe, "S BESS installations 'surged' in 2023 with 96% increase in cumulative capacity, ACP says," Energy Storage News, 18 March 2024. [Online]. Available: https://www.energy-storage.news/us-bess-installations-surged-in-2023-with-96-increase-in-cumulative-capacity-acp-says/. [Accessed 8 May 2024].

[138] M. Jacobson, No Miracles Needed: How Today's Technology Can Save Our Climate and Clean Our Air, Cambridge: Cambridge University Press, 2023.

[139] E. Fox, "Tesla Giga Texas Installs BESS Consisting of 68 Megapacks," Tesmanian, 6 May 2023. [Online]. Available: https://www.tesmanian.com/blogs/tesmanian-blog/tesla-giga-texas-installs-bess-consisting-of-68-megapacks. [Accessed 18 May 2024].

[140] Powin Energy, "Leading the Energy Transition," Powin, [Online]. Available: https://powin.com/company/. [Accessed 6 January 2024].

[141] Moss Construction, "HI 6," Moss Construction, 1 March 2024. [Online]. Available: https://moss.com/projects/hi-6/?pm=&pt=Solar. [Accessed 28 May 2024].

[142] Ulstein Group, "Energy Management System," Ulsten, [Online] Available: https://ulstein.com/system-integration/automation-solutions/energy-management-system. [Accessed 18 May 2024].

APPENDIX A - DIVISION OF RESPONSIBILITIES EXAMPLE

BESS Division of Responsibility (DoR)		
Project Name		
Project Location		
Developer		
Battery Integrator		
BoP Contractor		
Financier		
Owner's Engineer		

Item	Activity	BESS OEM	BoP Contractor	OE	Developer	Remarks
Part I: Engineering - BoP						
1.01	Product Drawings	Rev	R	I	Rev	
1.02	Power Single Line Diagram	Rev	R	I	Rev	
1.03	Protection & Control Single Line Diagram	Rev	R	I	Rev	
1.04	General Arrangement and Elevations	Rev	R	I	Rev	

Item	Activity	BESS OEM	BoP Contractor	OE	Developer	Remarks
1.06	Substation design		R			
1.07	Civil Works, Foundations, and Fencing (BESS enclosure foundation, transformer foundation, etc.)	Rev	R	I	Rev	
1.08	30%, 60%, 90%, IFP, and IFC packages	Rev	R	I	Rev	
Part II: Engineering - Controls						
2.01	Monitoring system design & implementation		R	I	Rev	
2.02	History query features design and implementation		R	I	Rev	
2.03	Host and network design and integration into operator EMS		R	I	I	
2.04	Power generation and load forecasting, optimal scheduling control design and implementation		R	I	I	
Part III: Engineering - BESS						
3.01	Product Drawings: Outline, Electrical and Physical Interface Points including Auxiliary Power Consumption, Weights, and Dimensions	R	I	I	Rev	

Item	Activity	BESS OEM	BoP Contractor	OE	Developer	Remarks
3.02	Protection & Control Single Line Diagram	n/a		I		
3.03	General Arrangement and Elevations	R		I		
3.05	Controls Architecture Diagram	R		I		
3.07	Cable Schedule and Wiring Diagrams	R		I		
3.08	Grounding Plan, Details, and Calculations	I		I	Rev	
3.09	Conduit, Trench, and Tray Plan, Details, and Calculations	I		I	Rev	
3.12	Noise Study	I		I	R	
3.13	Site Security System			I	R	
3.14	Site studies and surveys			I	Rev	
3.15	Foundation Plan, Details, and Calculations	I		I	Rev	
3.22	Ministerial and discretionary permits				Rev	
3.23	Lightning protection			I	Rev	
Part IV: Procurement						
4.01	Battery cells + Battery Enclosures	R				
4.02	PCS	R				

Item	Activity	BESS OEM	BoP Contractor	OE	Developer	Remarks
4.03	Energy Storage Plant Controller	R	I			
4.03	SCADA / Controls Panel		R	Rev	Rev	
4.04	High and Medium Voltage Switchgear including Protection Relays and Metering		R		Rev	
4.05	Control Enclosure		R		Rev	
4.06	Auxiliary Power Transformer		R		Rev	
4.07	Auxiliary Panel boards		R		Rev	
4.20	Battery substation equipment procurement		R			
Part IV: Construction						
5.01	Overall Safety Program			I	R	
5.02	Site Security During Construction		R	I	Rev	
5.03	Temporary facilities: Construction Trailers, Toilets, Temporary Fencing, Dumpsters		R	I	Rev	
5.04	Power and Water Supply During Construction (Customer to provide construction power source and water supply outlet at the site).		R	I	Rev	

Item	Activity	BESS OEM	BoP Contractor	OE	Developer	Remarks
5.05	Site Preparation (Clear and Grub, Grading, Roadways, Drainage)		R	I	Rev	
5.06	Civil Works, Foundations, and Fencing		R	I	Rev	
5.07	Install Ground Grid, Connectors, and Fittings and Related Materials		R	I	Rev	
5.08	Install underground piping/trenching between battery substation and HV substation		R	I	Rev	
5.09	Install conduit, tray between electrical components		R	I	Rev	
5.10	Install and Terminate All AC/DC Power cables		R	I	Rev	
5.11	Install and terminate control, and communications cables		R	I	Rev	
5.12	Install Transmission line		R	I	Rev	
5.13	Install utility meter			I	Rev	
5.14	Construct battery substation		R	I	Rev	
5.15	System Commissioning (For BESS OEM supplied equipment)	R				

Item	Activity	BESS OEM	BoP Contractor	OE	Developer	Remarks
5.16	Supply Recommended Spare Parts (For BESS OEM supplied equipment)	R			Rev	
5.17	Customer Operator Training for BESS OEM Supplied Equipment	R			Rev	
5.18	Documentation for BESS OEM Supplied Equipment/Design: O&M Manuals, As Built Drawings, Final Test Reports, Material Safety Data Sheets	R			Rev	
5.19	Disposal at End-of-Life of Batteries	I			R	

ABBREVIATIONS

A - Amps

AC - Alternating current

AGC - Automatic generator current

Ah - Amp-hours

AHJ - Authority having jurisdiction

Anode - The positively charged electrode from which electrons leave the cell

BCS - Battery control system

BESA - Battery energy supply agreement

BESS - Battery energy storage system. The term BESS will be used throughout the book to refer to the complete energy storage system, including the batteries themselves, PCSs, and controls.

BMS - Battery management system

BOL - Beginning-of-life

BoP - Balance of Plant

BTM - Behind the meter

C&I - Commercial and Industrial

CAISO - California ISO

Cathode - The negatively charged electrode through which the electrons enter the cell

CFD - Computational fluid dynamics

COD - Commercial operations date or commercial online date

CPUC - California Public Utilities Commission

C-rate - Charge rate

CSA - Canada Standards Association Group

DC - Direct current

DDP - Delivery duty paid

DOD - Depth of discharge

DoR - Division of Responsibilities

ECS - Energy control system

EIA - Environmental Impact Assessment

Electrolyte - The medium in a cell through which the ions flow

EMS - Energy management system

Energy - Power over a set amount of time. One megawatt of power expended over one hour is one megawatt hour

EOL - End of life

EoR - Engineer of Record

EPC - Engineering, procurement, and construction

ER - Employer's requirements

FE - Fundamentals of engineering

FERC - Federal Energy Regulatory Commission

FTM - Front-of-the-meter

GSU - Generator step up

HV - High voltage

HVAC - Heating, ventilation, and air-conditioning

HVT - High-voltage transformer. Sometimes also referred to as main power transformer (MPT) or the generator step-up (GSU)

IBC - International Building Code

IDLH - Immediately Dangerous to Life or Health

IE - Independent engineer

IFC - International Fire Code

IOU - Investor owned utility

IPP - Independent power producer

IRA - Inflation Reduction Act

IRC - International Residential Code

ISO - Independent System Operator

ITC - Investment Tax Credit

kV - Kilovolts

kW - Kilowatt

kWh - Kilowatt hour

LCO - Lithium cobalt oxide

LDs - Liquidated damages

LFL - Lower flammable limit

LFP - Lithium iron phosphate

Li-ion - Lithium-ion

LCOE - Levelized cost of energy

LMO - Lithium manganese oxide

LTO - Lithium titanate

LTSA - Long-term service agreement

LV - Low voltage

MPT - Main power transformer

MVA - Megavolt-amperes

MVT - Medium voltage transformer

MW - Megawatt

MWh - Megawatt hour

NCA - Nickel cobalt aluminum oxide

NEC - National Electrical Code

NERC - National Energy Reliability Commission

NFPA - National Fire Protection Association

NICd - Nickel-cadmium

NIMH - Nickel-metal hydride

NMC - Nickel manganese cobalt oxide

NRTL - Nationally Recognized Testing Laboratory

NRTL - Nationally Recognized Testing Laboratory

NYCFD - New York City Fire Department

OEM - Original equipment manufacturer

OpEx - Operating expenses

OSHA - US Occupational Safety and Health Administration

PCS - Power conversion system

POI - Point of interconnection

PPA - Power purchase agreement

PPC - Power plant controller

PPE - Personal protective equipment

RFP - Request for proposal

RTE - Round trip efficiency

RTO - Regional Transmission Organization

RTU - Ring terminal unit

SC - Substantial completion

SCADA - Supervisory control and data acquisition

SIPS - System Integration Protection Scheme

SOC - State of charge

SOH - State of health

TE - Tax equity

UL - Underwriters Laboratories

UPS - Uninterruptible power system

V - Volts

VPP - Virtual power plant

VT - Voltage transformer

W - Watt (power)

Wh - Watt-hour (energy)

WSB - Waratah Super Battery (Akaysha Energy project)

INDEX

E

F

G

H

I

L

T

U

V

W

Made in the USA
Las Vegas, NV
11 October 2024

96476008R00285